loIgG p235-6 p238-9
overall immunoregulatory dy

p206 allergies and EB

CHRONIC FATIGUE SYNDROME

A Biological Approach

CHRONIC FATIGUE SYNDROME

A Biological Approach

Edited by Patrick Englebienne, Ph.D.
Kenny DeMeirleir, M.D., Ph.D.

CRC PRESS

Boca Raton London New York Washington, D.C.

Library of Congress Cataloging-in-Publication Data

Chronic fatigue syndrome : a biological approach / editor, Patrick Englebienne.
p. cm.
Includes bibliographical references and index.
ISBN 0-8493-1046-6 (alk. paper)
1. Chronic fatigue syndrome. I. Englebienne, Patrick.
[DNLM: 1. Fatigue Syndrome, Chronic. WB 146 C55664 2002]
RB150.F37 C483 2002
616′.0478—dc21 2001052424

Direct all inquiries to CRC Press LLC, 2000 N.W. Corporate Blvd., Boca Raton, Florida 33431.

International Standard Book Number 0-8493-1046-6
Library of Congress Card Number 2001052424
Printed in the United States of America 1 2 3 4 5 6 7 8 9 0
Printed on acid-free paper

Foreword

Chronic fatigue syndrome (CFS) is now recognized as a common and extremely debilitating disease, which affects patients worldwide without respect to their socioeconomic status, race, sex, or premorbid psychological status. Early CFS research focused on the epidemiology and consensus of clinical description.

Efforts to determine the pathogenesis of the varied symptoms were frustrated by the heterogeneity of the patient population and seeming lack of measurable pathological findings. Eventually, researchers from around the world simultaneously described consistent patterns of immune activation, viral reactivation, and endocrine dysfunction. This suggested the existence of a common biochemical or molecular dysfunction, which could be induced in a genetically susceptible host by many different triggering events.

Whether for intellectual curiosity or for the clinical challenge, basic scientists and bedside and laboratory clinicians combined their observations to produce a plausible and powerful model for the production of chronic fatigue syndrome.

Chronic Fatigue Syndrome: A Biological Approach represents a monumental step in the journey to a unified understanding of CFS and establishes a scientific basis for treatment. This book represents a rare and much appreciated treatise on the current state of the art with respect to the worldwide scientifically documented basis of CFS and acknowledges the many as yet undiscovered or undefined pathogenetic mechanisms involved in the production of symptoms.

The authors, representing their clinical or basic research backgrounds, further outline future research imperatives. The clinician likewise is directed toward appropriate diagnostic and therapeutic strategies.

Great care was taken by the foremost researchers in the field of CFS not only to create a scholarly treatise, but also to clarify and describe these findings in a concise and well-illustrated format. This effort is particularly appreciated by clinicians unfamiliar with these highly technically and newly described areas of medicine.

In Chapter 1, Professor Lebleu and colleagues, review with clarity the interferon/2-5A RNase L pathway. They expand on what is known and postulate with respect to dysregulation of the antiviral pathway, control of cell growth, and apoptosis. Relevance to the pathophysiology of CFS and implications for future research are clearly outlined.

Chapter 2 presents an in-depth review of RNase L and reveals new information on the abnormal low molecular weight variants of RNase L seen specifically and perhaps uniquely in CFS patients. The Th2 cytokine shift commonly observed in these patients is explained and potential treatments suggested.

In Chapter 3, Dr. Suhadolnik and colleagues review their extensive work in characterizing the structure and function of the unique low molecular weight RNase L discussed in the previous chapter. They hypothesize the roles for persistent viral infection and immune activation in perpetuating the chronicity of the illness. Their work is based on a thoroughly studied subset of CFS patients. Work in progress to determine the amino acid sequence of the novel RNase L will further enlighten this basic research and may provide an avenue for rapid testing of this subset of patients.

Belgian and French collaborators in Chapter 4 discuss in depth the regulation of RNase L by RNase L inhibitor (RLI). Most practicing physicians are not familiar with RLI or the ATP-binding cassette (ABC) superfamily to which it belongs, and will find this chapter stimulating and challenging to their overall understanding of CFS.

Perhaps even more esoterically, Chapter 5 discusses the potential role of receptors and signal transduction cascades in the production of endocrine dysfunction. These abnormalities of the hypothalamic–pituitary–adrenal axis are commonly demonstrated to be present in CFS patients.

In Chapter 6, Dr. Englebienne and associates expand with clarity and depth the exceedingly complex phenomenon of apoptosis. They review for the first time in publication the altered PKR-mediated apoptosis demonstrated in CFS patients. It is further postulated that this immune perturbation may mechanistically explain many of the varied clinical manifestations of CFS.

From Australia, Dr. Neil McGregor, in Chapter 7, expands and builds upon the previous biochemical and molecular abnormalities and outlines (using a statistical model) how these processes may result in clustering of clinical signs and symptoms. Based on these models, the primary care physician is able to construct rational, and hopefully effective, empiric therapeutic strategies.

Using the previously established and postulated biochemical and molecular models of CFS, Dr. De Meirleir and colleagues develop a working clinical model in Chapters 8 and 10 for the etiology and pathogenesis of CFS. The gap between the disabled patient and the bench researcher is thus skillfully bridged. The physician or basic researcher is led logically to the detailed review of currently existent and potential therapies of CFS outlined in Chapter 9.

Chronic Fatigue Syndrome: A Biological Approach is a monumental and landmark publication. It represents years of clinical and basic research and will form the firm foundation for future research and understanding of chronic fatigue syndrome and related disorders.

Daniel L. Peterson, M.D.

The Editors

Patrick Englebienne, Ph.D., has been actively researching ligand receptor interactions for over 20 years. His current research involves unraveling the possible biological abnormalities in the immune systems of patients suffering from CFS and other immune dysfunctions using his knowledge of receptor and signal transduction.

Dr. Englebienne is a senior research associate with the Faculty of Medicine at the Free University of Brussels and consults with pharmaceutical companies.

Kenny De Meirleir, M.D., Ph.D., is professor of physiology and medicine at the Vrije Universiteit Brussel where he is director of the Human Performance Laboratory and Fatigue Clinic.

Since 1990, Dr. De Meirleir has seen approximately 8000 patients complaining of chronic fatigue at the university-based clinic in Brussels. He is a member of the board of directors of the American Association for Chronic Fatigue Syndrome, and is board certified in internal medicine (since 1982) and cardiac rehabilitation (since 1986) in Belgium.

Dr. De Meirleir serves as editor of the *Journal of Chronic Fatigue Syndrome* (2002). He has authored or coauthored several hundred journal articles and books on internal medicine, cardiology, exercise physiology, and chronic fatigue syndrome.

Currently Dr. De Meirleir conducts research on the etiology, diagnosis, and treatment of chronic fatigue syndrome. In addition to his research and teaching, he serves as president of a medical disciplinary committee and on a provincial medical regulatory commission.

Contributors

Lionel Bastide, Ph.D.
Unité de Génétique Moléculaire
Montpellier, France

Pascale De Becker, Ph.D.
Vrije Universiteit Brussels
Academic Hospital
Brussels, Belgium

Kenny De Meirleir, M.D., Ph.D.
Vrije Universiteit Brussels
Academic Hospital
Brussels, Belgium

Edith Demettre, M.S.
Unité de Génétique Moléculaire
Montpellier, France

Karen De Smet, Ph.D.
RED Laboratories, N.V.
Zellik, Belgium

Anne D'Haese, M.S.
RED Laboratories, N.V.
Zellik, Belgium

Karim El Bakkouri, Ph.D.
RED Laboratories, N.V.
Zellik, Belgium

Patrick Englebienne, Ph.D.
Department of Nuclear Medicine
Université Libre de Bruxelles,
Brugmann Hospital, Brussels
and
RED Laboratories, N.V.
Zellik, Belgium

Marc Frémont, Ph.D.
RED Laboratories, N.V.
Zellik, Belgium

C. Vincent Herst, Ph.D.
RED Laboratories, N.V.
Zellik, Belgium

Bernard Lebleu, Ph.D.
CNRS, Unité de Génétique Moléculaire
Montpellier, France

Camille Martinand-Mari, Ph.D.
Department of Biochemistry
Temple University School of Medicine
Philadelphia, Pennsylvania

Neil R. McGregor, B.D.Sc., M.D.Sc., Ph.D.
Department of Biological Sciences
University of Newcastle
Callaghan, NSW
and
Neurobiology Research Unit
Centre for Oral Health
University of Sydney
Westmead, Australia

Garth L. Nicolson, Ph.D.
Institute for Molecular Medicine
Huntington Beach, California

Jo Nijs, P.T.
Department of Sport Medicine
Vrije Universiteit Brussels, Academic Hospital
Brussels, Belgium

Roberto Patarca-Montero, M.D., Ph.D.
University of Miami School of Medicine
Miami, Florida

Daniel L. Peterson, M.D.
Sierra Internal Medicine Associates
Incline Village, Nevada

Simon Roelens, M.S.
RED Laboratories, N.V.
Zellik, Belgium

Nancy L. Reichenbach, B.S.
Department of Biochemistry
Temple University School of Medicine
Philadelphia, Pennsylvania

Susan E. Shetzline, Ph.D.
Temple University School of Medicine
Philadelphia, Pennsylvania

Robert, J. Suhadolnik, Ph.D.
Temple University School of Medicine
Philadelphia, Pennsylvania

Thierry Verbinnen
RED Laboratories N.V.
Zellik, Belgium

Michel Verhas, M.D., Ph.D.
Université Libre de Bruxelles
Brugmann Hospital, Department of Nuclear Medicine
Brussels, Belgium

Introduction

Among the many patients who seek medical care for the complaint of fatigue, a small number suffer from chronic fatigue syndrome (CFS). The syndrome is distinguished from chronic fatigue by a variety of symptoms present in addition to the fatigue, and by the severity and chronicity of both the fatigue and its accompanying symptoms.[1] CFS is clinically defined by persistent or relapsing fatigue lasting for over 6 months in the absence of any definable medical diagnosis. The fatigue is accompanied by a series of other symptoms: neurocognitive impairment, muscle pain, multijoint pain, headaches, unrefreshing sleep, postexertional malaise, adenopathy, and sore throat.[2,3] CFS sufferers exhibit diminished physical and mental health status, even when compared to patients with other chronic diseases, such as heart disease, diabetes mellitus, multiple sclerosis, and major depression.[4] CFS runs a relapsing–remitting course over many months or years without abnormalities detectable by routine laboratory testing. Patients may experience periods of feeling almost completely healthy and symptoms may vary from day to day. Recurrence of symptoms is unpredictable and while too strenuous exercise or emotional stress can trigger relapses, they sometimes occur for no apparent reason. The illness is more and more common and can be devastating to afflicted sufferers who may be bedridden or confined to a wheelchair; the social consequences can be very severe.[5] Half of CFS patients are unable to exert their professional activities, participate in household tasks, or attend school.[6]

Although the interest in CFS has grown since the mid-1980s, a number of medical publications suggest that fatigue syndromes are not new and have been reported for several centuries. CFS occurs worldwide, with sporadic outbreaks reported in the United Kingdom, New Zealand, Australia, Canada, and Japan as far back as the 1930s. Two epidemics that attracted considerable attention occurred respectively in 1934 at the Los Angeles County Hospital and in 1955 at the Royal Free Hospital in London. These outbreaks concerned the medical staff rather than the hospitalized patients.[7] More recent clusters have been reported respectively in Nevada in 1984,[8-10] in a female residential facility in 1990,[11] in two state buildings in California in 1991,[12,13] and in Michigan in 1992.[13] Various names have been used to describe these fatiguing illnesses. Some of these names refer to the geographical sites where an epidemic occurred such as Akueryi disease, Iceland disease, and Royal Free disease. Other names refer to a presumed etiology such as atypical poliomyelitis, myalgic encephalomyelitis (ME), neuromyasthenia, postviral fatigue syndrome, chronic candidiasis, chronic brucellosis, chronic Epstein–Barr virus syndrome, and chronic fatigue and immune dysfunction syndrome (CFIDS). ME is commonly used in British and New Zealand medical and patient literature in preference to CFS. The moniker "chronic fatigue syndrome" was recommended by the

1988 CDC research committee, as it was free from unproven etiological implication and instead described the cardinal symptom.[2]

Several case definitions of CFS have been developed throughout the years, each with its specific features. In order to reduce the diagnostic confusion surrounding CFS and to produce a rational basis for the evaluation of patients, a group of epidemiologists, researchers, and clinicians (from the U.S. Centers for Disease Control, CDC) developed a consensual case definition in 1988.[2] To meet the CDC case definition, a patient must fulfill both major and minor criteria. The minor criteria are divided into symptom criteria and physical signs (see Appendix for criteria). The next case definition, which is used by many investigators in the U.S., was later proposed by Fukuda and co-authors[3] and is the result of conceptual framework and guidelines for the diagnosis prepared by an international study group of the CDC (see Appendix). In this new "relaxed" definition of CFS, fewer of the minor symptoms need to be present. The relaxed definition excludes fewer patients, while on the other hand, a stricter definition may reduce heterogeneity in the patient population. A third set of criteria comprises the British definition proposed by Sharpe and colleagues in 1991.[14] This case definition was developed in 1990, when researchers and clinicians convened in Oxford. In the Oxford case definition, less emphasis is placed on the somatic symptoms and more emphasis is placed on acute onset. None of these definitions can be considered definitive,[3] and definitions will continue to be modified. In the interim, the diagnosis of CFS should be understood to be provisional, not final.[8]

Chronic fatigue syndrome, despite the absence of an apparent organic explanation, presents with common symptom clusters which persist indefinitely.[5] Many of the physical symptoms of CFS resemble classical physical symptoms of other diseases (e.g., fever, sore throat, enlarged lymph nodes, joint pains, night sweats, headaches). However, CFS also manifests more universal physical symptoms (e.g., visual disturbances, exaggerated allergic reactions, skin rash, intolerance to alcohol). Neuropsychiatric symptoms include confusion, memory loss, dizziness, anxiety, depression, difficulties in concentrating, spatial disorientation, and emotional lability. The chief complaint of CFS patients, though, is prolonged fatigue. The fatigue may be more profound after exercise (which was previously tolerated well).[5]

The physical examination of CFS patients is usually unremarkable with a few exceptions, e.g., inflamed pharynx, macular rash, hepatomegaly and splenomegaly, low body temperature, palpable posterior cervical/axillary adenopathy (typically nontender), impairment of tandem gait, or abnormal Romberg testing. A large number of CFS patients report atopy or allergic illness, documented both by history and skin testing, as well as inhalant, food, or drug allergies.[15,16] There is also a large prevalence of nasal, sinus, and other complaints that suggest allergic syndromes. A small subset of patients report sensitivities to perfumes, solvents, cosmetics, and other substances.[16] Patients with CFS often complain of symptoms that suggest central nervous system (CNS) dysfunction including: difficulty with concentration, attention, and memory; photophobia; paresthesias, paresis, visual loss, ataxia, or confusion.[17]

A number of diagnostic studies of the CNS have shown abnormalities in CFS patients. Different studies using magnetic resonance imaging (MRI) brain scans have documented CNS abnormalities in a large percentage of CFS patients, including punctuate areas of high signal intensity in the brain stem and subcortical regions.[17] A recent study showed that these abnormalities are predominantly found in the frontal lobes.[18] Disturbances in cerebral blood flow have been identified in several regions of the brain, including frontal area temporal lobes and basal ganglia.[19] Orthostatic intolerance, neurally mediated hypotension (NMH), and increased sympathetic and decreased parasympathetic tone are frequently observed in CFS patients.[20] Several studies have identified abnormalities of hypothalamic function in CFS and disruption of both serotonergic and noradrenergic pathways.[21] Typically, these abnormalities are opposite to those seen in depression.

Given the large number of well-recognized organic diseases that can produce chronic fatigue, laboratory tests play a major role in the differential diagnosis of CFS. Standard hematologic tests revealed leukocytosis, lymphocytosis, and leukopenia in approximately 20% of the patients.[22] Other blood parameters that may show abnormalities are, respectively, elevated transaminases and alkaline phosphatase,[23] increased levels of thyroid-stimulating hormone,[24] low red blood cell magnesium,[25] low serum acylcarnitine,[26] and increased serum angiotensin-converting enzyme activity.[27] Immunologic and cytokine abnormalities are found in most patients.

CFS occurs in men and women of all age, ethnic, and socio-economic groups, including young children. Depending on the studies, however, females are prevalent over males among patient populations by factors ranging from 1.5 to 4. The true incidence and prevalence of CFS in the general population are unknown. Estimates vary depending on the case definition used, the methods employed, and the population surveyed. A health maintenance organization estimates that there are 75 to 267 cases per 100,000, based on a study of 4000 randomly selected individuals.[28] More recent studies report a much higher prevalence of up to 0.74%.[29] The prognosis of CFS is hard to predict, although cases occurring as part of clusters appear to have a better prognosis than sporadic cases. Those with an acute onset also have a better prognosis than those with a gradual onset.[30,31] A recent study found that recovery was more likely to be reported in the earlier years of illness.[32] The improvement rates in prospective studies vary from 8%[33] to a maximum of 63 to 64% after a period of 31 to 39 months.[34,35] Severely disabled patients have a poorer prognosis for recovery. In this group, the majority showed no symptom improvement and only 4% of the patients ultimately recovered.[36] Although the prognosis for complete recovery in adults remains poor,[37] children have a far better prognosis. In a tertiary care pediatric infectious disease clinic, 76% of the children were reported to be completely cured.[38] Importantly, a subgroup of patients may be prone to develop cancer. In the years following the outbreak in Lake Tahoe, an increased occurrence of brain tumors, non-Hodgkin's lymphomas, breast carcinomas, basal cell carcinomas, and uterine, bladder, and prostate cancer has been observed.[39]

Like other chronic diseases such as multiple sclerosis, systemic lupus erythematosus, or rheumatoid arthritis, for which in the past no objective organic abnormalities could be identified,[40] the etiology and pathogenesis of CFS remain controversial. The long-lasting debate persists as to whether the origin of CFS is physical or psychological, and CFS remains a clinically defined medical condition. Nevertheless, evidence-based organic (cellular) abnormalities are being reported consistently in recent studies and the authors of this book were able to select subgroups of CFS patients in which they identified significant dysregulations of the innate cellular immunity system. These abnormalities suggest sound biological explanations for chronic fatigue and many of its associated symptoms, and shed a new light on the etiology, pathogenesis, and evolution of CFS. These results are detailed in the various chapters and confronted to the available clinical and biological literature. Ultimately, these recent findings may pave new avenues for an objective drug therapy. The editors consider this publication as timely and welcome comments from interested readers.

Pascale De Becker, Ph.D.
Kenny De Meirleir, M.D., Ph.D.
Patrick Englebienne, Ph.D.
Daniel L. Peterson, M.D.

REFERENCES

1. Bates, D. W. et al., Prevalence of fatigue and chronic fatigue syndrome in a primary care practice, *Arch. Intern. Med.,* 1993; 153: 2759–65.
2. Holmes, G. P. et al., Chronic fatigue syndrome, a working case definition, *Annu. Intern. Med.,* 1988; 108: 387–9.
3. Fukuda, K. et al., The chronic fatigue syndrome, a comprehensive approach to its definition and study, *Annu. Intern. Med.,* 1994; 121: 953–9.
4. Komaroff, A. et al., Health status in patients with chronic fatigue syndrome and in general population and disease comparison groups, *Am. J. Med.,* 1996; 101: 281–90
5. Lane, R. M., Aetiology, diagnosis and treatment of chronic fatigue syndrome, *J. Serotonin Res.,* 1994; 1: 47–60.
6. Lloyd, A. R. et al., The prevalence of chronic fatigue syndrome in an Australian population, *Med. J. Aust.,* 1990; 153: 522–528.
7. Dillon, M. J. et al., Epidemic neuromyasthenia. Outbreak among nurses at a children's hospital, *Br. Med. J.,* 1974; 1: 301–5.
8. Holmes, G. et al., A cluster of patients with a chronic mononucleosis-like syndrome, *JAMA,* 1987; 257: 2297–302.
9. Daugherty, S. A. et al., Chronic fatigue syndrome in Northern Nevada, *Rev. Infect. Dis.,* 1991; 13 (Suppl. 1): S39–S44.
10. Levine, P. H. et al., Clinical, epidemiological, and virological studies in four clusters of the chronic fatigue syndrome, *Arch. Intern. Med.,* 1992; 152: 1611–16.
11. Levine, P. H. et al., A cluster of cases of chronic fatigue and chronic fatigue syndrome: clinical and immunologic studies, *Clin. Infect. Dis.,* 1996; 23: 408–9.
12. Shefer, A. et al., Fatiguing illness among employees in three large state office buildings, California, 1993; was there an outbreak? *J. Psychiatr. Res.,* 1997; 31: 31–43.
13. Fukuda, K. et al., An epidemiological study of fatigue with relevance for the chronic fatigue syndrome, *J. Psychiatr. Res.,* 1997; 31: 19–29.
14. Sharpe, M. C. et al., A report on chronic fatigue syndrome. Guidelines for research, *J. R. Soc. Med.,* 1991; 84: 118–21.
15. Komaroff, A. L., *Chronic Fatigue Syndrome,* New York, Marcel Dekker, 1994.
16. Straus, S. E. et al., Allergy and the chronic fatigue syndrome, *J. Allergy Clin. Immunol.,* 1988; 81: 791–5.

17. Komaroff, A. L. and Buchwald, D., Chronic fatigue syndrome: an update, *Annu. Rev. Med.,* 1998; 49: 1–13.
18. Lange, G. et al., Brain MRI abnormalities exist in a subset of patients with chronic fatigue syndrome, *J. Neurol. Sci.,* 1999; 171: 3–7.
19. Schwartz, R. B. et al., SPECT imaging of the brain, comparison of findings in patients with chronic fatigue syndrome, AIDS dementia complex, and major unipolar depression, *Am. J. Roentgenology,* 1994; 162: 943–51.
20. De Becker, P. et al., Autonomic testing in patients with chronic fatigue syndrome, *Am. J. Med.,* 1998; 105: 22S–26S.
21. Scott L. V. and Dinan, T. G., The neuroendocrinology of chronic fatigue syndrome, focus on the hypothalamic-pituitary-adrenal axis, *Funct. Neurol.,* 1999; 14: 3–11.
22. Buchwald, D. and Komaroff, A. L., Review of laboratory findings for patients with chronic fatigue syndrome, *Rev. Infect. Dis.,* 1991; 13: S73–S83.
23. Tobi, M. et al., Prolonged atypical illness associated with serological evidence of persistent Epstein–Barr virus infection, *Lancet,* 1982; 1: 61–4.
24. Kroenke, K. et al., Chronic fatigue in primary care: prevalence, patient characteristics, and outcome, *JAMA,* 1988; 260: 929–34.
25. Cox, M., Campbell, M. J., and Dowson, D., Red blood cell magnesium and chronic fatigue syndrome, *Lancet,* 1991; 337: 757–60.
26. Kuratsune, H. et al., Acylcarnitine deficiency in chronic fatigue syndrome, *Clin. Infect. Dis.,* 1994; 18 (Suppl. 1): S62–S67.
27. Lieberman, J. and Bell, D. S., Serum angiotensin-converting enzyme as a marker for the chronic fatigue-immune dysfunction syndrome, a comparison to serum angiotensin-converting enzyme in sarcoidosis, *Am. J. Med.,* 1993; 95: 407–12.
28. Buchwald, D. et al., Chronic fatigue and the chronic fatigue syndrome: prevalence in a Pacific Northwest health care system, *Annu. Intern. Med.,* 1995; 123: 81–8.
29. Lawrie, S. M. et al., A population-based incidence study of chronic fatigue, *Psychol. Med.,* 1997; 27: 343–53.
30. Levine, P. H., Epidemiologic advances in chronic fatigue syndrome, *J. Psychiatr. Res.,* 1997; 31: 7–18.
31. Levine, P. H. et al., Epidemic neuromyasthenia and chronic fatigue syndrome in West Otago, New Zealand, a 10-year follow-up, *Arch. Intern. Med.,* 1997; 157: 750–54.
32. Reyes, M. et al., Chronic fatigue syndrome progression and self-defined recovery, evidence from the CDC surveillance system, *J. Chronic Fatigue Syndrome,* 1999; 5: 17–27.
33. Tirelli, U. et al., Immunological abnormalities in patients with chronic fatigue syndrome, *Scand. J. Immunol.,* 1994; 40: 601–8.
34. Wilson, A. et al., Longitudinal study of outcome of chronic fatigue syndrome, *BMJ,* 1994; 308: 756–9.
35. Bombardier, C. H. and Buchwald, D., Outcome and prognosis of patients with chronic fatigue syndrome, *Arch. Intern. Med.,* 1995; 155: 2105–10.
36. Hill, N. F. et al., Natural history of severe chronic fatigue syndrome, *Arch. Phys. Med. Rehab.,* 1999; 80: 1091–4.
37. Joyce, J., Hotopf, M., and Wessely, S., The prognosis of chronic fatigue and chronic fatigue syndrome; a systematic review, *Q. J. Med.,* 1997; 90: 223–33.
38. Marshall, G. S. et al., Chronic fatigue in children, clinical features, Epstein–Barr virus and human herpesvirus 6 serology and long term follow-up, *Pediat. Infect. Dis. J.,* 1991; 10: 287–90.
39. Levine, P. H. et al., Chronic fatigue syndrome and cancer: is there a relationship? *Proc. Sec. World Congr. Chronic Fatigue Syndrome and Related Disorders,* 1999, 18.
40. Komaroff, A. L., *Chronic Fatigue Syndrome,* John Wiley & Sons, Chichester, 1993.

CDC Case Definitions for CFS

A. Holmes et al., 1988 definition[2]

Major criteria:

- New onset of persistent or relapsing, debilitating fatigue in a person without a previous history of such symptoms does not resolve with bedrest and is severe enough to reduce or impair average daily activity to less than 50% of the patient's premorbid activity level for at least 6 months.
- The fatigue is not explained by the presence of other obvious medical or psychiatric illness.

In addition, eight of the minor criteria have to be present (eight symptom criteria or two physical criteria in combination with six symptom criteria).

Minor criteria:

1. Symptoms (patient's description of initial onset of symptoms as acute or subacute):
   - Mild fever (37.5 to 38.6°C orally) or chills
   - Sore throat
   - Posterior cervical, anterior cervical, or axillary lymph node pain
   - Unexplained generalized muscle weakness
   - Muscle discomfort or myalgia
   - Prolonged (at least 24 h) generalized fatigue following previously tolerable levels of exercise
   - New, generalized headaches
   - Migratory noninflammatory arthralgias
   - Neuropsychiatric symptoms, photophobia, transient visual scotoma, forgetfulness, excessive irritability, confusion, difficulty thinking, inability to concentrate, depression
   - Sleep disturbances (hypersomnia or insomnia)

2. Physical examination criteria (documented by a physician on at least two occasions, at least 1 month apart):
   - Low-grade fever (37.6 to 38.6°C oral or 37.8 to 38.8°C rectal)
   - Nonexudative pharyngitis
   - Palpable or tender anterior cervical, posterior cervical, or axillary lymph nodes (less than 2 cm in diameter)

B. "Relaxed" definition (Fukuda et al., 1994)[3]

1. CFS is clinically evaluated as unexplained, persistent, or relapsing chronic fatigue that is of new or definite onset (i.e., not lifelong). The fatigue is not the result of ongoing exertion, is not substantially alleviated by rest, and results in substantial reductions in previous levels of occupational, educational, social, or personal activities.
2. There must be concurrent occurrence of four or more of the following symptoms, and all must be persistent or recurrent during 6 or more months of the illness and not predate the fatigue:
    - Self-reported persistent or recurrent impairement in short-term memory or concentration severe enough to cause reductions in previous levels of occupational, educational, social, or personal activities
    - Sore throat
    - Tender cervical or axillary lymph nodes
    - Muscle pain
    - Multiple joint pain without joint swelling or redness
    - Headaches of a new type, pattern, or severity
    - Unrefreshing sleep
    - Postexertional malaise lasting more than 24 h

REFERENCES

2. Holmes, G. P. et al., Chronic fatigue syndrome, a working case definition, *Annu. Intern. Med.,* 1988; 108: 387–9.
3. Fukuda, K. et al., The chronic fatigue syndrome, a comprehensive approach to its definition and study, *Annu. Intern. Med.,* 1994; 121: 953–9.

Table of Contents

1 Interferon and the 2-5A/Pathway

Lionel Bastide, Edith Demettre, Camille Martinand-Mari, and Bernard Lebleu

CONTENTS

1.1 OVERVIEW AND HISTORICAL PERSPECTIVE

Interferons (IFNs) were discovered through studies of viral interference mechanisms and extensively studied for their remarkable capacity to confer large spectrum antiviral protection in mammalian cells.[1] IFNs act as transcriptional activators of IFN-stimulated genes (ISGs) whose products mediate a wide panel of biological responses.

The IFNs can be classified into two groups; namely the type I (IFNα/β essentially) gene family clustered on the short arm of chromosome 9, and type II or γ IFN. Almost all cell types are capable of producing IFNα/β upon response to various pathogens (including viruses, bacteria, or mycoplasma) or to other cytokines (as, for example, tumor necrosis factor). IFNγ, on the contrary, is produced essentially in the context of T lymphocyte activation; hence the alternative terminology of immune IFN.

The regulation of IFNα/β and IFNγ gene expression is tightly regulated at both transcriptional and posttranscriptional levels, and has been extensively studied (see Young and Ghosh[2] for a recent review). Type I and type II IFNs make use of different sets of cellular membrane receptors and signaling pathways, as outlined in Figure 1.1.

0-8493-1046-6/02/$0.00+$1.50

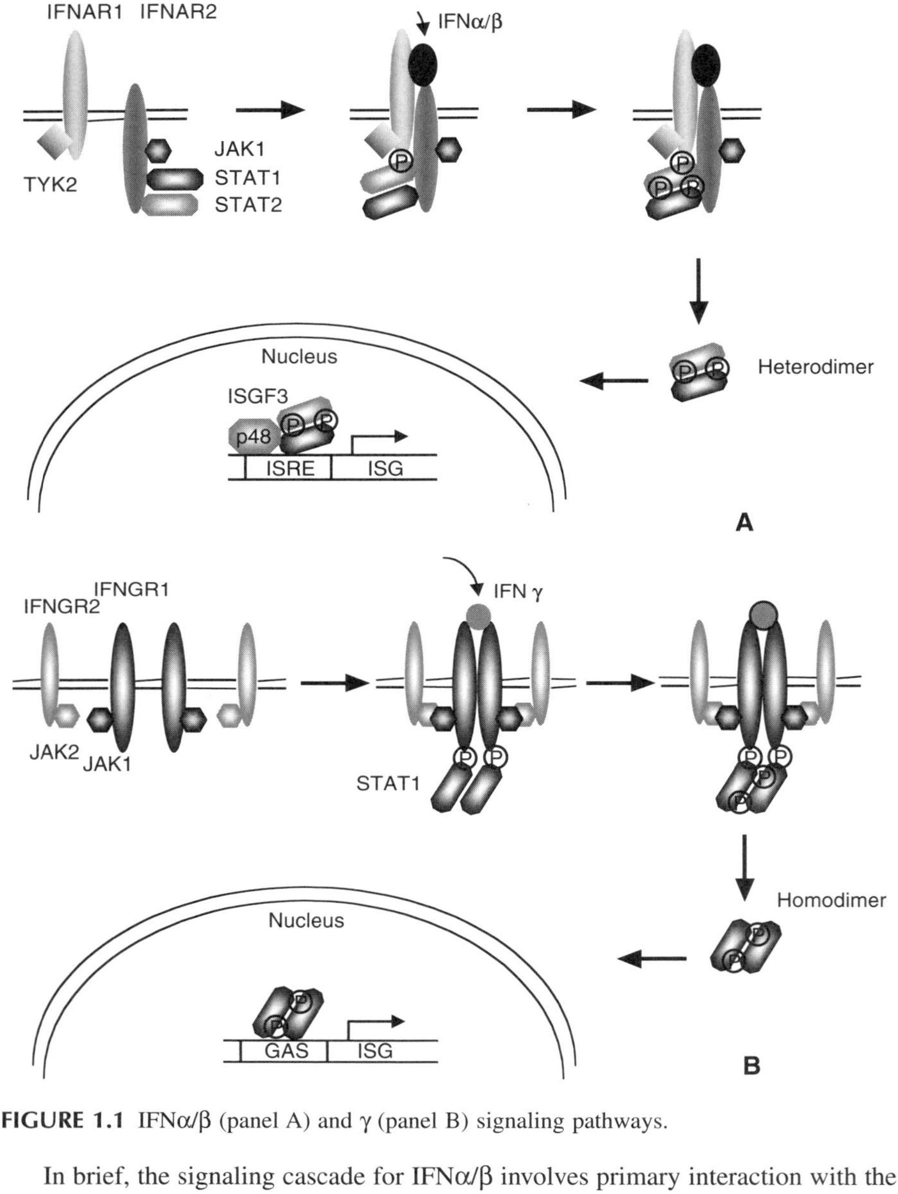

FIGURE 1.1 IFNα/β (panel A) and γ (panel B) signaling pathways.

In brief, the signaling cascade for IFNα/β involves primary interaction with the IFNAR-1 and IFNAR-2 components of the receptor. Ligand-driven receptor dimerization triggers a complex intracellular cascade of tyrosine phosphorylations involving JAKs (Janus kinase) and STATs (for signal transducers and activators of transcription). Activation of STATs leads to the assembly and nuclear translocation of an ISGF3 transcription factor. ISGF3 in turn activates the transcription of the genes (ISG = interferon stimulated gene) comprising the appropriate ISRE sequence (interferon stimulated regulatory element) in their promoter. Likewise, cellular responses

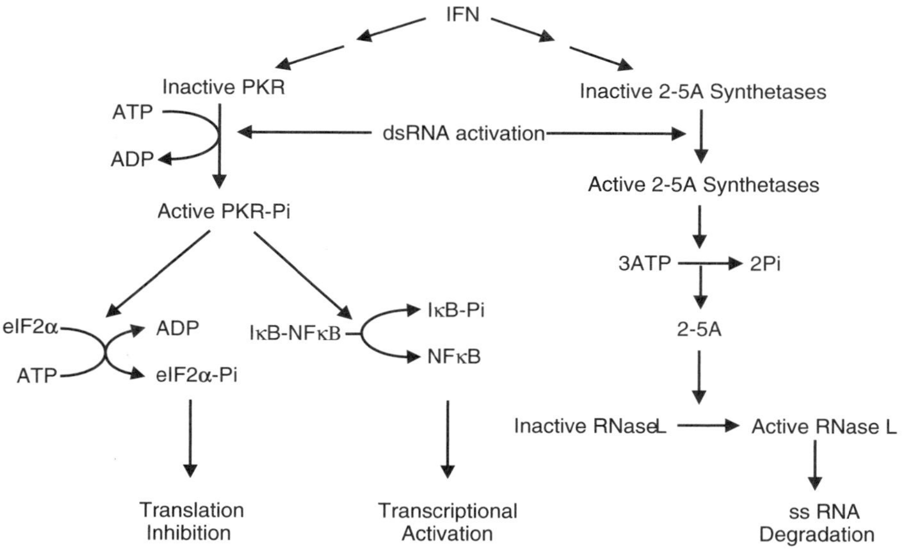

FIGURE 1.2 Outline of the IFN-induced dsRNA-dependent pathways.

to IFNγ require binding to the IFNGR I and IFNGR 2 subunits of the IFNγ receptor, signaling, through JAK1 and STAT1, homodimerization of STAT1, nuclear translocation of the active phosphorylated STAT1 homodimer, and binding to the specific GAS response elements of IFNγ inducible genes.

Alternative signaling pathways, for instance the MAPK (mitogen-activated protein kinase) pathway, can be activated as well, but the physiological relevance of these observations is not yet known. Negative regulation of ISG transcription to limit expression in the absence of IFN or to down-regulate IFN induction has also been reported. Several comprehensive recent reviews can be consulted for details.[3,4]

The most studied ISGs are the Mx gene, known to play a pivotal role in the control of influenza virus replication, and the double-stranded RNA (dsRNA) activated pathways, namely PKR (dsRNA-dependent protein kinase) and RNase L (Figure 1.2). The Mx pathway has been found to be a key player in the genetic susceptibility of certain murine strains to infection by influenza virus, as reviewed by Haller et al.[5]

The dsRNA-activated pathways were both discovered in the mid-1970s through studies of the mechanisms by which IFN down-regulates viral mRNA translation in cell extracts. IFN treatment of cell cultures induces the synthesis of a 68kDa protein kinase (known as dsRNA-activated protein kinase or PKR) which requires activation by dsRNA to phosphorylate its substrates.[6] The transcriptional activation of the PKR gene involves IFN-responsive elements in its promoter,[7] but other stimuli, such as TNF-α, are capable of promoting PKR transcriptional activation.[8] Activation of the protein kinase itself requires binding of dsRNA to a double-stranded RNA binding motif (ds RBM) and autophosphorylation[9] (Figure 1.3). The activated form of PKR phosphorylates several substrates, among which the

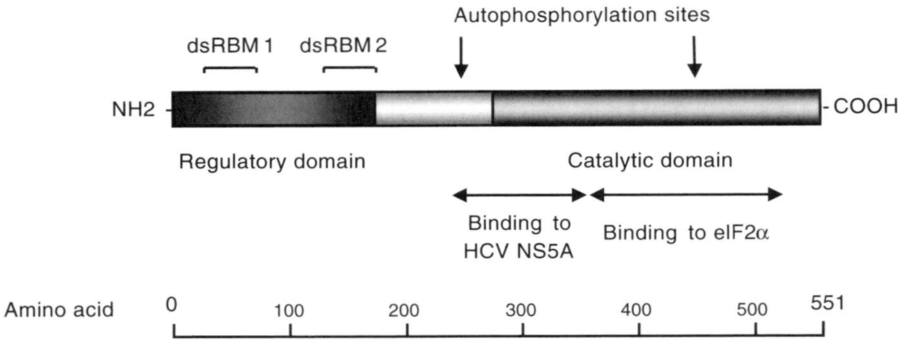

FIGURE 1.3 Structure–function of PKR.

α-subunit of the eIF2 translation initiation factor and the NF kappa B-associated I kappa B inhibitor are the best characterized. The *in vitro* expression of a dominant negative PKR mutant and the availability of PKR null (–/–) transgenic mice have provided clear evidence for the implication of the PKR pathway in the antiviral activity of IFNs on picornaviruses[10] and in cell apoptosis.[11] Interestingly, PKR-related eIF2α-specific kinases are activated in other responses to stress, such as heme deprivation in reticulocytes[12] or amino acid starvation.[13] Not surprisingly, many viruses have engineered strategies to counteract PKR-associated antiviral activity. As an example, the HCV NS5A protein binds PKR, thereby inhibiting its antiviral and apoptosis-inducing activities.[14]

The 2-5A/RNase L pathway was discovered in studies from Kerr's and Lengyel's groups aiming at understanding protein synthesis inhibition by dsRNAs in extracts from IFN-treated cells.[15,16] The characterization of 2-5A as a protein synthesis inhibitor[17] and the direct demonstration of a 2-5A dependent RNase activity[18] rapidly followed, leading to the general scheme presented in Figure 1.2.

Cloning of RNase L by the group of Silverman[19] has given insights into the structure and function of this unique endoribonuclease (see Section 1.2 and Chapter 2). Likewise, the expression of the cloned RNase L cDNA has provided important tools to ascertain the role of RNase L in the control of virus multiplication and in cellular physiology as detailed in Sections 1.3 and 1.4. Excellent and comprehensive reviews by Silverman[20] and by Player[21] and Torrence can also be consulted. More than 100 genes have been identified as significantly induced by IFNα/β, by IFNγ, or by both IFN types, and this number is likely to increase with the availability of increasingly sophisticated technologies for transcriptome analysis (see Table 1.1 for a few representative examples of ISGs).

Differential screening of a cDNA library of IFN-treated human lymphoblastoïd (Daudi) cell line by Mechti in our group has led to the discovery of two unknown ISGs that code for a transcription factor named Staf-50[22] and for a nuclear PML-associated protein named ISG-20.[23] More recently, application of gene-array technology has allowed the discovery of additional ISGs, including a phospholipid scramblase.[24] A recent survey of IFN-β-treated WM9 melanoma cells using a probe set corresponding to around 5700 human genes[25] estimated the number of genes of

TABLE 1.1
IFN-Induced Genes: A Few Examples

Name	Function	Induced by IFN
2-5A synthetases	2-5A synthesis	α, β, γ
PKR	Protein kinase	α, β, γ
Mx	GTPase	α, β, γ
MHC I, heavy chain	Antigen presentation	α, β, γ
MHC II, heavy chain	Antigen presentation	γ
IRF-1	Transcription factor	α, β, γ
IRF-2	Transcription factor	α, β
calpain	Protein degradation	γ
LMP2, LMP7	Units of proteasome	γ
cathepsins H, B, and L	Lysosomial proteins	γ
phospholipid scramblase	Antitumoral	α, β, γ
TNF receptor	Membrane receptor/apoptosis	γ
ICE	Caspase/apoptosis	γ

which expression is increased more than twice upon IFN influence to be 105. The total number of ISGs could thus be extrapolated to several hundreds. It is worth pointing out here that a fair proportion of these ISGs have not yet been identified and that many of them have been poorly characterized functionally. Attempts to relate entirely biological functions and therapeutic efficacy of IFNs to the expression of PKR and of RNase L should therefore be considered with caution. As underlined by Borden in a recent survey of IFNs as anticancer agents,[25] we still ignore which properties of IFNs are important for their antitumoral activities.

1.2 THE 2-5A/RNase L PATHWAY: AN OVERVIEW

IFNα/β and γ induce the transcription of several 2-5A synthetase isozymes. In human cells, the low molecular weight (40 and 46 kDa) 2-5A synthetases are produced by alternative splicing of a single gene.[26] The medium-size (69 kDa) isoforms[27] and the high molecular weight (100 kDa) form[28] are synthesized from separate genes. The 2-5A synthetase isoforms differ in intracellular localization, in enzymatic properties (for instance, in their requirements for dsRNA activation), in oligomeric structure, and in length of the synthesized 2-5A oligomers. However, the physiological relevance of these differences is not understood. Activated 2-5A synthetases polymerize ATP into 2'-5' linked oligomers of various lengths (Figure 1.4), except for the p100 isoform which primarily synthesizes 2-5A dimers[29] which do not activate RNase L. It is worth mentioning that 2-5A immunoreactive material (2-5A dimer cores) has been detected in nonmammalian tissues using a 2-5A core-specific radioimmunoassay,[30] thus suggesting alternative roles for 2-5A synthetases.

The only known protein binding 2-5A oligomers is RNase L, which migrates in PAGE-SDS gels as an 83 kDa polypeptide. 2-5A binding and 2-5A affinity can

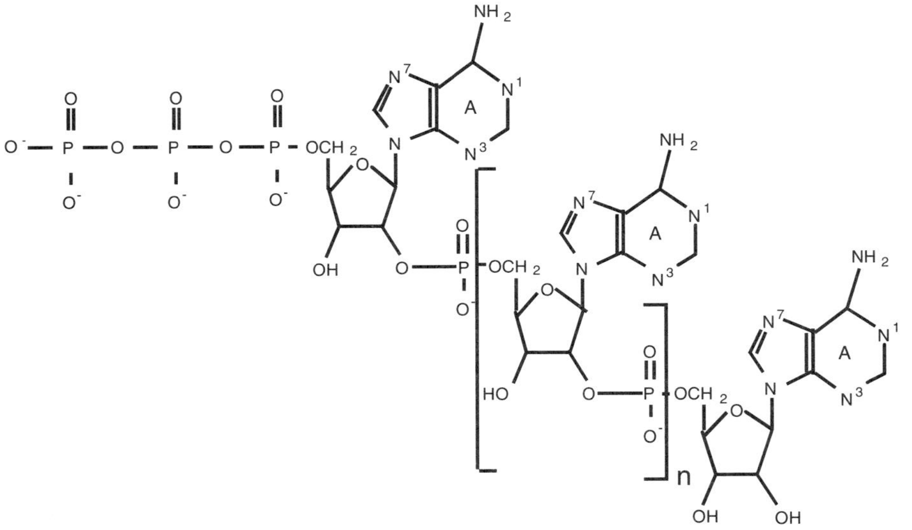

ppp (A2'p5')$_n$A

FIGURE 1.4 Structure of triphosphorylated 2-5A.

be monitored with several assays, as outlined in Figure 1.5. The cloning and expression of the human and murine forms of RNase L[19] have allowed a detailed analysis of its structure and function (Figure 1.6). The N-terminal part of RNase L (aa 1-340) contains nine ankyrin domains often involved in protein–protein interactions. Ankyrin domains 7 and 8 contain P-loop motifs which are required for 2-5A binding[31] with high affinity (Kd = 5×10^{-11}M). Intriguingly, a large part of the C-terminal portion of RNase L (aa 365-741) has significant homology with the yeast IRE-1 protein kinase, but no kinase activity has ever been detected in RNase L.[32] A Cys-rich domain extending from amino acids 401 to 436 bears a zinc finger-like motif characteristic of nucleic acid-binding proteins and might be involved in RNA binding.[19] The catalytic domain is still very poorly characterized; it includes a stretch of 89 amino acids at the C-terminal end, since truncation of this peptide gives rise to a dominant negative (ZB1) mutant of RNase L.[33]

Binding of 2-5A is required for RNase L homodimerization and activation.[34] Although 2-5A synthetases synthesize 5'-triphosphorylated 2-5A derivatives of various lengths, the monophosphorylated 2-5A trimer is sufficient for the activation of human RNase L. Numerous 2-5A analogs have been synthesized as potential antiviral or antitumoral agents. They turned out as useful tools to define requirements for RNase L binding and activation, or to probe RNase L biological functions (see Reference 21 for a comprehensive review). As an example, some phosphorothioate analogs of 2-5A behaved as RNase L agonists (e.g., pRpRpR), while others were RNase L antagonists (e.g., pSpSpSp)[35] when microinjected in intact cells. RNase L is a poorly specific endoribonuclease with preferential cleavage sites on the 3' side

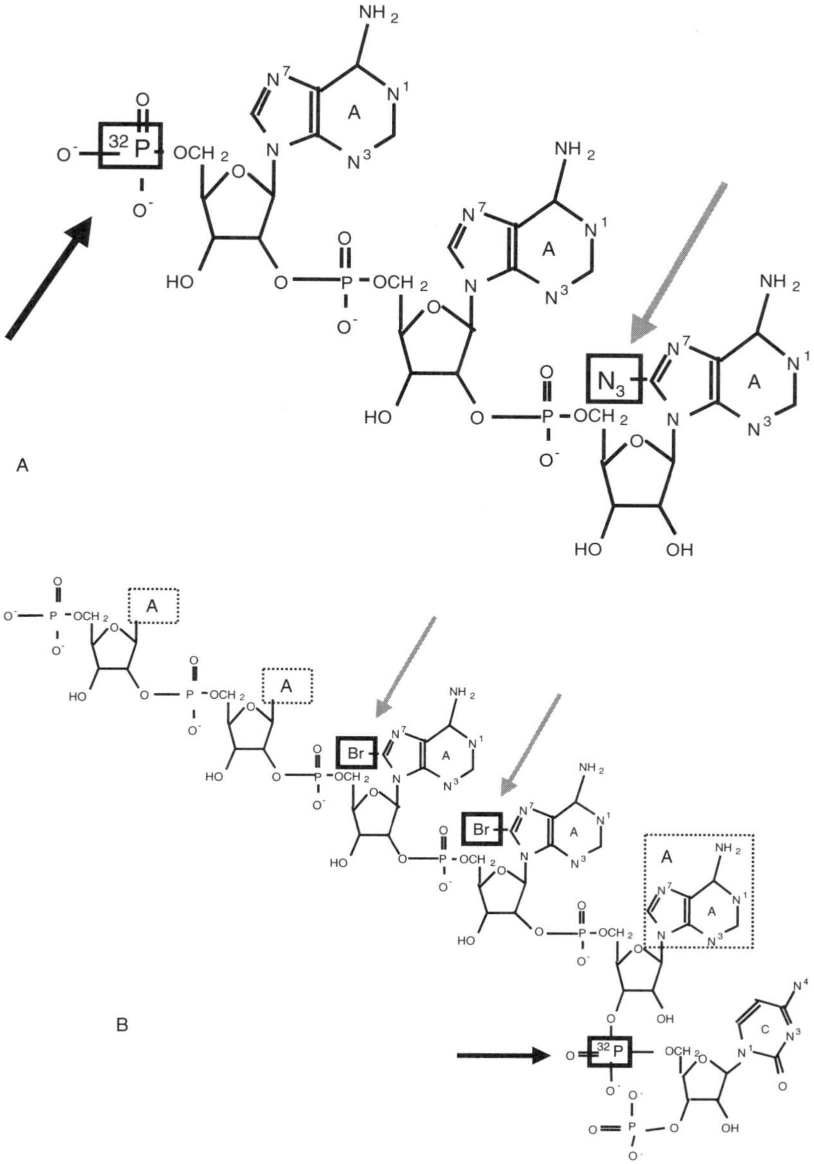

FIGURE 1.5 Affinity probing of 2-5A binding proteins (RNase L). Covalent labeling of 2-5A binding proteins can be achieved with an azido 2-5A derivative[72] (panel A), with a bromo 2-5A derivative[73] (panel B), or with a 3'-oxydized 2-5A pCp derivative[68] (panel C). Black arrows indicate the [^{32}P] labeling sites and gray arrows indicate the covalent fixation sites.

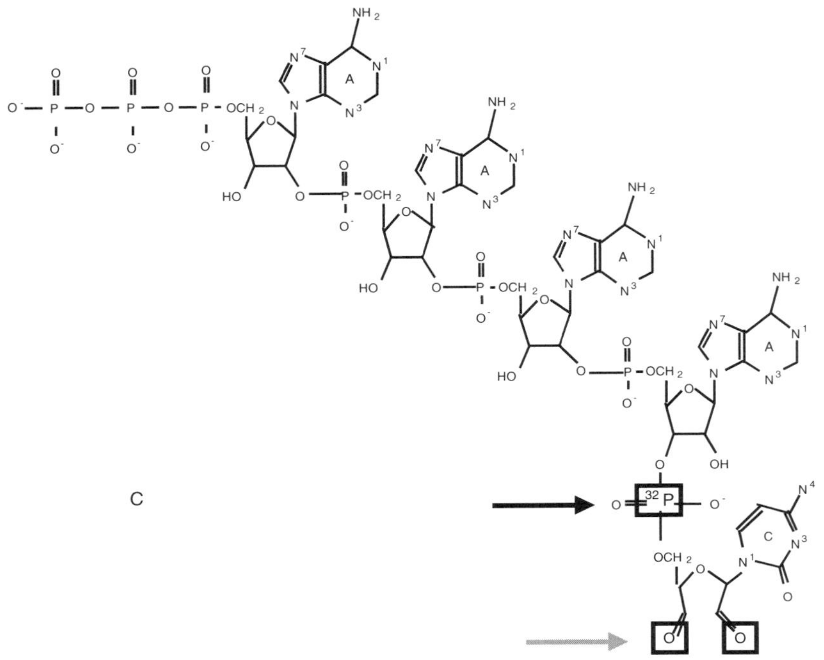

FIGURE 1.5 *Continued.*

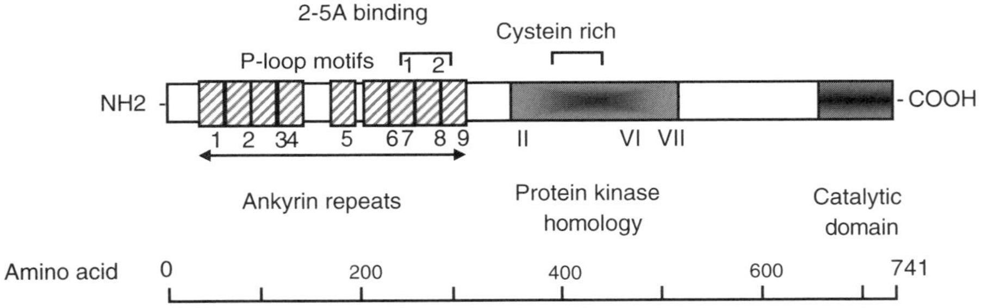

FIGURE 1.6 Structure–function of RNase L.

of UpU and UpA sequences, whether the endogenous enzyme in cell extracts[36] or the purified recombinant protein[37] is used. Explaining how RNase L activation in IFN-treated cells could lead to the preferential degradation of some viral (or cellular) RNAs is difficult and will be discussed in Section 1.4.

Although homodimerization has been clearly established as a prerequisite for RNase L activation, the following observations are worth mentioning. A truncated RNase L form lacking ankyrin domains 1 to 9 retains nuclease activity independently of 2-5A binding.[31] An attractive hypothesis would thus be that 2-5A-induced dimerization simply relieves a negative control imposed by the N-terminal ankyrin-rich region. Along the same lines, our group has obtained evidence for heterodimer formation with an 80 kDa polypeptide (named RNA-BP), which could clearly be distinguished from the 2-5A binding 83 kDa polypeptide (Reference 38 and unpublished observations). RNA-BP has now been identified as a component of the translation machinery (Le Roy et al., unpublished observations), but the physiological relevance of these observations cannot yet be fully appreciated.

Not surprisingly, the 2-5A/RNase L pathway is tightly regulated at several steps. The activity of 2-5A synthetases is controlled at the transcriptional (through promoter activation) and posttranscriptional levels (through dsRNA availability and subcellular localization). Moreover, 2-5A oligomers have a short (5 to 30 min) functional half-life since they are degraded by phosphodiesterases[39] and dephosphorylated to inactive core 2-5A by phosphatases.[40] Finally, a protein inhibitor of RNase L (RLI or RNase L inhibitor) whose association with RNase L prevents 2-5A binding has been described.[41]

1.3 INVOLVEMENT OF THE 2-5A/RNase L PATHWAY IN THE CONTROL OF VIRUS MULTIPLICATION

The most direct evidence of the broad spectrum antiviral mechanism triggered by IFN production comes from the increased sensitivity of transgenic mice lacking IFN α/β and γ receptors to virus infections.[42] The role played by the 2-5A/RNase L pathway in the control of virus multiplication and in the antiviral activity of IFNs has been reasonably well established for vaccinia virus, retrovirus, HIV-1, and picornaviruses. The most convincing and complete demonstration has been gathered for picornaviruses such as encephalomyocarditis virus (EMCV) and Mengo virus and will be discussed here. The transcriptional induction of 2-5A synthetase genes upon IFN treatment and the activation of the enzyme by virus-associated dsRNA structures (It should be recalled here that picornaviruses have a single-stranded RNA genome whose replication requires a partially double-stranded replicative intermediate.) should lead to 2-5A accumulation. Increased levels of 2-5A were indeed found in cell culture experiments[43] and in infected mice[44] upon IFN treatment and picornavirus infection. Direct evidence for an association of the 40 kDa form 2-5A synthetase to EMVC RNA has also been obtained.[45]

The introduction of 2-5A analog inhibitors of RNase L in intact cells inhibits RNase L activity and partially antagonizes the antiviral activity of IFNs.[46] Cloning of the 2-5A/RNase L pathway genes has allowed the direct manipulation of the system and provided the most definitive arguments for its antiviral role. Overexpression of a cDNA encoding the 40/46 kDa informs of 2-5A synthetase in cells (which were not treated with IFN) severely inhibits the replication of picornavirus, but not of vesicular stomatitis virus (VSV).[47] Although overexpression of a gene does not necessarily reflect cell physiology, these data are in line with the activation of 2-5A synthetases by dsRNA structures produced during the course of picornavirus infection, while it is generally accepted that the replication of rhabdoviruses does not give rise to dsRNA accumulation. The strongest arguments for a role of the 2-5A/RNase L pathway in the antiviral activity of IFNs against picornaviruses has come from the manipulation of RNase L level *per se*. The stable expression of the

ZB1 trans-dominant negative mutant of RNase L in various cell lines significantly antagonizes (about 250-fold) the antiviral activity of IFN against EMCV but not against VSV.[33] Along the same lines, the overexpression of RNase L inhibitor cDNA or of RNase L antisense constructions partially antagonizes the antiviral activity of IFN against EMCV but not against VSV.[48]

Finally, transgenic mice in which the RNase L gene had been inactivated by homologous recombination were more sensitive to EMCV infection than normal mice.[49] The protective effect of IFN was moderately reduced, however, in these knock-out mice which means that several IFN-activated biochemical pathways cooperate in the antiviral activity. It should be noted here that transgenic mice in which the Mx gene, PKR, and RNase L have been completely inactivated still exhibited a significant antiviral activity against picornaviruses when treated with IFN,[50] thus pointing to the pleiotropic character of this defense mechanism.

1.4 INVOLVEMENT OF THE 2-5A/RNase L PATHWAY IN OTHER BIOLOGICAL RESPONSES

An implication of the 2-5A/RNase L pathway in the control of cell growth and differentiation has long been suspected, but unequivocal experimental evidence has been provided only recently. An inverse correlation between 2-5A synthetase levels and cell growth has been documented in several biological models. As an example, the levels of RNase L[51] and of 2-5A synthetase[52] increase at the transition from exponential cell growth to growth arrest. Likewise, a decrease in RNase L and in 2-5A synthetase levels has been observed *in vivo* in regenerating rat liver after partial hepatectomy.[53] The transfection of metabolically stable 2-5A analogs inhibits cell proliferation.[54] Along the same line, overexpression of a 2-5A synthetase cDNA[55] or of RNase L cDNA[56] inhibits cell growth. Taken together these data indicate that the activation of the 2-5A/RNase L pathway in the absence of IFN treatment leads to cell growth inhibition.

Definitive arguments for the implication of RNase L in the control of cell proliferation again come from experiments in which the 2-5A/RNase L pathway has been inactivated. IFN treatment of murine fibroblasts in which the ZB1 transdominant negative RNase L mutant has been overexpressed does not inhibit cell growth.[33] Likewise, we have shown that the overexpression of RLI cDNA or a RNase L antisense construction antagonizes the antiproliferative activity of IFNs (Le Roy et al., unpublished observations). In none of these experiments has the cellular target of RNase L been demonstrated.

Several studies have pointed to the implication of the IFN system in apoptosis, and several genes involved in the negative control of IFN-induced apoptosis have been identified.[57] Again, direct evidence for an implication of the RNase L pathway has come from genetic manipulation of the RNase L. Expression of RNase L from a vaccinia virus vector enhances apoptosis,[58] while, on the contrary, the transfection of a transdominant negative RNase L mutant prevents apoptosis.[59] Murine embryonic fibroblasts isolated from RNase L-null mice were resistant to apotosis induced by staurosporine, or by IFN and 2-5A treatment.[49] Interestingly, these RNase L null transgenic mice had enlarged thymus and are deficient in spontaneous splenic and thymic cell apoptosis.[49] Apoptosis in response to RNase L activation leads to cytochrome c release from the mitochondria and to caspase activation.[60] The pathway from RNase L activation to cytochrome c release remains unknown. It should be pointed out, in this context, that we have recently demonstrated an RNase L-dependent increased rate of degradation of mitochondrial mRNAs in IFN-treated cells (Le Roy et al., unpublished observations).

As pointed out above, the cellular targets of RNase L have not yet been identified with certainty, even in cells undergoing RNase L-associated growth arrest or apoptosis. Two recent publications have documented RNase L-mediated degradation of cellular mRNAs. Differential display analysis in RNase L-deficient and competent cell lines has revealed a decline of ISG43 mRNA expression upon IFN treatment and an increased half-life in an RNase L-deficient cell line.[61] Along the same lines, our laboratory has demonstrated an implication of the 2-5A/RNase L pathway in terminal myogenic differentiation through selective degradation of Myo D mRNA, a transcription factor controlling muscle differentiation.[62]

The implication of the 2-5A/RNase L pathway in the control of cellular mRNA stability is particularly appealing since it sheds light on possible mechanisms for IFN effects on cell physiology and because very few nucleases involved in the modulation of mRNA decay rates have been identified.[63] The mechanism allowing selective degradation of defined mRNAs (or groups of mRNAs) by such a poorly specific endoribonuclease as RNase L is not known. mRNA structure, intracellular compartmentalization, and protein binding obviously restrict RNase L accessibility. This might explain preferential degradation of viral mRNAs in reovirus-infected cells.[64] An attractive hypothesis of localized activation of the 2-5A/RNase L pathway has been proposed by Nilsen and Baglioni[65] but still lacks direct demonstration. According to their model, double-stranded regions in viral or cellular RNAs will activate 2-5A synthetase, temporarily and locally increase 2-5A concentration, activate RNase L at the site of 2-5A production, and therefore lead to preferential degradation of this particular RNA. Selective degradation of a complementary RNA by 2-5A[66] or 2-5A analogs[67] conjugated to an antisense oligonucleotide does indeed indicate that RNase L will cleave the RNA to which it is addressed, whatever its sequence.

1.5 CONCLUSION AND PERSPECTIVES: POSSIBLE RELEVANCE OF THE RNase L PATHWAY TO CFS PHYSIOPATHOLOGY

As extensively discussed in this issue, infection by viruses (or other pathogens) and chronic immune activation are frequently associated with chronic fatigue syndrome (CFS).

In keeping with these observations, altered IFN responses and dysfunctioning of the 2-5A/RNase L pathway were reported by several laboratories, including our own[68] (see Chapters 2 and 3 for extensive coverage). In summary, increased proteolytic activity in PBMC from CFS patients leads to the cleavage of the native 83 kDa RNase L and to the accumulation of a 37 kDa 2-5A binding polypeptide

(References 69, 70, and unpublished observations) a potential biochemical marker for CFS[68]. Whether this 37 kDa 2-5A binding polypeptide is a deregulated form of RNase L accounting for increased RNase L activity[71] is still a matter of debate, and awaits complete purification and biochemical characterization. Whatever the case, an altered RNase L activity (whether decreased, increased, or deregulated) will obviously profoundly affect cellular physiology. Indeed, as reported in Section 1.4 of this chapter, there is now clear evidence for an involvement of the 2-5A/RNase L pathway in apoptosis in response to various stimuli, in cell growth and differentiation, and in the metabolic stability of several cellular mRNA.

ACKNOWLEDGMENTS

Research on the 2-5A/RNase L pathway and on its dysfunctioning in CFS in the author's laboratory has been sponsored by the Centre National de la Recherche Scientifique, the Association pour la Recherche sur le Cancer, the Chronic Fatigue Association of America, and R. E. D. Laboratories. The authors wish to thank Dr. I. Robbins for a critical reading of the manuscript.

REFERENCES

1. Isaacs, A. and Lindenmann, J., Virus interference. I. The interferon, *Proc. R. Soc. London B, Biol. Sci.*, 147, 258, 1957.
2. Young, H.A. and Ghosh, P., Molecular regulation of cytokine gene expression: interferon-gamma as a model system, *Prog. Nucleic Acid Res. Mol. Biol.*, 56, 109, 1997.
3. Bach, E.A., Aguet, M., and Schreiber, R.D., The IFN gamma receptor: a paradigm for cytokine receptor signaling, *Annu. Rev. Immunol.*, 15, 563, 1997.
4. Stark, G.R. et al., How cells respond to interferons, *Annu. Rev. Biochem.*, 67, 227, 1998.
5. Haller, O., Frese, M., and Kochs, G., Mx proteins: mediators of innate resistance to RNA viruses, *Rev. Sci. Tech.*, 17, 220, 1998.
6. Lebleu, B. et al., Interferon, double-stranded RNA, and protein phosphorylation, *Proc. Natl. Acad. Sci. U.S.A.*, 73, 3107, 1976.
7. Tanaka, H. and Samuel, C.E., Mechanism of interferon action: structure of the mouse PKR gene encoding the interferon-inducible RNA-dependent protein kinase, *Proc. Natl. Acad. Sci. U.S.A.*, 91, 7995, 1994.
8. Yeung, M.C., Liu, J., and Lau, A.S., An essential role for the interferon-inducible, double-stranded RNA-activated protein kinase PKR in the tumor necrosis factor-induced apoptosis in U937 cells, *Proc. Natl. Acad. Sci. U.S.A.*, 93, 12451, 1996.
9. Meurs, E. et al., Molecular cloning and characterization of the human double-stranded RNA- activated protein kinase induced by interferon, *Cell*, 62, 379, 1990.
10. Yang, Y.L. et al., Deficient signaling in mice devoid of double-stranded RNA-dependent protein kinase, *Embo. J.*, 14, 6095, 1995.
11. Tan, S.L. and Katze, M.G., The emerging role of the interferon-induced PKR protein kinase as an apoptotic effector: a new face of death? *J. Interferon. Cytokine Res.*, 19, 543, 1999.
12. Chen, J.J. and London, I.M., Regulation of protein synthesis by heme-regulated eIF-2 alpha kinase, *Trends Biochem. Sci.*, 20, 105, 1995.

13. Sood, R. et al., A mammalian homologue of GCN2 protein kinase important for translational control by phosphorylation of eukaryotic initiation factor-2alpha, *Genetics*, 154, 787, 2000.
14. Gale, M., Jr. et al., Antiapoptotic and oncogenic potentials of hepatitis C virus are linked to interferon resistance by viral repression of the PKR protein kinase, *J. Virol.*, 73, 6506, 1999.
15. Kerr, I.M., Brown, R.E., and Ball, L.A., Increased sensitivity of cell-free protein synthesis to double-stranded RNA after interferon treatment, *Nature*, 250, 57, 1974.
16. Sen, G.C. et al., Interferon, double-stranded RNA, and mRNA degradation, *Nature*, 264, 370, 1976.
17. Kerr, I.M. and Brown, R.E., pppA2'p5'A2'p5'A: an inhibitor of protein synthesis synthesized with an enzyme fraction from interferon-treated cells, *Proc. Natl. Acad. Sci. U.S.A.*, 75, 256, 1978.
18. Clemens, M.J. and Williams, B.R., Inhibition of cell-free protein synthesis by pppA2'p5'A2'p5'A: a novel oligonucleotide synthesized by interferon-treated L cell extracts, *Cell*, 13, 565, 1978.
19. Zhou, A., Hassel, B.A., and Silverman, R.H., Expression cloning of 2-5A-dependent RNAase: a uniquely regulated mediator of interferon action, *Cell*, 72, 753, 1993.
20. Silverman, R.H., 2-5A-Dependant RNase L: a regulated endoribonuclease in the interferon system, *Ribonucleases: Structures and Functions*, D'Alessio, G. and Riordan, J.F., Academic Press, New York, 1997, 515.
21. Player, M.R. and Torrence, P.F., The 2-5A system: modulation of viral and cellular processes through acceleration of RNA degradation, *Pharmacol. Ther.*, 78, 55, 1998.
22. Tissot, C. and Mechti, N., Molecular cloning of a new interferon-induced factor that represses human immunodeficiency virus type 1 long terminal repeat expression, *J. Biol. Chem.*, 270, 14891, 1995.
23. Gongora, C. et al., Molecular cloning of a new interferon-induced PML nuclear body-associated protein, *J. Biol. Chem.*, 272, 19457, 1997.
24. Der, S.D. et al., Identification of genes differentially regulated by interferon alpha, beta, or gamma using oligonucleotide arrays, *Proc. Natl. Acad. Sci. U.S.A.*, 95, 15623, 1998.
25. Borden, E.C. et al., Novel interferon stimulated genes (ISGs) potently induced by IFN-β in WM9 melanoma cells, *Eur. Cytokine Netw.*, 11, 101, 2000.
26. Marié, I. et al., Differential expression and distinct structure of 69- and 100-kDa forms of 2-5A synthetase in human cells treated with interferon, *J. Biol. Chem.*, 265, 18601, 1990.
27. Marié, I. and Hovanessian, A.G., The 69-kDa 2-5A synthetase is composed of two homologous and adjacent functional domains, *J. Biol. Chem.*, 267, 9933, 1992.
28. Rebouillat, D. et al., The 100-kDa 2',5'-oligoadenylate synthetase catalyzing preferentially the synthesis of dimeric pppA2'p5'A molecules is composed of three homologous domains, *J. Biol. Chem.*, 274, 1557, 1999.
29. Hovanessian, A.G. et al., Characterization of 69- and 100-kDa forms of 2-5A-synthetase from interferon-treated human cells, *J. Biol. Chem.*, 263, 4959, 1988.
30. Laurence, L. et al., Immunological evidence for the *in vivo* occurrence of (2'-5')adenylyladenosine oligonucleotides in eukaryotes and prokaryotes, *Proc. Natl. Acad. Sci. U.S.A.*, 81, 2322, 1984.
31. Dong, B. and Silverman, R.H., A bipartite model of 2-5A-dependent RNase L, *J. Biol. Chem.*, 272, 22236, 1997.
32. Dong, B. and Silverman, R.H., Alternative function of a protein kinase homology domain in 2',5'-oligoadenylate dependent RNase L, *Nucl. Acids Res.*, 27, 439, 1999.

33. Hassel, B.A. et al., A dominant negative mutant of 2-5A-dependent RNase suppresses antiproliferative and antiviral effects of interferon, *EMBO J.*, 12, 3297, 1993.
34. Dong, B. and Silverman, R.H., 2-5A-dependent RNase molecules dimerize during activation by 2-5A, *J. Biol. Chem.*, 270, 4133, 1995.
35. Charachon, G. et al., Phosphorothioate analogues of (2'5')(A)4: agonist and antagonist activities in intact cells, *Biochemistry*, 29, 2550, 1990.
36. Floyd-Smith, G., Slattery, E., and Lengyel, P., Interferon action: RNA cleavage pattern of a (2'-5')oligoadenylate-dependent endonuclease, *Science*, 212, 1030, 1981.
37. Dong, B. et al., Intrinsic molecular activities of the interferon-induced 2-5A-dependent RNase, *J. Biol. Chem.*, 269, 14153, 1994.
38. Salehzada, T. et al., 2',5'-Oligoadenylate-dependent RNase L is a dimer of regulatory and catalytic subunits, *J. Biol. Chem.*, 268, 7733, 1993.
39. Schmidt, A. et al., An interferon-induced phosphodiesterase degrading (2'-5') oligoisoadenylate and the C-C-A terminus of tRNA, *Proc. Natl. Acad. Sci. U.S.A.*, 76, 4788, 1979.
40. Bayard, B. et al., Increased stability and antiviral activity of 2'-O-phosphoglyceryl derivatives of (2'-5')oligo(adenylate), *Eur. J. Biochem.*, 142, 291, 1984.
41. Bisbal, C. et al., Cloning and characterization of a RNAse L inhibitor. A new component of the interferon-regulated 2-5A pathway, *J. Biol. Chem.*, 270, 13308, 1995.
42. Van den Broek, M.F. et al., Antiviral defense in mice lacking both alpha/beta and gamma interferon receptors, *J. Virol.*, 69, 4792, 1995.
43. Williams, B.R. et al., Natural occurrence of 2-5A in interferon-treated EMC virus-infected L cells, *Nature*, 282, 582, 1979.
44. Hearl, W.G. and Johnston, M.I., Accumulation of 2',5'-oligoadenylates in encephalomyocarditis virus-infected mice, *J. Virol.*, 61, 1586, 1987.
45. Gribaudo, G. et al., Interferon action: binding of viral RNA to the 40-kilodalton 2'-5'-oligoadenylate synthetase in interferon-treated HeLa cells infected with encephalomyocarditis virus, *J. Virol.*, 65, 1748, 1991.
46. Watling, D. et al., Analogue inhibitor of 2-5A action: effect on the interferon-mediated inhibition of encephalomyocarditis virus replication, *EMBO J.*, 4, 431, 1985.
47. Chebath, J. et al., Constitutive expression of (2'-5') oligo A synthetase confers resistance to picornavirus infection, *Nature*, 330, 587, 1987.
48. Martinand, C. et al., RNase L inhibitor (RLI) antisense constructions block partially the down regulation of the 2-5A/RNase L pathway in encephalomyocarditis-virus-(EMCV)-infected cells, *Eur. J. Biochem.*, 254, 248, 1998.
49. Zhou, A. et al., Interferon action and apoptosis are defective in mice devoid of 2',5'-oligoadenylate-dependent RNase L, *EMBO J.*, 16, 6355, 1997.
50. Zhou, A. et al., Interferon action in triply deficient mice reveals the existence of alternative antiviral pathways, *Virology*, 258, 435, 1999.
51. Jacobsen, H. et al., Induction of ppp(A2'p)nA-dependent RNase in murine JLS-V9R cells during growth inhibition, *Proc. Natl. Acad. Sci. U.S.A.*, 80, 4954, 1983.
52. Stark, G.R. et al., 2-5A synthetase: assay, distribution, and variation with growth or hormone status, *Nature*, 278, 471, 1979.
53. Etienne-Smekens, M. et al., (2'-5')Oligoadenylate in rat liver: modulation after partial hepatectomy, *Proc. Natl. Acad. Sci. U.S.A.*, 80, 4609, 1983.
54. Hovanessian, A.G. and Wood, J.N., Anticellular and antiviral effects of pppA(2'p5'A)n, *Virology*, 101, 81, 1980.
55. Rysiecki, G., Gewert, D.R., and Williams, B.R., Constitutive expression of a 2',5'-oligoadenylate synthetase cDNA results in increased antiviral activity and growth suppression, *J. Interferon Res.*, 9, 649, 1989.

56. Zhou, A. et al., Impact of RNase L overexpression on viral and cellular growth and death, *J. Interferon Cytokine Res.*, 18, 953, 1998.
57. Cohen, O., Feinstein, E., and Kimchi, A., DAP-kinase is a Ca2+/calmodulin-dependent, cytoskeletal-associated protein kinase, with cell death-inducing functions that depend on its catalytic activity, *EMBO J.*, 16, 998, 1997.
58. Diaz-Guerra, M., Rivas, C., and Esteban, M., Activation of the IFN-inducible enzyme RNase L causes apoptosis of animal cells, *Virology*, 236, 354, 1997.
59. Castelli, J.C. et al., A study of the interferon antiviral mechanism: apoptosis activation by the 2-5A system, *J. Exp. Med.*, 186, 967, 1997.
60. Rusch, L., Zhou, A., and Silverman, R.H., Caspase-dependent apoptosis by 2',5'-oligoadenylate activation of RNase L is enhanced by IFN-beta, *J. Interferon Cytokine Res.*, 20, 1091, 2000.
61. Li, X.L. et al., RNase-L-dependent destabilization of interferon-induced mRNAs. A role for the 2-5A system in attenuation of the interferon response, *J. Biol. Chem.*, 275, 8880, 2000.
62. Bisbal, C. et al., The 2'-5' oligoadenylate/RNase L/RNase L inhibitor pathway regulates both MyoD mRNA stability and muscle cell differentiation, *Mol. Cell Biol.*, 20, 4959, 2000.
63. Tharun, S. and Parker, R., Mechanisms of mRNA turnover in eukaryotic cells, in *mRNA Metabolism and Post-Transcriptional Gene Regulation*, Wiley-Liss, 1997, 181.
64. Baglioni, C., De Benedetti, A., and Williams, G.J., Cleavage of nascent reovirus mRNA by localized activation of the 2'-5'-oligoadenylate-dependent endoribonuclease, *J. Virol.*, 52, 865, 1984.
65. Nilsen, T.W. and Baglioni, C., Mechanism for discrimination between viral and host mRNA in interferon- treated cells, *Proc. Natl. Acad. Sci. U.S.A.*, 76, 2600, 1979.
66. Torrence, P.F. et al., Targeting RNA for degradation with a (2'-5')oligoadenylate-antisense chimera, *Proc. Natl. Acad. Sci. U.S.A.*, 90, 1300, 1993.
67. Robbins, I. et al., Selective mRNA degradation by antisense oligonucleotide-2,5A chimeras: involvement of RNase H and RNase L, *Biochimie*, 80, 711, 1998.
68. De Meirleir, K. et al., A 37 kDa 2-5A binding protein as a potential biochemical marker for chronic fatigue syndrome, *Am. J. Med.*, 108, 99, 2000.
69. Lebleu, B. et al., A truncated form of RNaseL accumulates in PBMC of chronic fatigue syndrome patients, *3rd Joint Meet. ICS/ISICR*, Amsterdam, The Netherlands, 2000.
70. Roelens, S. et al., G-actin cleavage parallels 2-5A-dependent RNase L cleavage in peripheral blood mononuclear cells. Relevance to a possible serum-based screening test for dysregulation in the 2-5A pathway, *J. Chronic Fatigue Syndrome*, 8, 63, 2001.
71. Suhadolnik, R.J. et al., Biochemical evidence for a novel low molecular weight 2-5A-dependent RNase L in chronic fatigue syndrome, *J. Interferon Cytokine Res.*, 17, 377, 1997.
72. Shetzline, S.E. and Suhadolnik, R.J., Characterization of a 2-5A dependent 37-kDa RNase L 2. Azido photoaffinity labeling and 2-5A-dependent activation, *J. Biol. Chem.*, 276, 23707, 2001.
73. Nolan-Sorden, N.L. et al., Photochemical crosslinking in oligonucleotide-protein complexes between a bromine-substituted 2-5A analog and 2-5A-dependent RNase by ultraviolet lamp or laser, *Anal. Biochem.*, 184, 298, 1990.

2 Ribonuclease L: Overview of a Multifaceted Protein

Patrick Englebienne, C. Vincent Herst, Simon Roelens, Anne D'Haese, Karim El Bakkouri, Karen De Smet, Marc Frémont, Lionel Bastide, Edith Demettre, and Bernard Lebleu

CONTENTS

2.1 INTRODUCTION

The 2',5'-oligoadenylate (2-5A)-dependent ribonuclease (RNase) is known as RNase L, a terminology which underlines the latency of this protein. RNase L is an endonuclease central to the 2-5A antiviral pathway.[1] This enzyme is dormant in nearly every mammalian cell type and is activated by binding small adenylate oligomers linked in 2',5', of which the general structure is shown in Figure 2.1. The activation is accompanied by a homodimerization of the enzyme. This activation occurs in the presence of double-stranded RNAs (ds-RNA) of viral origin, a process mediated by type I interferons.[1] Homodimerization confers the catalytic activity to the protein which cleaves single-stranded RNA (ss-RNA).[2] The enzymatic activity

0-8493-1046-6/02/$0.00+$1.50

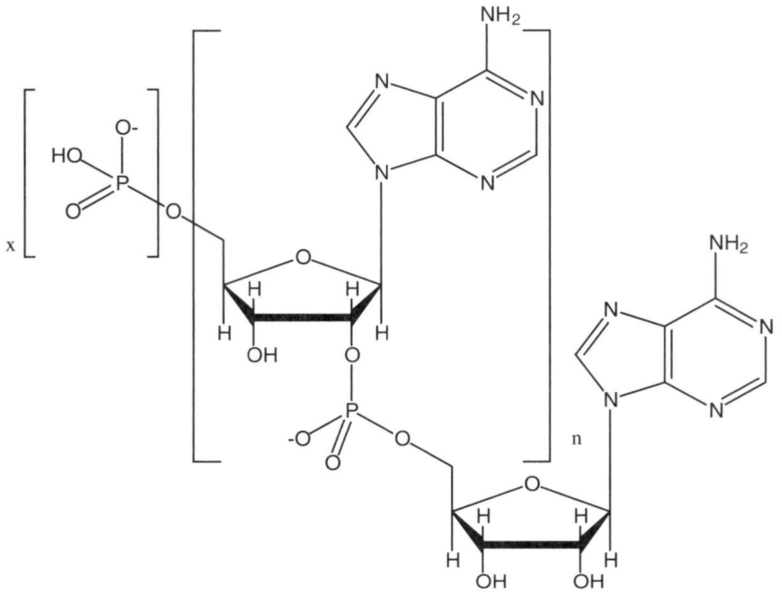

FIGURE 2.1 General structure of the 2',5'-oligoadenylates, the activators of RNase L. The oligomers contain from one to three phosphates in 5' (x = 1 to 3) and several adenylate repeats (n ≥ 1).

of RNase L is further regulated by a natural inhibitor (RNase L inhibitor, RLI).[3] In chronic fatigue syndrome (CFS), the 2-5A pathway is dysregulated and peripheral blood mononuclear cell (PBMC) extracts are characterized by an upregulated RNase L activity, a down-regulated RLI, and the presence of low molecular weight forms of RNase L, including a 37-kDa 2-5A-binding protein, (37-kDa RNase L).[4-6] Before entering into the details of the biological significance of these abnormalities, it is important to understand the implications of the protein structure and characteristics within its normal cellular activation, function, origin, and distribution, respectively.

2.2 CHARACTERISTICS OF THE PRIMARY STRUCTURE OF RNase L

Human RNase L is a polypeptide comprised of 741 aminoacids.[7] The physical characteristics of the protein are summarized in Table 2.1. The data provided concern the monomeric enzyme of 83-kDa, which homodimerizes upon activation by 2-5A.[2] The protein is composed by close to one-third (28.7%) of charged aminoacid residues, of which 113 are negatively and 100 positively charged, respectively. A further statistical analysis of the protein sequence[8] reveals neither specific charge clusters nor significant charge patterns. The hydrophobicity profile of the protein displayed in Figure 2.2 reveals only a small significant hydrophobic segment (AA 95-113) which could be indicative of the presence of a transmembrane region (spanning helix). However, such observation is not necessarily conclusive[9] and the protein is

TABLE 2.1
Physical Characteristics of Monomeric Human RNase L

Number of amino acids	741
Molecular weight	83,532.8 Da
pI	6.2
Extinction coefficient ($\varepsilon_{280\ nm}$)	63,290 ± 594 mol^{-1} cm^{-1}
Number of negatively charged residues (Asp + Glu)	113
Number of positively charged residues (Arg + Lys)	100

Amino acid composition:	**n**	**%**
Ala (A)	50	6.7
Arg (R)	38	5.1
Asn (N)	31	4.2
Asp (D)	51	6.9
Cys (C)	14	1.9
Gln (Q)	27	3.6
Glu (E)	62	8.4
Gly (G)	52	7.0
His (H)	31	4.2
Ile (I)	29	3.9
Leu (L)	85	11.5
Lys (K)	62	8.4
Met (M)	11	1.5
Phe (F)	28	3.8
Pro (P)	24	3.2
Ser (S)	45	6.1
Thr (T)	29	3.9
Trp (W)	7	0.9
Tyr (Y)	18	2.4
Val (V)	47	6.3

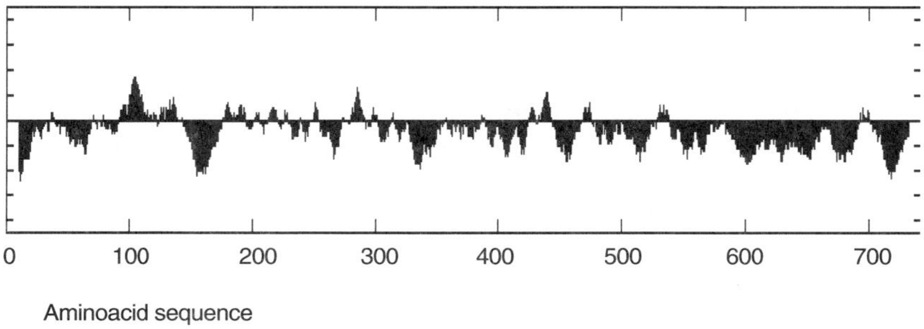

FIGURE 2.2 Hydrophobicity (hydropathy) profile of human RNase L.

TABLE 2.2
N-glycosylation, N-myristoylation and Amidation Sites in the Human RNase L Sequence

	Motif	AA Sequence
N-glycosylation	NDTD	267–270
	NESD	601–604
N-myristoylation	GANVNF	48–53
	GADVNE	115–120
	GANVNL	148–153
	GGATAL	167–172
	GADVNA	192–197
	GADVNV	229–234
	GIYLGF	378–383
	GSPRAQ	397–402
	GAGGAS	729–734
	GGASGL	731–736
Amidation	SGRR	17–20

most probably a soluble protein. The protein sequence contains also a four-amino-acids KHKK pattern (AA 676-679) indicative of nuclear targeting. The protein has been identified not only in cytosolic cell fractions, more precisely in polysomes,[10] but also in the nucleus.[11] The RNase L sequence contains two N-glycosylation sites, 10 N-myristoylation sites and one amidation site, which are identified in Table 2.2. The protein sequence also contains several possible phosphorylation sites, summarized in Table 2.3.

The analysis of the aminoacid sequence of RNase L allows us to identify several features, some of which share homologies with known protein families. These are illustrated and summarized in the diagram shown in Figure 2.3. First, starting from the N-terminal end, the sequence of the first 330 aminoacids presents nine sequence repeats sharing a high degree of homology with the ankyrin repeat motif. This homology is further enlightened by the comparative alignment with the ankyrin consensus sequence shown in Figure 2.4. The identities of these repeats with the ankyrin consensus range from 53 to 84%. Ankyrins constitute a family of proteins containing a series of specific repeat motifs of about 33 aminoacids that control interactions between integral membrane components and cytoskeletal elements. Ankyrin proteins have been isolated from erythrocytes and brain tissue, but there are hints that ankyrin exist in many other and perhaps all cells.[12] The integral membrane proteins known to associate *in vitro* with ankyrins are the anion–exchanger of erythrocytes, the Na^+/K^+-ATPase of kidney distal tubules, the voltage-dependent Na^+ channel at Ranvier's nodes and the cluster determinant (CD) 44 of lymphocytes.[13] The ankyrins link these integral membrane proteins to the cytoskeleton so as to possibly organize their presence in specialized cell areas such as basolateral domains of epithelial tissues, caps of lymphocytes, and specific

TABLE 2.3
Possible Phosphorylation Sites Present in the Amino Acid Sequence of RNase L

Kinase	Motif	RNase L AA Sequence
c-AMP- and c-GMP-dependent protein kinase	RRKT	154-157
	KKGS	533-536
Protein kinase C	SGR	17-19
	SVK	104-106
	TAR	307-309
	SPR	398-400
	SSR	410-412
	SHR	425-427
	SKK	495-497
	SIK	507-509
	TLR	595-597
	TRK	607-609
	TTK	633-635
Tyrosine kinase	KFFIDEKY	362-369
	KIGDPSLY	684-691
	KTFPDLVIY	694-702
Caseine kinase II	SRED	70-73
	TKED	157-160
	SSDD	212-215
	SDVE	216-219
	TDSD	269-272
	TFCE	393-396
	SSRE	410-413
	TLCE	435-438
	THQD	482-485
	SFED	538-541
	SNEE	546-549
	SPDE	554-557
	SESE	610-613
	TVGD	658-661
	TFPD	695-698

domains of neurons.[13,14] Besides its ankyrin homology, this aminoterminal region of the protein also contains several tandem and periodic aminoacid repeats (gray-boxed in Figure 2.4): GANVN (AA 48-52 and 148-152), GADVN (AA 115-119, 192-196 and 229-233), and GADVNXCD (AA 115-122 and 192-199) respectively, which are parts of the possible myristoylation sites (Table 2.2.). The ankyrin repeat region of RNase L also contains the binding site for 2-5A which the protein binds with very high affinity (K_A 10^{10}–10^{11} mol^{-1}).[15] This part of the amino-terminal region of RNase L contains two conserved phosphate-binding loop motifs (P-loops) between residues 229-241 (GADVNVRGER<u>GKT</u>) and 253-275 (GLVQRLLEQEHIE-INDTDSD<u>GKT</u>), respectively.

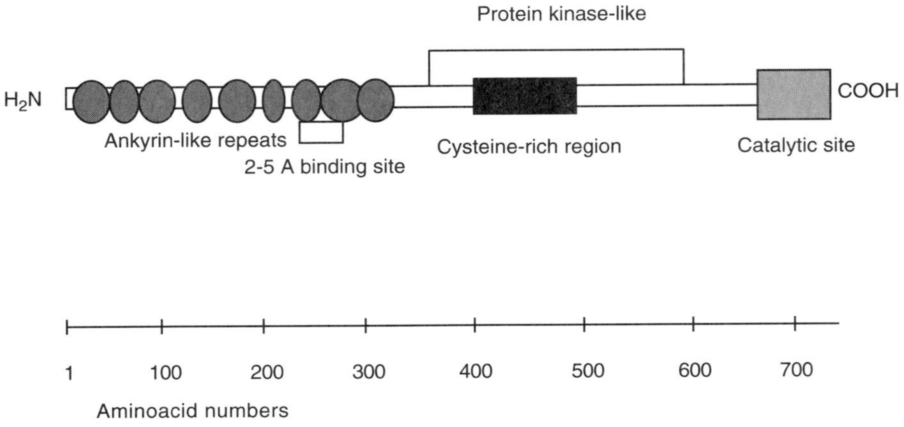

FIGURE 2.3 Relevant features of human RNase L.

Ankyrin-repeat consensus:

```
XGXTP-------------  LHXAAXXXX---XXX  VXXLLXX-GAXXXXXX DX
XGXTP-------------  AHXAAXXXX---XXX  VXXLLXX-GAXXXXXX NX
```

RNase L sequence:

```
11   EGPTSSSGRRAAVEDNHL  LIKAVQNED---VDLVQQLLEG-GANVNFQEEE
58   GGWTP-------------  LHNAVQMSR---EDIVELLRH-GADPVLRKK-
91   NGATP-------------  FILAAIAGS---VKLLKLFLSK-GADVNECDF-
124  YGFTA-------------  FMEAAVYGK---VKALKFLYKR-GANVNLRRKT
167  GGATA-------------  LMDAAEKGH---VEVLKILLDEMGADVNACDN-
201  MGRNA-------------  LIHALLSSDDSDVEAITHLLDHGADVNVRGER
238  RGKTP-------------  LILAVEKKH---LGLVQRLLEQEHIEINDTDSD
272  DGKTA-------------  LLLAVELKL----KKIAELLCKRGASTDCGDL-
301  CGDLV-------------  MTARRNYD---HSLVKVLLSH-GAKEDFHPPA
```

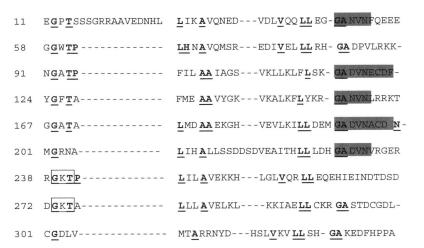

FIGURE 2.4 Comparative sequence alignment (ClustalW) between the ankyrin consensus sequence and the nine NH_2-terminal repeats of RNase L. Identical amino acids are bolded and underlined. The two P-loops are boxed and other periodic repeats are gray-boxed.

```
          ....|....| ....|....| ....|....| ....|....| ....|....| ....|....|
              5         15         25         35         45         55
RNase L   VSCLQSSREN SHLVTFYGSE SHRGHLFVCV TLCEQTLEAC LDVHRGEDVE NEEDEFARNV
hGR       KLCLVCSDE- -------ASG CHYG-VLTCG S-CKVFFKRA VEGQ--HNYL CAGR--NDCI
hMR       KICLVCSDE- -------ASG CHYG-VVTCG S-CKVFFKRA VEGQ--HNYL CAGR--NDCI
hPr       KICLICGDE- -------ASG CHYGVVLTCG S-CKVFFKRA MEGQ--HNYL CAGR--NDCI
hAR       LTCLICGDE- -------ASG CHYG-ALTCG S-CKVFFKRA AEGK--QKYL CASR--NDCT
hER       TRCAVCNDY- -------ASG YHYG-VWSCE G-CKAFFKRS IQGH--NDYM CPAT--NQCT
hTR       EQCVVCGDK- -------ATG YHYR-CITCE G-CKSFFRRT IQKNLHPSYS CTYD--GCCV
VDR       RICGVCGDR- -------ATG FHFN-AMTCE G-CKGFFRRS MQKR-KAMFT CPFD--GCCK
RAR       --CFVCQDK- -------STG YHYG-VSACE G-CKGFFRRS VQKN--MVYT CHRD--KNCI
erbA      ELCVVCGDK- -------ATG YHYR-CATCE G-CKSFFRRT IQKNLHPSYS CTYD--GCCV
                    Finger 1                P-Box Intermediate region    D-Box
          ....|....| ....|....| ....|....|
              65         75         85
RNase L   LSSIFKAVQE LHLSCGYTHQ DLQPQNILID
hGR       ID---KIRRK NCPACRY--R KCLQAGMNL-
hMR       ID---KIRRK NCPACRL--Q KCLQAGMNL-
hPr       VD---KIRRK NCPACRL--R KCCQAGMVL-
hAR       ID---KFRRK NCPSCRL--R KCYEAGMTL-
hER       ID---KNRRK NCPACRY--R KCYEVGMMK-
hTR       ID---KITRN QCQLCRF--K KCISVGMAM-
VDR       IT---KDNRR HCEACRL--K RCVDIGMMK-
RAR       IN---KVTRN RCQYCRL--Q KCFEVGM---
erbA      ID---KITRN QCQLCRF--K KCISVGMAM-
          D-Box  Finger 2
```

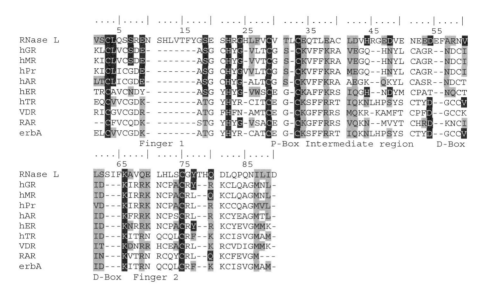

FIGURE 2.5 Comparative sequence alignment (ClustalW) between the cysteine-rich region of RNase L and the finger-like domains of human glucocorticoid (hGR), mineralocorticoid (hMR), progesterone (hPr), androgen (hAR), estrogen (hER), thyroid (hTR), vitamin D (VDR), retinoic acid (RAR) receptors, and the *v-erbA* oncogene (erbA). Identical amino acids are in a black background and conservative replacements are gray-boxed.

Farther along the sequence, the protein contains a protein kinase-like domain (AA 365-586) which includes a cysteine-rich segment (AA 395-444) reminding us of the finger-like sequence of the steroid and thyroid receptor superfamily. These finger-like domains interact with hormone-responsive elements on DNA and are constrained by two Zn coordination centers made of four cysteine residues separated by an amphipathic α-helix. The fingers are made of antiparallel β-sheets which help orient the residues that contact the phosphate backbone of DNA (P-box). The second finger module is more important (D-box) for phosphate contacts and receptor dimerization.[16] By analyzing the sequence homology of this region of RNase L with the finger-like sequence of steroid receptors (Figure 2.5), it is already apparent that the homology is quite high, particularly in the first finger sequence. However, one of the receptor cys residues (cys-6 of Figure 2.5) required for the Zn coordination center is replaced by a serine in RNase L. The same conclusion can be drawn for the second Zn coordination center where the first two cys residues (AA 51 and 59, Figure 2.5) are replaced by two asparagines in RNase L. Most striking, however, is the high homology within the D-box of most steroid receptors. Since this region of the receptors is central to dimerization contacts, we can suspect by analogy that this region of the sequence of RNase L is involved in the dimerization process required for activation of the catalytic activity of the enzyme. This suggestion is supported by the fact that the conserved lysine residue 392, situated three aminoacids upstream of the first cys residue of the cys-rich domain, has been demonstrated to play a critical role in the homodimerization of the protein.[17]

```
                 ....|....| ....|....| ....|....| ....|....| ....|....| ....|....|
                     5          15         25         35         45         55
RNase L   531    VVKKGSISFE DLKAQSN--- EEVVQLSPDE E-------TK DLIHRLFHPG EHVRDCLSDL
Ire1β     716    VLSGGSHPFG DSLYRQANIL TGVPCLAHLE EEVHDKVVAR DLVAAMLSLL PQARPSAPQV

                 ....|....| ....|....| ....|....| ....|....| ....|....| ....|....|
                     65         75         85         95         105        115
RNase L          LGHPFFWTWE SRYRTLRNVG NESDIKTRKS ESE-ILRLLQ PGPSEHSKSF DKWTTKINEC
Ire1β            LAHPFFWSRA KQLQFFQDVS ---DWLEKES EQEPLMRALE AGG--CTVVR DNWHEHISMP

                 ....|....| ....|....| ....|....| ....|....| ....|....| ....|....|
                    125        135        145        155        165        175
RNase L          VMKKMNKFYE KRGNFYQNTV GDLLKFIRNL GEHIDEEKHK KMKLKIGDPS ---LYFQKTF
Ire1β            LQIDLRKFRS YKG----TSV RDLLRAVRNK KHHYRELPVE VRQALGQVPD GFVQYFTNRF

                 ....|....| ....|....| ....|....| ....|....| ....|
                    185        195        205        215        225
RNase L          PDLVIYVYTK LQNTEYRKHF PQTHSPNKPQ CDGAGGASGL ASPGC
Ire1β            PQLLLHTHRV MRSCASESLF LPYYPPDSEA RGPCPGAAGR -----
```

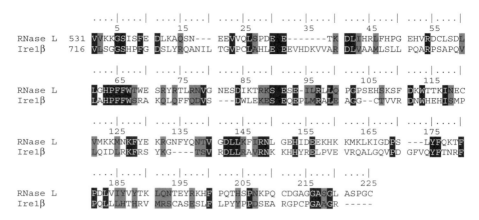

FIGURE 2.6 Comparative sequence alignment (ClustalW) of the C-terminal domains of human RNase L and human Ire1β. Identical amino acids are in a black background and conservative replacements are gray-boxed.

Finally, the carboxy-terminal part of the protein includes the enzymatic catalytic site.[18] However, this C-terminal part of RNase L does not share much relatedness with any sequence signature of the ribonucleases superfamily.[15] The only significant resemblance between RNase L and the superfamily is the fact that the catalytically active proteins are dimers.[19] A sequence similarity search indicates that human RNase L is highly homologous to human Ire1β, a protein which shares the protein kinase-like structure of RNase L, as well as its endoribonuclease activity.[20] The homology is particularly striking in the C-terminal part of both human proteins as shown in Figure 2.6. Ire1 is a protein conserved in eukaryotic cells from the budding yeast *Saccharomyces cerevisiae*.[21] This protein is an endoplasmic reticulum (ER) resident kinase which acts as a sensor of ER stress. Ire1 transduces a lumenal signal imparted by the presence of incorrectly folded proteins (the unfolded protein response, UPR) in the ER to a nuclear event that results in the increased transcription of genes which encode ER resident proteins promoting the folding of newly synthesized peptides in the ER lumen.[22] When unfolded proteins accumulate in the ER, Ire1, which is both a kinase and an endonuclease,[20] probably oligomerizes, self-phosphorylates, and cuts an intron out of the precursor mRNA for the transcription factor of the folding protein genes.[23] A futher study of the specificity of the endonucleolytic activity of these proteins will probably shed some light on the active site of RNase L.

2.3 PHYLOGENETIC ORIGIN OF RNase L

The story starts with the striking homology between human RNase L and Ire1β, and continues with other players of the ER stress response. Ire1β pertains to the UPR arm of the cell response to ER stress. A second component of this response consists of a profound and rapid repression of protein synthesis.[24] Both responses are aimed at relieving the ER stress by increasing the capacity of the ER to actively fold proteins and to decrease the demand made on the organelle by decreasing protein synthesis.

The second component of the ER-stress response is mediated by the phosphorylation of the eukaryotic translation initiation factor eIF2α. Besides the heme-regulated eIF2α kinase (HRI)[25] and the ds-RNA-dependent protein kinase (PKR),[26] the two known kinases which phosphorylate eIF2α in mammalian cells, two new related kinases linked to the ER have been recently identified. These have been respectively named PERK (for PKR-like ER kinase)[27] and PEK (for pancreatic eIF2α kinase).[28] These proteins share important homologies not only with PKR, but also with Ire1β.[27] It has been suggested[23] that HRI, PERK/PEK, PKR, Ire1, and RNase L were proteins evolved from common ancestors. In order to verify this possibility, we aligned the sequences of PERK, PKR, RNase L, and Ire1β. The results are shown in Figure 2.7. All these four-protein sequences contain important homologies, which are particularly significant between PERK and PKR on the one hand, and Ire1β and RNase L on the other hand. Overall, however, the four proteins share consensus segments which in some regions of the sequences represent more than 20% similarity.

Besides their sequence homologies, all these proteins share a functional link: they are involved in mediating apoptosis or programmed cell death. Ire1β and RNase L induce the process by degrading mRNA,[23,29] while PKR and PERK act at decreasing the initiation of translation by phosphorylating eIF-2α.[27,30] In view of these singular similarities, we attempted the computation of a distance matrix from the four protein sequences and obtained the tree shown in Figure 2.8. According to this tree, the four proteins would originate from a common ancestor gene that evolved into two daughters (1 and 2 in Figure 2.8), each having respectively evolved into either RNase L and Ire1β, or PERK and PKR. As diagrammatically shown in Figure 2.8, the distance between 2 and RNase L/Ire1β is null, while the distance between 1 and PERK/PKR is more significant. These results are consistent with the suggestion[23] that the daughters 1 and 2 referred to above are respectively the yeast proteins GCN2 (the yeast eIF2α kinase[31]) and Ire1.

2.4 MOLECULAR STRUCTURE AND ACTIVATION OF RNase L BY 2-5A

Viral infection and replication within eukaryotic cells results in the production of ds-RNA, which in turn stimulates the production of type I interferons (IFNs α and β). IFNs are a class of proteins and glycoproteins which interact with membrane receptors and trigger a wide range of biological effects including inhibition of virus replication, changes in cell membrane, inhibition of cell growth, and modification of the immune response.[32] Upon binding to their receptors, IFNs induce signaling cascades which start by the catalytic activation of the receptor-associated Janus kinases (JAKs), which in turn phosphorylate the cytoplasmic region of the receptor. This effects the recruitment of SH2-containing signal transducers and activators of transcription (STATs) on the membrane protein phosphorylated motifs, and the further phosphorylation of these transcription factors.[33] Phosphorylation activates STATs, which translocate into the nucleus where they induce the transcription of target genes.[34] These genes include those encoding the 2-5A synthetase (2-5OAS).[35]

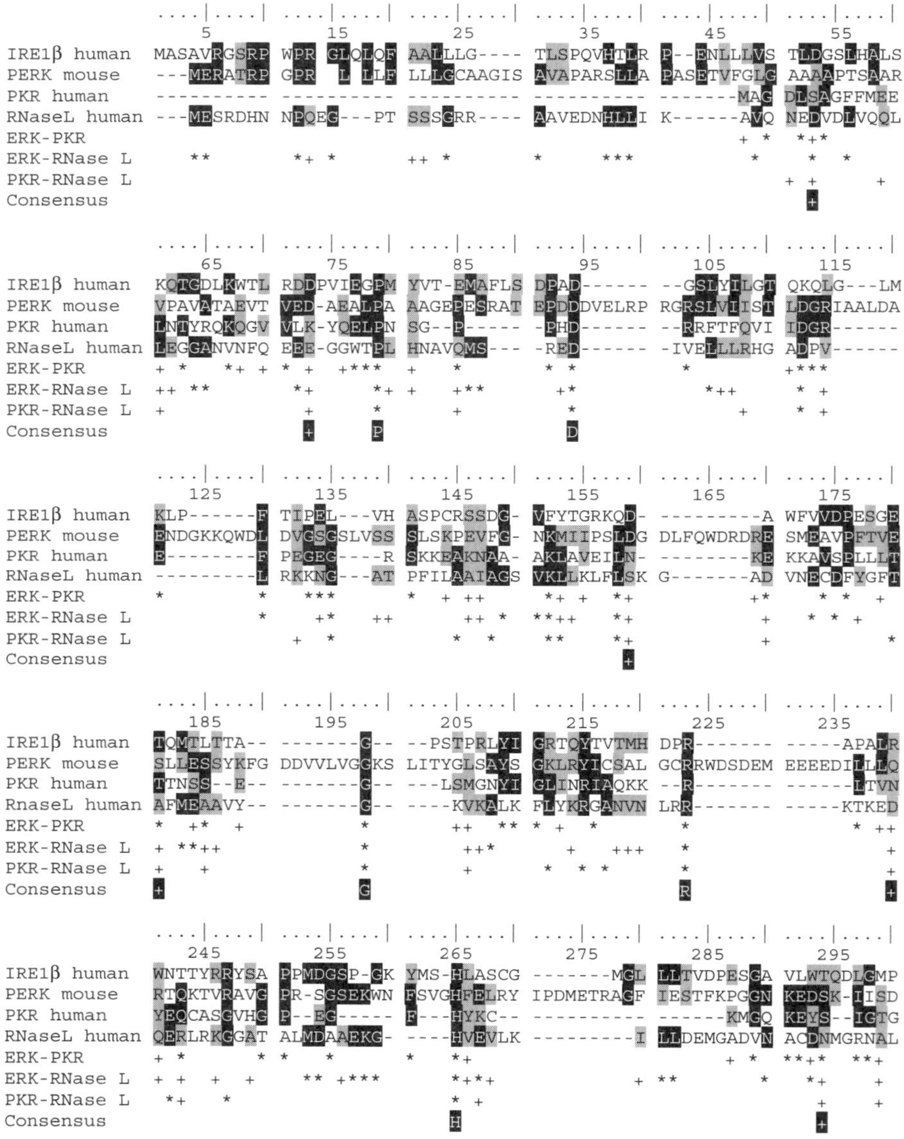

FIGURE 2.7 Comparative sequence alignment (ClustalW) of human RNase L, Ire1β, PKR, and mouse PERK, respectively. Identical amino acids are in a black background and conservative replacements are gray-boxed. Identities and conservations between the ER kinases (ERK), PKR, and RNase L are respectively indicated below the sequences (* and + signs). The consensus among all four sequences is in a black background.

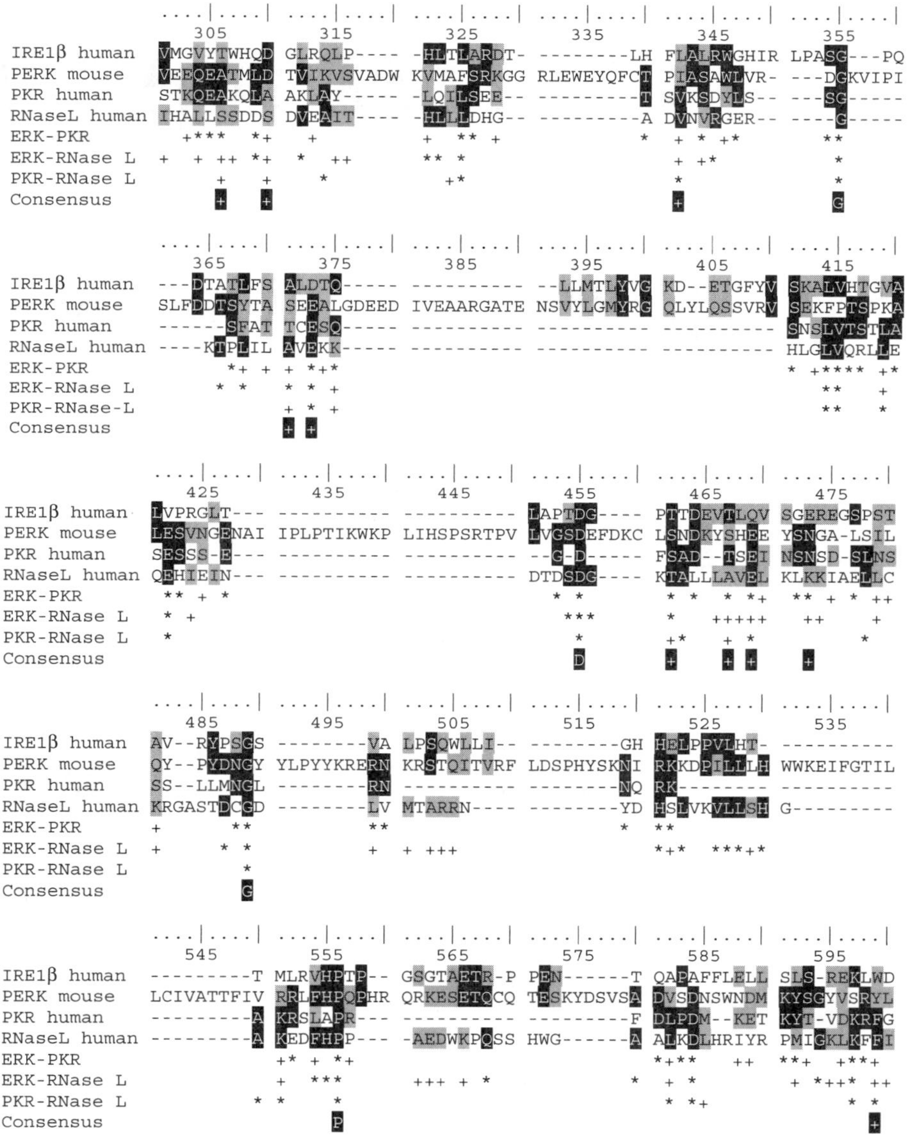

FIGURE 2.7 *Continued.*

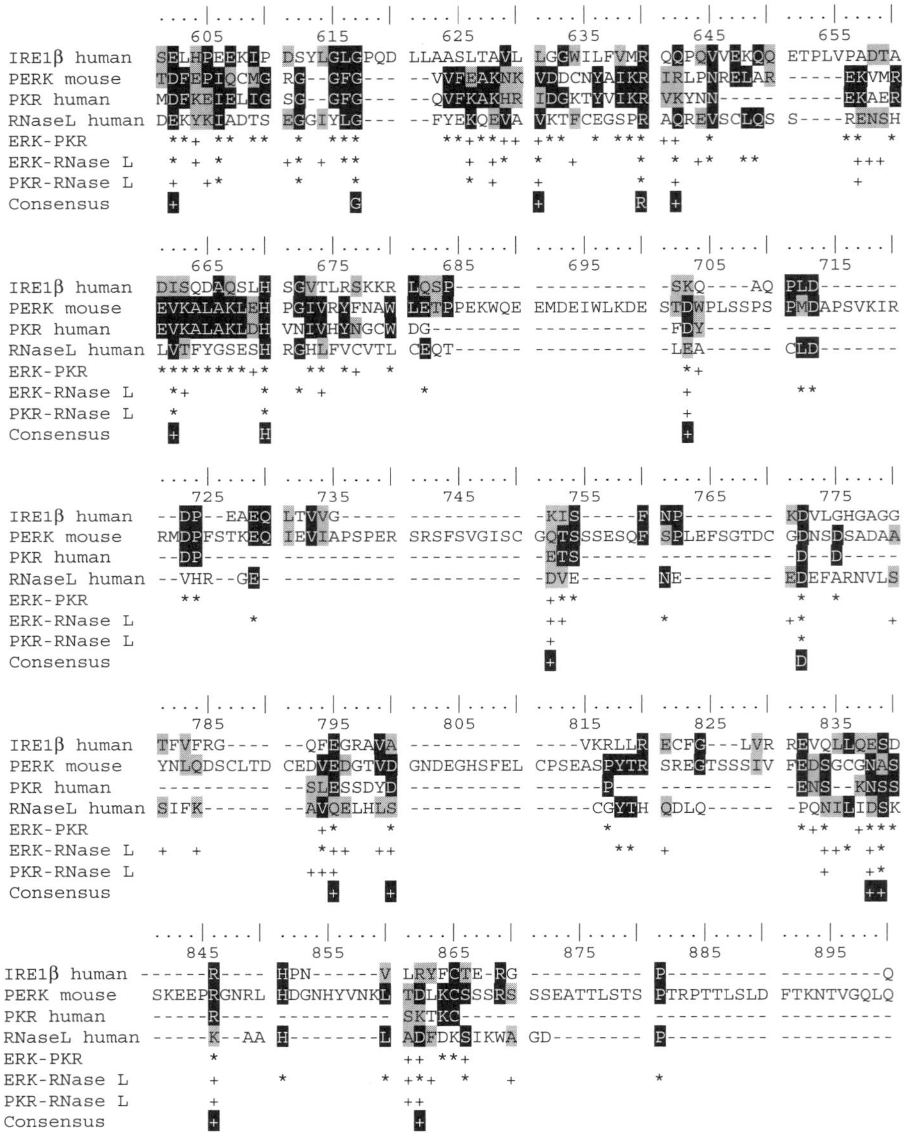

FIGURE 2.7 *Continued.*

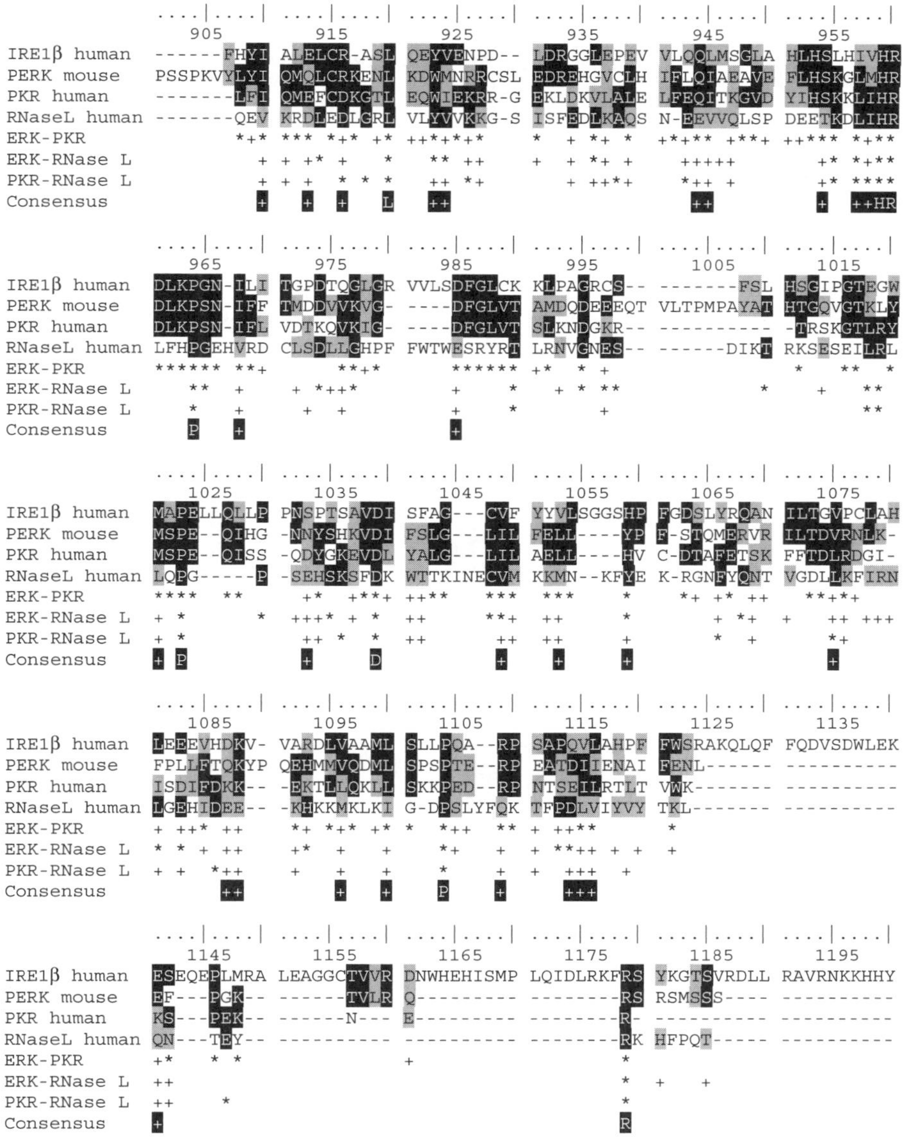

FIGURE 2.7 *Continued.*

```
              ....|....| ....|....| ....|....| ....|....| ....|....| ....|....|
                   1205       1215       1225       1235       1245       1255
IRE1β human   RELPVEVRQA LGQVPDGFVQ YFTNRFPQLL LHTHRVMRSC ASESLFLPYY PPDSEARGPC
PERK mouse    ---------- ---------- --------GT KHSR--QPSC S--------- ------YSPL
PKR human     ---------- ---------- ---------- -HT------C ---------- ----------
RNaseL human  ---------- ---------- ---------- -HSPN-KPQC DGAG------ ---GASGLAS
ERK-PKR                                         **      *
ERK-RNase L                                     **   +* *                 +
PKR-RNase L                                     *+      *
Consensus                                       H+      C
```

```
              ....|.
                   1265
IRE1β human   PGAAGR
PERK mouse    PGN---
PKR human     ------
RNaseL human  PGC---
ERK-PKR
ERK-RNase L   **
PKR-RNase L
Consensus
```

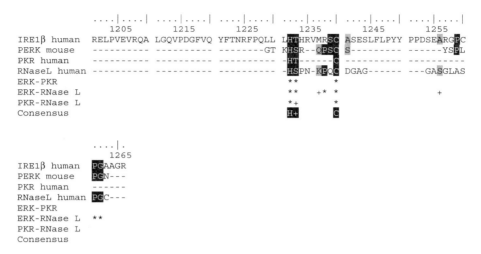

FIGURE 2.7 *Continued.*

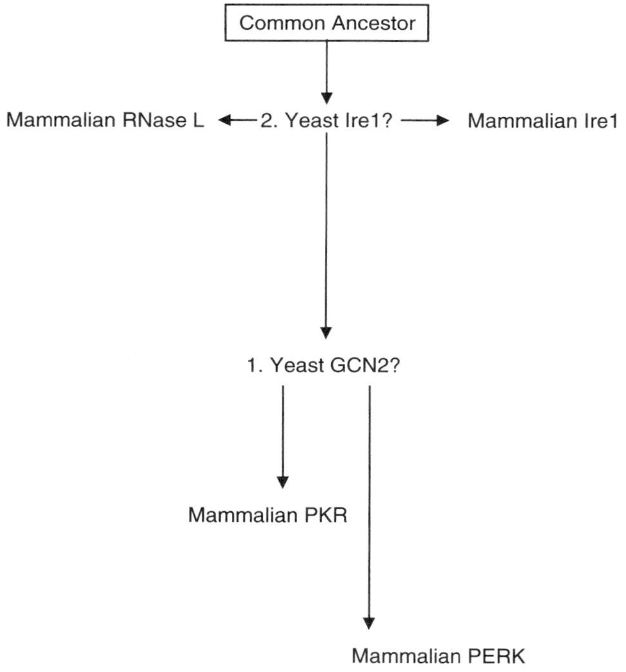

FIGURE 2.8 Phylogenetic tree computed by distance matrix from the amino acid sequence homologies shown in Figure 2.7.

The enzyme synthesizes 2-5A oligomers of the general formula shown in Figure 2.1 by polymerizing ATP. The 2-5OAS is activated not only by ds-RNA, but also by single-stranded RNA (ss-RNA).[36] The basic human enzyme unit is a 400 amino acid-long protein. This enzyme unit (p40) pertains to a larger family of 2-5A synthetases. Three different isoforms of the enzyme exist, respectively small (p40), medium (p69), and large (p100), which contain: one, two, and three basic units and are each encoded by distinct genes clustered on chromosome 12.[37] Activation of the enzyme induces its oligomerization and tetramers of the small isoenzymes are required for enzymatic activity,[38] while dimers of the larger forms are active.[39] Among these isoforms, p69 and p100 display different catalytic activities, p69 synthesizing higher 2-5A oligomers whereas p100 synthesizes preferentially 2-5A dimers (n = 1; see Figure 2.1).[40] The medium size 2-5OAS form (p69) has a minimal size requirement of 25bp ds-RNA for proper activation. Indeed, lower size ds-RNA have been shown to induce the enzyme to produce dimeric instead of trimeric or higher length 2-5As.[41] The enzyme is associated with various subcellular fractions including mitochondria, nucleus, and rough and smooth microsomes.

The fact that some isoforms are likely to manifest differential catalytic activities favors the hypothesis that these enzymes might have other functions in the cell besides those in the 2-5A system.[40] This is further supported by the fact that a 56-kDa 2-5A synthetase-like protein has been identified. The gene encoding for this protein maps to the same chromosome as 2-5A synthetase (12q24.22) in the proximity of the 2-5A synthetase locus.[42] This protein, like 2-5A synthetase, is induced by interferons, and its amino acid sequence shares 41% identity with p40 in the first 350 out of the 514 amino acid residues. This 56-kDa protein has been identified as the thyroid receptor interacting protein 14 (TRIP14),[43] which interacts with the ligand-binding domain of the thyroid receptor, but does not require the presence of thyroid hormones for its interaction (Chapter 5). TRIP14 does not have 2-5A synthetase catalytic activity and is expressed in most tissues, with the highest levels in primary blood leukocytes and other hematopoietic system tissues.[44] This element further exemplifies the complexity of the possible interactions in the 2-5A pathway covered in Chapter 1 of this volume.

As mentioned earlier, RNase L requires 2-5A binding for homodimerization and activation.[2] For a better understanding of the 2-5A binding and activation process of the enzyme, we have constructed a three-dimensional (3D) model of the protein[45] which is presented in Figure 2.9A.* The model is colored according to the protein domains presented in Figure 2.3, namely ankyrin (red), protein kinase-like (blue), cysteine-rich (blue-green), and catalytic (green). The 2-5A-binding site location has been assigned within ankyrin repeats 6 to 9 (AA 212-339).[46] This region contains the two conserved phosphate-binding loop motifs situated respectively between residues 229-241 and 253-275 (yellow in Figure 2.9.A and the two conserved GKT residue triads of the P-loops are in magenta). The ankyrin domain and the 2-5A-binding site are likely to act as key modulators of the endonuclease activity of the protein, because a truncated protein containing only the protein kinase and catalytic domains (AA 340-741) produces a fully active enzyme, irre-

---

* Color insert figures follow page 76.

A

B

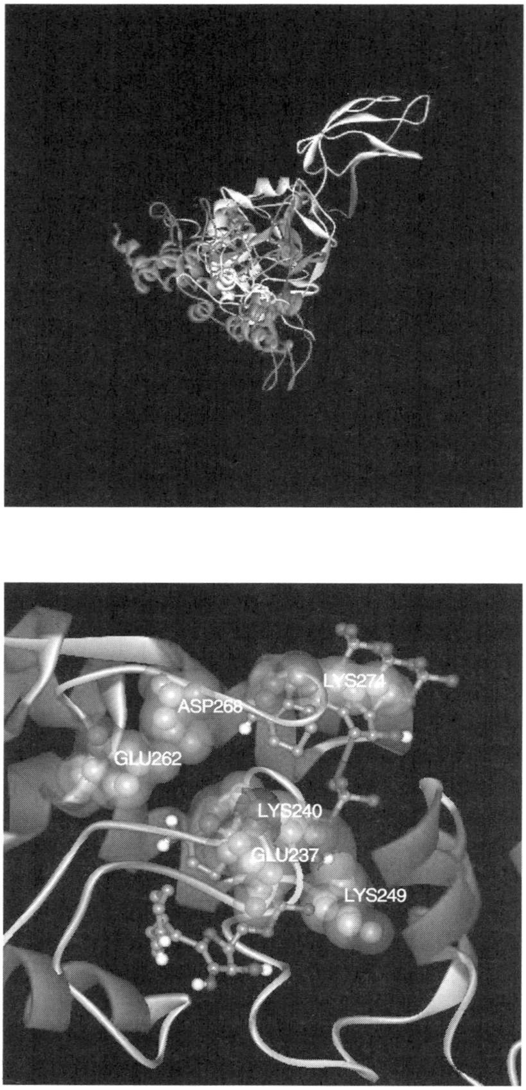

FIGURE 2.9 Part A. 3D-model of human RNase L. The protein is presented in ribbon colored according to the parts presented in Figure 2.3. The ankyrin domain is in red and includes the 2-5A binding site (yellow) and the two P-loops (GKT, magenta). The protein kinase-like domain is in blue and includes the cysteine-rich region (blue-green). The catalytic domain is in green. Part B. Structure of the 2-5A-binding site filled with the 2-5A trimer (ball and stick). The interacting residues are numbered and represented by their surface atoms. (See color figures 2.9A and 2.9B.)

sponsive to excess 2-5A.[46] Binding of 2-5A to the ankyrin domain of RNase L probably releases an internal clamp imposed on the catalytic domain by intrasteric regulation,[47] which permits dimerization and activation, respectively.

The clamping effect of the ankyrin domain on the catalytic domain is clearly sketched on the model presented in Figure 2.9.A. The two P-loops of the 2-5A-binding site of RNase L contain two conserved lysine residues (K240 and K274) which interact with the phosphate groups of the oligoadenylates.[15] The region is also particularly rich in aspartic and glutamic acid residues that could interact with either the N^1-nitrogen or exocyclic amino groups of adenine, which have been shown to be critical for 2-5A binding to the endonuclease.[48] Noteworthy also are the two lysine residues (K249 and K250) present between the P-loops and which could interact with the phosphodiester linkages of the oligoadenylates. We further identified these residues on our 3D-model and tethered those capable of interacting with a 5'-triphosphorylated 2-5A trimer (x = 3 and n = 2; Figure 2.1) previously docked in the P-loops of the enzyme. The result is shown in Figure 2.9B.* This model of the 2-5A binding site shown in Figure 2.9.B is further consistent with the facts that the 5' phosphates (at least one) are required for enzymatic activation,[49,50] and that an oligomer chain longer than the trimer is not an important determinant of the activating capacity.[50] Indeed, as can be clearly seen in Figure 2.9, a chain longer than the trimer would position the extra adenylyl groups out of the protein binding pocket. However, the 5'-triphosporylated 2-5A dimer is most probably able to bind to the monomeric enzyme, but unable to induce the homodimerization and the activation of the enzyme.[2]

Tethering the oligoadenylate trimer in the binding site of the 3D-model of RNase L produces a tremendous change in the conformation of the protein, which not only releases the ankyrin clamp on the catalytic site, but makes the cysteine-rich region used for protein–protein interactions during dimerization[17] to protrude out of the bulk protein structure. This further allows the interaction between the 2-5A-modified 83-kDa enzyme species in order to form the homodimer, the structure of which (stereo view) is shown in Figure 2.10.* The two catalytic carboxy-terminal ends of both monomers are clearly detached in the upper-left part of the model. This model of association between two 2-5A-liganded monomers of the enzyme to form the homodimer is in complete agreement with the stoechiometric and biophysical characterization of RNase L activation.[51,52] The model of interaction presented in Figure 2.9.B is also consistent with the fact that the 2-5A dimer is unable to induce the homodimerization of RNase L because the N^1 group of the last adenylyl residue of the trimer interacts with the first P-loop of the protein sequence (E237) and the last 2',5' phosphate linkage interacts with K249, a residue intermediate of the P-loops. These residues are respectively situated on the loop between ankyrin repeats 6 and 7 and on the helix of ankyrin repeat 7 (Figures 2.4 and 2.9). This interaction could consequently exert the leverage required for inducing the change in protein conformation observed with our simulations.

---

* Color insert figures follow page 76.

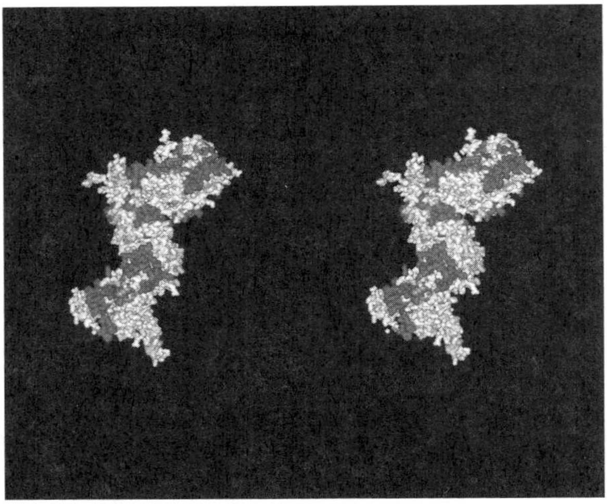

FIGURE 2.10 Stereo view of the 3D-model of RNase L homodimer. The proteins are shown as surfaces colored according to the secondary structure. The atoms of the 2-5A trimers docked in the respective binding sites are presented as balls. (See color figure 2.10.)

2.5 CATALYTIC ACTIVITY OF RNase L

The activated enzyme is capable of cleaving synthetic oligoribonucleotides with poly(U) and to a lesser extent poly(A) sequences,[53] providing fragments ranging from between 4 and 22 nucleotides in length. The enzyme is, however, unable to cleave poly(C), but capable of cleaving dyads of the form UU, UA, AU, AA, and UG inserted between poly(C).[54] The enzyme has a strong cleavage preference on the 3' side of the dimers, leaving cleavage products with a 3'-phosphoryl and 5'-hydroxyl groups.[54] RNase L cleaves neither oligoribonucleotides with poly(G) sequences nor oligodeoxyribonucleotides with poly(A) or poly(T) sequences.[53] The general specificity of the enzyme can consequently be attributed to -UpNp- sites with a preference for UU and UA. Manganese, magnesium ions, and ATP[53] enhance the nuclease activity of RNase L, but ATP does not bind to the enzyme.[15]

It has been suggested that the specificity of RNase L for its substrate could be conveyed by the structure of the 2-5A ligand.[15] This suggestion is supported by the use of antisense-2-5A chimeras allowed to exploit the catalytic activity of RNase L for the directed degradation of a targeted mRNA sequence. The antisense portion of the chimera addresses the specific nucleotide sequence of the targeted RNA, and the 2-5A portion attracts and activates the enzyme which degrades the selected RNA.[55,56]

2.6 THE RNS4 GENE, LOCALIZATION, AND BIOLOGICAL ROLES OF RNase L

The gene coding for RNase L has been localized in human chromosome 1q25.[57] Although no direct evidence has been provided linking the gene defects with any known disease, several abnormalities colocalized with the RNS4 gene have been identified in human chromosome 1q. Of particular interest are the abnormalities in this chromosome frequently observed in human cancers.[57] These observations are further likely to support the link between RNase L and active apoptotic cellular activity.[29,58]

RNase L is present in the cells of many different organs in mammals, including man. The protein has been identified in liver, kidney, lung, spleen, intestine, brain, testis, thymus, heart, lymphocytes, and reticulocytes.[15] The system of ss-RNA cleavage by activated RNase L has been claimed to be specific for the RNA of viral origin.[59] Some data were indeed likely to support the limited synthesis of 2-5A and activation of the endonuclease at the site of viral RNA synthesis. However, other data indicate that the 2-5A synthase is likely to be present in relatively high levels in cells not stimulated by IFNs such as rabbit reticulocyte lysates,[60] which suggests a wider significance of the 2-5A system. Further data support a role for the 2-5A-dependent RNase in the regression of the tubular gland of chicken oviduct upon estrogen withdrawal, by a rapid degradation of the ovalbumin mRNA[60] as well as rRNA.[61] More recent evidence[11] indicates that the monomeric endonuclease is present in the cytosol as well as the nucleus of human lymphocytic leukemia cells (CEM and Jurkat) and is likely to be inducible as a function of cell growth.

Furthermore, the nuclear fraction of the 83-kDa enzyme could not be detected by labeled 2-5A-binding unless freed of a natural ligand, suggesting its presence in an activated state. While the exact roles and functions of the enzyme in the absence of viral infection remain to be determined, several observations strongly suggest that the endonuclease is active during apoptosis.[29,58,62-64] More recently[65] a study showed direct evidence that the 2-5A pathway and RNase L were involved in mRNA regulation during muscle differentiation in mouse. During viral infection and upon cellular stimulation by type I IFNs, RNase L has also been suggested to play a dual role.[66] On the one hand, the enzyme degrades ss-RNA of viral origin as well as cellular mRNA and 18s rRNA (the effector arm of RNase L activity), which results in antiviral growth inhibition and apoptosis of the infected cells. Apoptosis induced by activated RNase L is characterized by cytochrome c release in the cytoplasm and activation of caspases 9 and 3.[67]

On the other hand, RNase L is also likely to degrade mRNAs of specific interferon stimulated genes (ISG), including ISG15 and ISG43.[66] ISG15 and ISG43 are directly induced as a primary response to interferons and encode respectively a 15-kDa ubiquitin-like protein and a 43-kDa ubiquitin-specific protease, a family of enzymes which remove ubiquitin adducts from their substrates and impair their eventual degradation by the proteasome.[68] Interestingly, these proteins seem to play a defensive role during acute infection loads, while RNase L is likely to mediate IFN action at lower viral loads. Consequently, RNase L is suspected to down-regulate part of the IFN-mediated activity (attenuator arm of RNase L activity). This further underlines the broad biological functions of RNase L in health and disease.

2.7 REGULATION OF RNase L ACTIVITY

The RNase L protein is capable of protein–protein interactions. Dimerization of the protein has been demonstrated upon activation by 2-5A.[2] Interestingly, RNase L is likely to be capable also of forming a heterodimer with a 83-kDa RNA-binding protein,[69] recently identified as a component of the translation machinery (Le Roy, Bisbal, Salehzada, Sihol, and Lebleu, personal communication). Moreover, the screening of a lambda-Zap Daudi cell c-DNA library for expression of 2-5A-binding proteins allowed to isolate a c-DNA encoding for the 68-kDa RNase L inhibitor (RLI),[3] of which part of the sequence matched a fragment of the distinct 83-kDa binding protein previously identified.[69] RLI is not induced by interferons, associates with RNase L, and inhibits its capacity to bind 2-5A. The RLI gene (RNS4I) has been localized on chromosome 4q31,[70,71] indicating that the genes of the 2-5A pathway are not organized in a cluster in the human genome. The RLI c-DNA seems to encode two mRNA of, respectively, 3.8 and 2.4kb, each coding for an active protein.[71] The amino acid sequence of RLI contains a ferredoxin-like cysteine rich domain (residues 55 to 67, $CX_2CX_2CX_3CP$), which has been suggested either to be involved in interactions between RLI and nucleic acids, or in mediating the formation of heterodimers with RNase L.[3] Alternatively, we have suggested (Reference 45, Chapter 4) that RLI could be susceptible to interact with the ankyrin domain of RNase L through a tetrapeptide sequence (AIIK, residues 166 to 169). This sequence is indeed strikingly homologous to the peptide motifs (respectively, ALLK and ALLLK) which allow the Na^+, K^+-ATPase[72] and the erythroid anion exchanger,[73] respectively, to interact with ankyrin proteins.

We have modeled the formation of the complex between the 83-kDa RNase L and RLI according to this proposal.[45] This resulted in the blockade of the 2-5A-binding site of RNase L by steric hindrance from RLI positioning, which may explain the capacity of RLI to inhibit the binding of 2-5A by RNase L in a noncompetitive manner.[3] Moreover, the molecular model shows that upon heterodimerization with RNase L, the cysteine-rich ferredoxin-like motif present in the RLI sequence lines up with the kinase homology domain of RNase L which serves to homodimerization. This ferredoxin-like motif of RLI might consequently play the role of a mock target for other 2-5A complexed RNase L monomers looking for dimerization partners.[45]

The role of RLI is the down-regulation of RNase L activity. The inhibition of RNase L activation by RLI depends on the ratio between these proteins in the cell. Since RLI is not induced by IFN, the induction of RNase L by IFN treatment modifies the ratio between the two proteins and shifts the balance toward RNase L activation.[3] Some viruses, however, are capable of inducing RLI so as to counter the effects of the 2-5A pathway. To date, two of such cases of viral counteraction of the RNase L activity through RLI have been reported with the encephalomyocarditis virus[74] and the human immunodeficiency virus type 1.[75]

2.8 RNase L IN CHRONIC FATIGUE SYNDROME

As early as the 1990s, a preliminary report[76] indicated the upregulation of RNase L activity in the PBMC of CFS patients. Later,[4,5] this upregulation was shown to result from the presence of low molecular weight variants of the enzyme. These studies pointed to the presence, besides the normal monomeric 83-kDa RNase L, of low molecular weight proteins capable of binding 2-5A, which migrated in polyacrylamide gel electrophoresis (PAGE) with apparent molecular weights of 42- and 37-kDa,

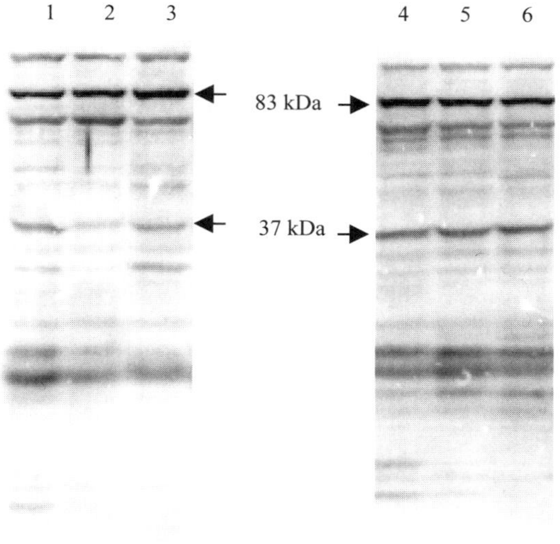

FIGURE 2.11 Detection of RNase L fragments in PBMC extracts by PAGE and immunoblotting using a polyclonal antibody raised against the recombinant enzyme. Lanes 1 to 3 are PBMC extracts from healthy controls. Lanes 4 to 6 are PBMC extracts from CFS patients. The 37-kDa fragment is recognized by the antibody, indicating its structure homology with the 83-kDa RNase L.

respectively. The 42-kDa truncated enzyme was present in the PBMC of both healthy and CFS individuals. The 37-kDa protein was detected preferentially in the PBMC of CFS patients. More recently, the presence of this 37-kDa in PBMC was validated as a biological marker for CFS.[6] Interestingly, the partially purified 37-kDa protein has retained the capacity to cleave poly(U), a characteristic currently under close scrutiny (see Chapter 3). The origin of these low molecular weight variants has remained elusive for some time and several hypotheses were proposed to account for the presence of these abnormal proteins, including proteolytic cleavage of the native 83-kDa protein and the constitutive presence of the low molecular weight variants. Recent evidence from our laboratories indicates that the low molecular weight variants are produced by proteolytic cleavage of the native 83-kDa monomer.[77,78]

The recent availability of the recombinant 83-kDa RNase L enzyme[45,77] allowed us to conduct several key experiments. First, we were able to raise polyclonal antibodies, which allow us to detect several fragments of the native enzyme in PBMC extracts by immunoblotting. This is shown in Figure 2.11. One can observe that, overall, the number and quantity of RNase L fragments detected by the antibody are much higher in the PBMC extracts of CFS patients than in the healthy controls. The 37-kDa fragment is also clearly detected by the antibody, confirming its structure homology with the 83-kDa RNase L.

Second, we digested the recombinant enzyme covalently labeled with ^{32}P-2-5A with PBMC extracts from healthy individuals and from CFS patients. As shown in Figure 2.12A, progressive incubation at 37°C during, respectively, 5, 15, and 30 min

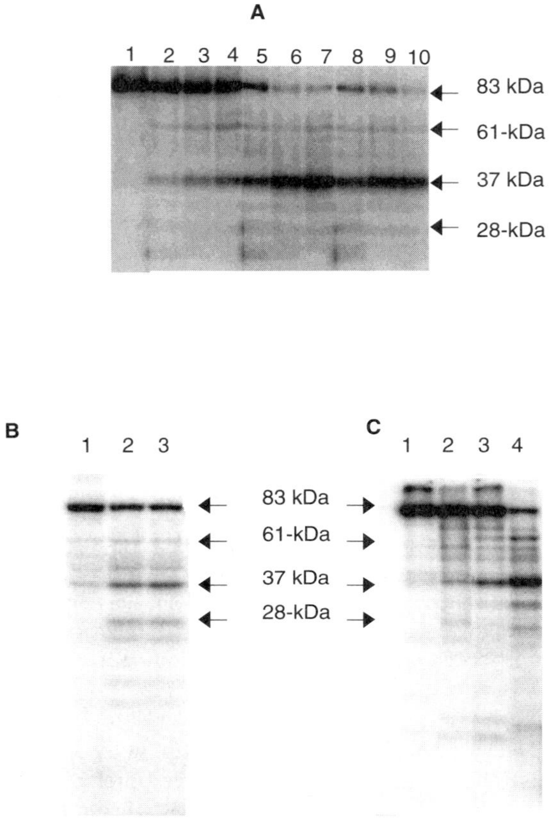

FIGURE 2.12 (A) Progressive incubation at 37°C during respectively 5, 15, and 30 min of ^{32}P-2-5A prelabeled recombinant RNase L (control lane 1) with PBMC extracts from either a healthy control (lanes 2 to 4), or two CFS patients (lanes 5 to 7 and 8 to 10, respectively) generates 2-5A binding fragments of 61-, 37-, and 28-kDa, of which the 37-kDa variant is the most abundant in the presence of CFS PBMC extracts (lanes 5 to 10). (B) The fragments generated by incubation of the recombinant RNase L with the PBMC extracts match exactly the sizes of the native RNase L and its fragments observed in three different PBMC extracts labeled with ^{32}P-2-5A (lanes 1 to 3). (C) Incubation at 37°C of the recombinant RNase L labeled with ^{32}-P-2-5A (control lane 1) with either m-calpain (5 and 30 min, lanes 2 and 3) and human leucocyte elastase (5 min, lane 4) generates fragments with sizes identical to those found in the PBMC extracts

of the prelabeled recombinant enzyme (control lane 1) with PBMC extracts from either a healthy control (lanes 2 to 4), or two CFS patients (lanes 5 to 7 and 8 to 10, respectively) generated 2-5A binding fragments of 61-, 37-, and 28-kDa, of which the 37-kDa variant was the most abundant. This latter fragment was generated rapidly by CFS PBMC extracts, much less by the control extract. The fragments generated by incubation of the recombinant RNase L with the PBMC extracts matched exactly the sizes of the native RNase L and its fragments observed in three different PBMC extracts labeled with ^{32}P-2-5A (Figure 2.12B, lanes 1 to 3). In an attempt at identifying the enzyme responsible for such cleavage, we incubated the recombinant RNase L labeled with ^{32}P-2-5A with several enzymes. Among the enzymes tested, m-calpain and human leukocyte elastase (HLE) were capable of cleaving the recombinant enzyme and generating fragments with sizes identical to those found in the PBMC extracts (Figure 2.12C). More recently, our laboratories also identified cathepsin G as a possible protease involved in the generation of the RNase L fragments. The specificity of these cleavages was verified by their inhibition in the presence of various inhibitors.

Calpain is a cysteine protease particularly active during apoptosis.[79] Increased apoptotic populations has been previously observed in the PBMC of CFS patients and an increased expression of protein kinase R has been held responsible for this dysfunction.[80] Consequently, it is not surprising to observed an enhanced calpain activity in PBMC of CFS patients. We tested other apoptotic proteases including caspase 3[81] for their possible proteolytic activity toward the recombinant RNase L, but were unable to detect any fragmentation.[77] Elastase and cathepsin G are serine proteases found in the azurophilic granules of polymorphonuclear leukocytes and are thought to be involved in the cleavage of a variety of coagulation proteins and in the degradation of connective tissue proteins which may lead to pulmonary emphysema, rheumatoid arthritis, and atherosclerosis.[82] Interestingly, these proteases are thus involved in processes of host defense, extracellular tissue remodeling, and inflammation, all processes involved in CFS. However, the physiological link is still tentative at this stage. As pointed out in a recent review,[83] protease inhibitors currently available usually lack complete specificity, which makes it hard to identify a physiologically relevant protease absolutely.

The observation that the 37-kDa truncated RNase L is generated by proteolytic cleavage of the 83-kDa enzyme raises interesting questions. As mentioned above, this fragment is likely to have both the 2-5A-binding and the catalytic activity toward poly(U) of the native RNase L. The first question that can be raised is how the catalytic activity can be exerted by this fragment since the monomeric enzyme is inactive and must homodimerize to gain the catalytic activity.[2] This would suggest that the fragment has lost the ankyrin domain suspected to exert a conformational clamp on the catalytic site of the monomer.[47] Recently, we reported the loss of at least a portion of the ankyrin domain in the 37-kDa fragment, which would also suggest that it could no longer be regulated by RLI.[45] However, it has been shown[6] that RLI prevented 2-5A-binding by the low molecular weight RNase L species, which indicates that these proteins have retained their capacity to interact with RLI.

A second question is how the 37-kDa fragment can retain both the 2-5A-binding and catalytic domains of RNase L. Indeed, a careful examination of the localization of these characteristic sites on the amino acid sequence of the monomeric RNase L (Figure 2.3) immediately draws attention to the fact that a fragment of that size cannot retain these domains in a continuous sequence. Consequently, the 37-kDa band detected by PAGE contains either two different proteolytic fragments, one containing the 2-5A-binding site and one containing the catalytic domain, or a single protein containing both activities generated by endoproteolytic cleavage of RNase L. In the latter case, the two domains could be held together after cleavage by a disulfide bond between either cys293 or cys301 (both close to the 2-5A-binding site), and cys639 (close to the catalytic domain). In an attempt to sort out this latter question, we examined all the possible cleavage sites in RNase L by both m-calpain and HLE[84,85] and calculated systematically the molecular weight of the fragments generated from the N- or C-terminal end of the enzyme. The results are summarized in Tables 2.4. and 2.5. for m-calpain and HLE, respectively. This exercise did not allow

TABLE 2.4
Possible Cleavage Motifs by m-calpain in the RNase L Molecule. The Theoretical Molecular Weight of the Fragments Generated from N- and C-Termini Is Also Given.

Motif at Cleavage	Amino Acids in Sequence	MW of Fragment from N-Terminus	MW of Fragment from C-Terminus
LR	79-80	8,906	74,437
	87-88	9,751	73,592
	153-154	16,937	66,406
	164-165	18,376	64,967
	596-597	66,839	16,504
	615-616	68,994	14,349
LK	108-109	11,853	71,490
	141-142	15,506	67,837
	184-185	20,381	62,962
	283-284/285-286	31,110/31,351	52,233/51,992
	361-362	39,939	43,404
	542-543	60,372	22,971
	663-664	74,650	8,693
	683-684	77,141	6,202
LY	144-145	15,929	67,414
	529-530	58,942	24,401
	690-691	77,868	5,475
LM	172-173	19,105	64,238
IK	508-509	56,490	26,853
	605-606	67,795	15,548
IR	666-667	75,066	8,277
IY	353-354	39,009	44,334
	379-380	42,008	41,335
	701-702	79,219	4,124
VK	105-106	11,517	71,826
	138-139	15,194	68,149
	317-318	34,894	48,499
	391-392	43,399	39,944
	517-518	57,500	25,843
	532-533	59,268	24,075
VR	573-574	63,961	19,382
VY	134-135	14,782	68,561
	703-704	79,481	3,862

TABLE 2.5
Possible Cleavage Motifs by Human Leukocyte Elastase in the RNase L Molecule. The Theoretical Molecular Weight of the Fragments Generated from N- and C-Termini Is Also Given.

Motif at Cleavage	Amino Acids in Sequence	MW of Fragment from N-Terminus	MW of Fragment from C-Terminus
PV	85-86	9,482	73,861
AV	133-134	14,596	68,747
	246-247	26,935	56,408
	280-281	30,740	52,603
	390-391	43,271	40,072
	471-472	52,306	31,037
IV	74-75	8,282	75,061
LV	40-41	5,227	78,116
	254-255	27,839	55,504
	304-305	33,323	50,020
	316-317	34,766	48,577
	417-418	46,299	37,044
	527-528	58,666	24,677
	699-700	78,943	4,400
FV	431-432	47,916	—
YV	530-531	59,041	24,302
	702-703	78,796	4,547
VV	531-532	59,140	24,203
	550-551	61,228	22,115
PM	356-357	39,393	43,950

us to identify any fragment of the proper size when compared to the size of the fragments detected by PAGE. This theoretical examination, supported by the experimental evidence showing that the 37-kDa fragment does not contain the N-terminal ankyrin domain,[45] suggests that the second possibility could prevail.

In order to further verify the validity of the disulfide bridge hypothesis, we used the recombinant RNase L, which contains a His6 tag at the N-terminus. We digested the protein with a PBMC extract, m-calpain, and HLE, submitted the digests to PAGE in reducing conditions, and detected the fragments generated by immunoblotting with a monoclonal antibody directed against the His tag.[45] We then analyzed the various fragments generated (all containing the ankyrin domain) for any possible match with the possible cleavage sites by m-calpain and HLE on RNase L sequence. The results are displayed in Figure 2.13. Once the fragments containing the ankyrin domain of RNase L were characterized in terms of the possible cleavage sites by both m-calpain and HLE, we ascribed these cleavage sites to the RNase L amino acid sequence. We then looked at all possible arrangements of the puzzle pieces in relationship with the fragments, including the 37-kDa protein, retaining the 2-5A-binding capacity as detected with ^{32}P-2-5A, and the catalytic activity, while involving

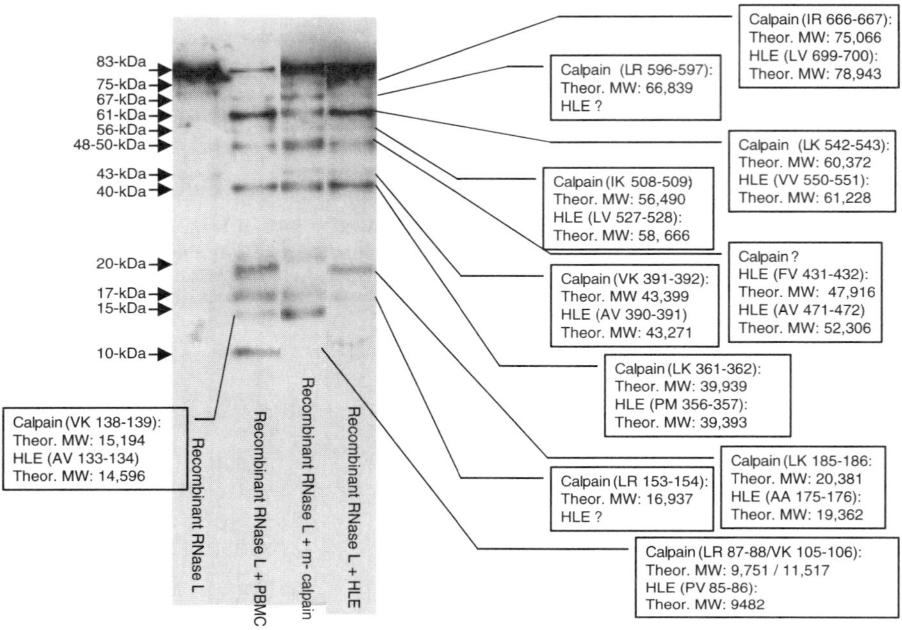

FIGURE 2.13 Analysis of the RNase L fragments containing the ankyrin domain for possible matches with cleavage sites by m-calpain and HLE on the protein sequence. The recombinant enzyme was digested as indicated by a PBMC extract, m-calpain, and HLE, submitted to PAGE, and the fragments were detected by immunoblotting using an anti-His antibody.[45] The amino acid pairs in the sequence at the cleavage sites are given by their single code letter and are localized by their respective numbers in the boxes. The theoretical molecular weights (MW) of the fragments are given in daltons. (With permission from Englebienne, P. et al., *J. Chronic Fatigue Syndrome,* 2001.)

the possible disulfide bridge in the process. The results of this analysis are displayed in Figure 2.14.

The first fragments likely to be produced by proteolytic cleavage are proteins of 61- and 56-kDa, respectively. The lower molecular weight fragments are likely to originate by further cleavage of these proteins. These conclusions can be drawn from kinetic experiments with the recombinant enzyme of which an example is shown in Figure 2.15. These kinetics indicate that the lower molecular weight proteins appear progressively and in parallel with the disappearance of the higher molecular weight bands. All these elements permit us to delineate the progressive cleavage of the 83-kDa RNase L in lower molecular weight fragments of 61-, 56-, 50-, and 37-kDa respectively, as schematically depicted in the right-hand side of Figure 2.14. In such a model, the generation of the lower size proteins of 28-, 25-, and 20-kDa could only be explained by a further reduction of the disulfide bridge, as shown in the right-hand side of Figure 2.14. However, the question can be raised as to how the disulfide bridge could be maintained in some proteins even in the strong reducing conditions already used during PAGE. And why in this context the

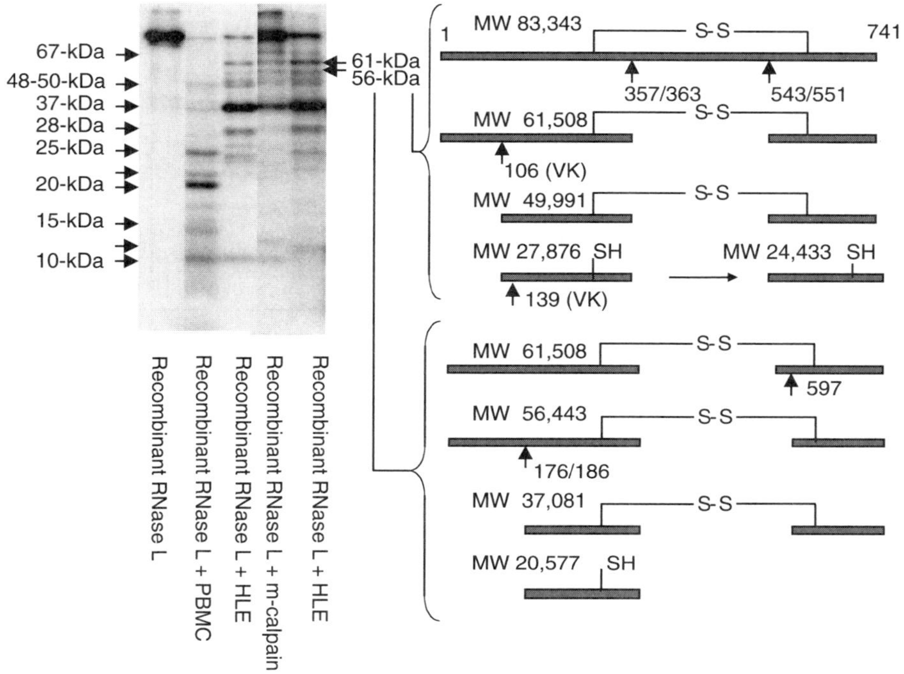

FIGURE 2.14 Schematic representation (right part) of the possible generation by proteolytic cleavage fragments of RNase L retaining the 2-5A-binding capacity, separated by PAGE in reducing conditions and detected by ^{32}P-2-5A labeling of the recombinant enzyme digested respectively with a PBMC extract, m-calpain, or HLE (left part). The presence of a 37-kDa fragment retaining both the 2-5A-binding and catalytic domains of the protein can only be explained by the presence of a disulfide bridge between cys293/301 and cys639. The lower molecular weight fragments (28-, 25-, and 20-kDa, respectively) could be produced by further reduction of the disulfide bridge.

lower molecular weight bands would be more apparent in some proteolytic conditions than in others (Figure 2.14). In order to further investigate these important questions, we reasoned that a lower pH during the reductive process could possibly favor the reduction of the disulfide bridge by a preferential hydrogenation of the sulfhydryl groups. As shown in Figure 2.16, the use of a lower pH during the reduction by either dithiothreitol or β-mercaptoethanol before PAGE indeed favors the production of the lower molecular weight fragments (compare right-hand side with left-hand side in Figure 2.16).

The 40-kDa RNase L-like protein, which has also been claimed to contain both the 2-5A-binding and catalytic domains of the enzyme and is detected on these blots, can be generated from the further cleavage of the 56-kDa band by calpain (motif VK 138-139). The need for very harsh reducing conditions to succeed at reducing only partly the disulfide bond likely to be present in the RNase L molecule is not surprising. Indeed, the reduction of internal disulfide bonds in proteins and,

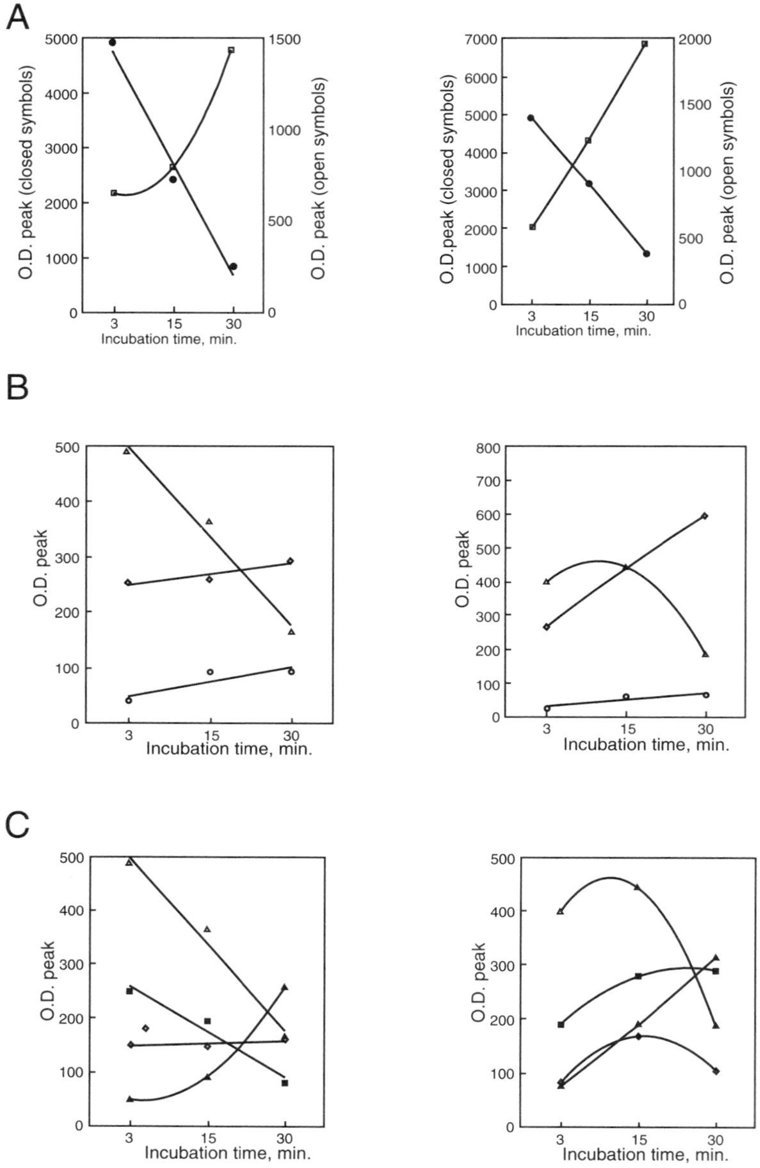

FIGURE 2.15 Kinetic analysis of the evolution of the major species of RNase L during proteolytic degradation of the recombinant enzyme prelabeled with ^{32}P-2-5A and incubated at 37°C with two different PBMC samples during the times indicated. Part A shows the parallel generation of the 37-kDa fragment (open squares) from cleavage of the 83-kDa enzyme (closed circles). Part B shows the progressive appearance of the 56- (diamonds) and 20-kDa (open circles) from the 61-kDa species (open triangles). Part C shows the progressive appearance of the 50- (closed squares), 28- (closed triangles), and 25-kDa (closed diamonds) fragments from the 61-kDa species.

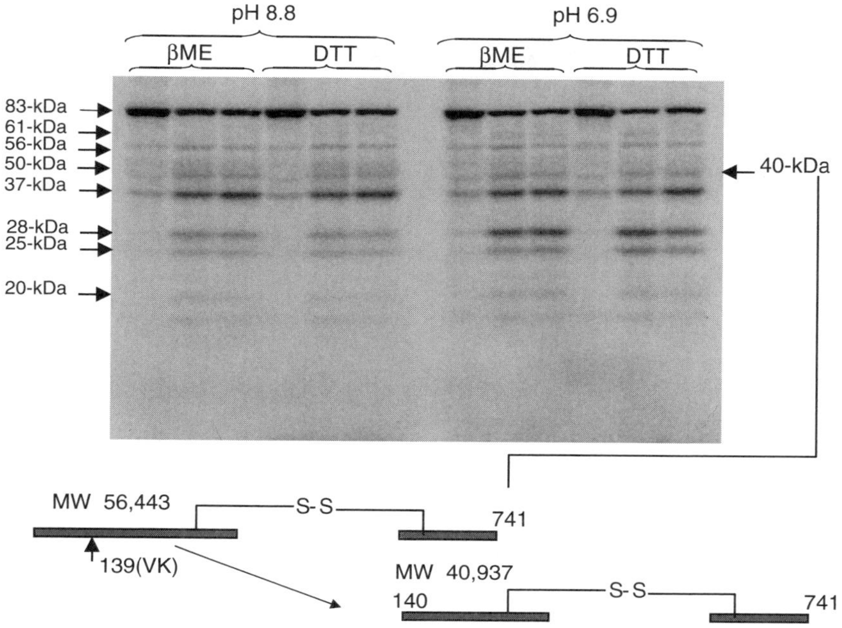

FIGURE 2.16 Comparative blots of three independent PBMC samples labeled with ^{32}P-2-5A separated by PAGE in reducing conditions. The reductive reactions before blotting were performed with either β-mercaptoethanol or dithiothreitol, respectively, in a buffer adjusted to either pH 8.8 (left) or 6.9 (right). The more acidic conditions during reduction generate more of the lower molecular weight fragments of 28-, 25-, and 20-kDa (compare right blot with left blot). The 40-kDa fragment (which possesses both 2-5A-binding and catalytic activity) is generated by further cleavage of the 56-kDa fragment, as depicted schematically at the bottom of the figure.

particularly, in ribonucleases is dependent on the protein conformation and the degree of bond internalization. In RNase A for instance, even in strongly reducing conditions, the reduction of the disulfide bonds must be preceded by a local or global unfolding step that exposes the bonds to the redox reagent in order to be successful.[86]

We also wanted to know if the different cells present in the PBMC contained different levels of the RNase L low molecular weight fragments. To this aim, we separated the cells by affinity chromatography according to their cluster determinants (CD) surface markers prior to extraction and examined the extracts by PAGE. We separated the cells according to either their CD3 (T-cells),[87] or CD14 (monocytes)[88] expression. As shown in Figure 2.17, the low molecular weight of RNase L was primarily retrieved in the cell fractions containing monocytes (CD14+/CD3–), not in the fractions containing T-cells (CD3+/CD14–). This has major implications in the context of innate and humoral immunity. Indeed, monocytes can give rise to either antigen presenting dendritic cells or scavenger macrophages.[88] The activation of dendritic cells allows them to acquire T-cell stimulatory capacity. Any failure from these dendritic cells to follow a specific kinetic process of activation in the

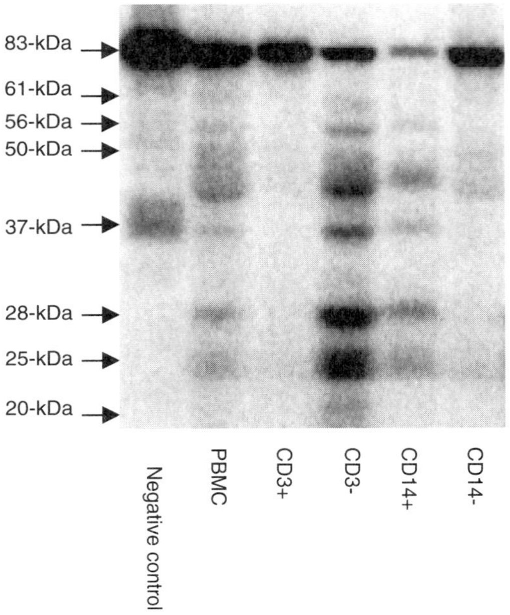

FIGURE 2.17 Localization of the RNase L low molecular weight fragments in PBMC cell populations. PBMC cells were separated by affinity chromatography on CD3 and CD14 beads prior to extraction and the 2-5A-binding proteins were then detected in the extracts of cells either bound (+) or unbound (–) by the beads, respectively, using PAGE and labeling with ^{32}P-2-5A. The low molecular weight fragments were retrieved mainly in the monocyte fractions (CD14+ and CD3–).

presence of infectious agents may lead to a distorted trigger of T-cells differentiation and polarization into T_H1 and T_H2 subtypes.[89]

Finally, we raised the question as to whether the RNase L monomer and dimers would be cleaved as easily by the proteases involved in the pathological process observed in PBMC of chronic fatigue patients. This issue is far from trivial as the dimerization process might involve changes in protein folding and conformation[90] which might limit or change the susceptibility to proteolytic cleavage.[91] Aiming to resolve this important question, we submitted the recombinant RNase L to proteolytic cleavage by various PBMC samples. To make the difference between their potency at cleaving the monomer and the dimer, prior to cleavage, we preincubated the RNase L with PBS devoid or containing an excess of 2-5A tetramer. The preincubation of 2-5A tetramer was intended at driving the formation of the homodimer. After incubation, we detected the 83-kDa RNase L by immunoblotting. Figure 2.18 displays the results of this study. As is clearly indicated, the PBMC samples positive for the presence of low molecular weight RNase L fragments cleaved the monomer rather than the dimer of the recombinant protein preferentially. This consequently has major implications for the possible physiological role of 2-5A in preventing pathological cleavage of the nuclease within the cells.

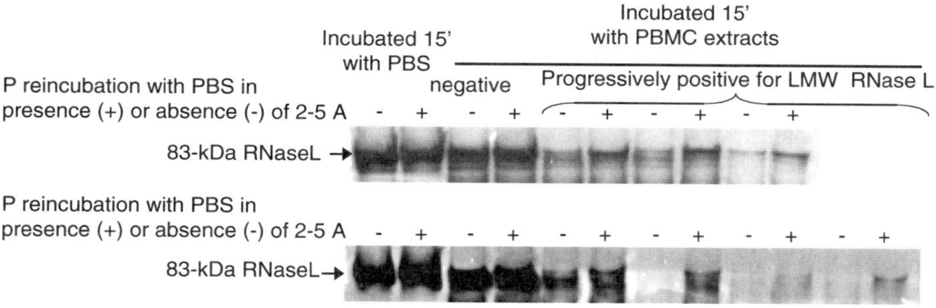

FIGURE 2.18 RNase L monomer rather than the dimer is preferentially cleaved by PBMC extracts. Recombinant RNase L was preincubated with PBS either devoid or containing an excess of 2-5A, respectively, prior to digestion with PBMC extracts and the 83-kDa protein content after incubation was detected by immunoblotting. A higher content of 83-kDa protein was systematically recovered when preincubation with 2-5A had taken place, driving the protein to homodimerization.

2.9 CONCLUSIONS AND PROSPECTS

In this chapter, we have reviewed the characteristic features of the 83-kDa RNase L and its implications in the 2-5A pathway. By doing that, we were able to delineate the molecular structure of the protein along with the possible ways for leading from the normal monomer to the abnormal low molecular weight variants observed in the PBMC of individuals suffering from CFS. The fact that apototic and inflammatory proteases such as m-calpain, elastase, and cathepsin-G are likely to be implicated in the cleavage of RNase L permits us to unravel the downstream effects of cellular stress and inflammatory processes that have been observed in CFS.[80,92] The fact that the cellular RNase L abnormalities are found in monocytes and not in T-cells sheds some new light in the biological process that leads to the T_H2 shift of CFS, i.e., an improper trigger of T-cell differentiation by dendritic cells leading to the reported incapacity of the T_H1 subsets at activating NK cells by nitric oxide mediation,[93] and to the improperly driven T_H2/T_H1 cytokine imbalance reported in the syndrome.[94-96] These observations lead to new therapeutic approaches of the immune dysfunctions associated with CFS using protease inhibitors and regulators of calcium homeostasis.

Our observation that the monomeric form of RNase L is more susceptible to proteolytic cleavage than the homodimer points to the central role of the 2',5'-oligoadenylate synthetase (2-5OAS) activation in the syndrome. Improper activation of 2-5OAS leads to the production of 2-5A dimers preferentially to higher oligomers,[97] dimers which are known to bind but do not activate (i.e., dimerize) RNase L.[2] In case of need, any failure of the 2-5OAS to proper activation would consequently lead to a lack of activation (homodimerization) of RNase L which would remain latent as the monomeric enzyme in the cell. In case of parallel apoptotic induction, i.e., by the protein kinase R (PKR), such a mechanism fuels apoptotic and inflammatory proteases such as m-calpain and elastase with their substrate, which leads to the profound dysregulation of the 2-5A pathway observed in CFS.[4,5] In normal cell homeostasis, the production of the RNase L dimer prevents this from occurring. Such an observation opens new therapeutic approaches using immunomodulators capable of activating the 2-5OAS system such as bile salt[98,99] or retinoic acid derivatives.[100-102]

Our studies have shown that the possible presence of an internal disulfide bridge in the RNase L molecule allows explanation of the simultaneous presence of the 2-5A-binding domain and of the catalytic site in some low molecular weight fragments, including the 37-kDa molecule, produced by the proteolytic cleavage of the endogenous RNase L. Lower molecular weight 2-5A-binding fragments, lacking the catalytic site, arise by the reduction of this disulfide bridge. An important question remaining unresolved is how, if any, the regulation of the catalytic activity of the monomer and of the low molecular weight variants occurs. Dong and Silverman[2] have shown that homodimerization of the protein was mandatory for the activation of the catalytic activity. They have also shown[18] that the ankyrin domain played a regulatory role on this activity, but that a RNase L mutant lacking the full ankyrin repeat region, including the 2-5A-binding site, was fully active independently of 2-5A. They further localized the catalytic site in the C-terminal end of the protein, more precisely in the 20 to 30 residues downstream of amino acid 725.

The low molecular weight variants of 40- and 37-kDa reported to share 2-5A-binding and catalytic activity[4] lack both the protein kinase homology domain and the first N-terminal ankyrin repeat. They are consequently unable to dimerize upon binding 2-5A and their catalytic activity is no longer regulated by the clamp imposed on the monomeric protein by this ankyrin region. The binding and catalytic sites of ribonucleases are made of alternating base- (B sites) and phosphate-interacting (P sites) amino acids. Among the amino acids interacting with bases are T, F, S, N, Q, and E;[103,104] among those interacting with the phosphodiester bonds, K and H are the most relevant,[103,104] but R and Y can be involved.[105,106] These residues are plentiful in the sequence identified as the catalytic site of the enzyme.

In the dimeric state, these structures are freed from the ankyrin clamp and could line up with one another to create the catalytic site, as schematically displayed in Figure 2.19A. In the active low molecular weight fragments (Figure 2.19B), the loss of the ankyrin domain could allow for the catalytic sequence of the original monomer to fold differently in order to create the catalytic site. A full verification of this proposed model will involve mutations in this domain of both the monomer and the truncated variants. A truncated variant unable to be regulated by the ankyrin domain would be suspected to inflict a lot of damage at the cellular level by splicing mRNA, thereby impairing the translation of key signal transduction factors.[66]

While our understanding of the RNase L structure and the way protein truncations occur at the cellular level in CFS has improved very much these last few years, a lot of questions remain unanswered. We expect that the answers to these questions will allow us to fully apprehend the central role played by the degradation of this protein in CFS etiology and pathogenesis.

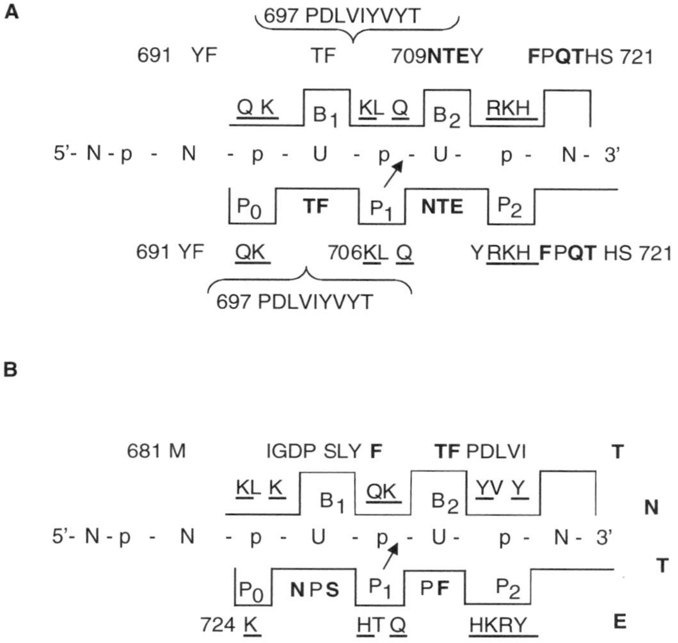

FIGURE 2.19 Tentative catalytic site structure of the dimeric (A) and low molecular weight variants (B) of RNase L. In the dimer (A), the identified catalytic motifs from both monomers are freed from the ankyrin clamp and line up to form the catalytic site. In the low molecular weight variants (B), the ankyrin domain has been lost and the catalytic sequence of the monomer is allowed to fold differently in order to create the catalytic site of the enzyme. The amino acids interacting with the bases are in bold and those interacting with the phosphate groups are underlined. The arrow shows the cleavage site.

REFERENCES

1. Baglioni, C., Minks, A., and Maroney, P.A., Interferon action may be mediated by activation of a nuclease by pppA2'p5'A2'p5'A, *Nature*, 1978; 273: 684–7.
2. Dong, B. and Silverman, R.H., 2-5A-dependent RNase molecules dimerize during activation by 2-5A, *J. Biol. Chem.,* 1995; 270: 4133–7.
3. Bisbal, C. et al., Cloning and characterization of a RNase L inhibitor: a new component of the interferon-regulated 2-5A pathway, *J. Biol. Chem.,* 1995; 270: 13308–17.
4. Suhadolnik, R.J. et al., Biochemical evidence for a novel low molecular weight 2-5A-dependent RNase L in chronic fatigue syndrome, *J. Interferon Cytokine Res.,* 1997; 17: 377–85.
5. Suhadolnik, R.J. et al., Biochemical dysregulation of the 2-5A synthetase/RNase L antiviral defense pathway in chronic fatigue syndrome, *J. Chronic Fatigue Syndrome,* 1999; 5: 224–41.
6. De Meirleir, K. et al., A 37kDa 2-5A binding protein as a potential biochemical marker for chronic fatigue syndrome, *Am. J. Med.,* 2000; 108: 99–105.

7. Zhou, A., Hassel, B.A., and Silverman, R.H., Expression cloning of 2-5A-dependent RNAase: a uniquely regulated mediator of interferon action, *Cell*, 1993; 72: 753–65.
8. Brendel, V. et al., Methods and algorithms for statistical analysis of protein sequences, *Proc. Natl. Acad. Sci. U.S.A.*, 1992; 89: 2002–6.
9. Jennings, M.L., Topography of membrane proteins, *Annu. Rev. Biochem.*, 1989; 58: 999–1027.
10. Salehzada, T. et al., Polyclonal antibodies against RNase L: subcellular localization of this enzyme in mouse cells, *J. Biol. Chem.*, 1991; 266: 5808–13.
11. Bayard, B.A. and Gabrion, J.B., 2',5'-Oligoadenylate-dependent RNAase located in nuclei: biochemical characterization and subcellular distribution of the nuclease in human and murine cells, *Biochem. J.*, 1993; 296: 155–60.
12. Lux, S.E., John, K.M., and Bennett, V., Analysis of cDNA for human erythrocyte ankyrin indicates a repeated structure with homology to tissue-differentiation and cell-cycle control proteins, *Nature*, 1990; 344: 36–42.
13. Bennett, V. et al., Diversity of ankyrins in the brain, *Biochem. Soc. Trans.*, 1991; 19: 1034–9.
14. Nelson, J.W., Generation of plasma membrane domains in polarized epithelial cells: role of cell–cell contacts and assembly of the membrane cytoskeleton, *Biochem. Soc. Trans.*, 1991; 19: 1055–9.
15. Silverman, R.H., 2-5A-Dependent RNase L: a regulated endoribonuclease in the interferon system, in *Ribonucleases: Structures and Functions,* Orlando, Academic Press, 1997: 515–42.
16. Tsai, M.-J. and O'Malley, B.W., Molecular mechanisms of action of steroid/thyroid receptor superfamily members, *Annu. Rev. Biochem.*, 1994; 63: 451–86.
17. Dong, B. and Silverman, R.H., Alternative function of a protein kinase homology domain in 2',5'-oligoadenylate dependent RNase L, *Nucleic Acids Res.*, 1999; 27: 439–45.
18. Dong, B. and Silverman, R.H., A bipartite model of 2-5A-dependent RNase L, *J. Biol. Chem.*, 1997; 272: 22236–42.
19. Ciglic, M.I. et al., Origin of dimeric structure in the ribonuclease superfamily, *Biochemistry,* 1998; 37: 4008–22.
20. Iwawaki, T. et al., Translational control by the ER transmembrane kinase/ribonuclease IRE1 under ER stress, *Nature Cell Biol.,* 2001; 3: 158–64.
21. Nikawa, J. and Yamashita, S., IRE1 encodes a putative protein kinase containing a membrane-spanning domain and is required for inositol phototropy in *Saccharomyces cerevisiae, Mol. Microbiol.,* 1992; 6: 1441–6.
22. Shamu, C.E., Splicing together the unfolded-protein response, *Curr. Biol.,* 1997; 7: R67–70.
23. Silverman, R.H. and Williams, B.R.G., Translational control perks up, *Nature*, 1999; 397: 208–11.
24. Wong, W.L. et al., Inhibition of protein synthesis and early protein processing by thapsigargin in cultured cells, *Biochem. J.,* 1993; 289: 71–9.
25. Chen, J.J. and London, I.M., Regulation of protein synthesis by heme-regulated eIF-2 alpha kinase, *Trends Biochem. Sci.,* 1995; 20: 105–8.
26. Thomis, D.C., Doohan, J.P., and Samuel, C.E., Mechanism of interferon action: cDNA structure, expression, and regulation of the interferon-induced, RNA-dependent P1/eIF-2 alpha protein kinase from human cells, *Virology,* 1992; 188: 33–46.
27. Harding, H.P., Zhang, Y., and Ron, D., Protein translation and folding are coupled by an endoplasmic-reticulum-resident kinase, *Nature*, 1999; 397: 271–4.

28. Sood, D. et al., Pancreatic eukaryotic initiation factor-2α kinase (PEK) homologues in human, *Drosophila melanogaster*, and *Caenorhabditis elegans* that mediate translational control in response to endoplasmic reticulum stress, *Biochem. J.*, 2000; 346: 281–93.
29. Castelli, J.C. et al., The role of 2'-5' oligoadenylate-activated ribonuclease L in apoptosis, *Cell Death Differ.*, 1998; 5: 313–20.
30. Srivastava, S.P., Kumar, K.U., and Kaufman, R.J., Phosphorylation of eukaryotic translation initiation factor 2 mediates apoptosis in response to activation of the double-stranded RNA-dependent protein kinase, *J. Biol. Chem.*, 1998; 273: 2416–23.
31. Romano, P.R. et al., Autophosphorylation in the activation loop is required for full kinase activity *in vivo* of human and yeast eukaryotic initiation factor 2α kinases PKR and GCN2, *Mol. Cell. Biol.*, 1998; 18: 2282–97.
32. Arai, K. et al., Cytokines: coordinators of immune and inflammatory responses, *Annu. Rev. Biochem.*, 1990; 59: 783–836.
33. Darnell, J.E., Jr., Kerr, I.M., and Stark, G.R., Jak-STAT pathways and transcriptional activation in response to IFNs and other extracellular signaling proteins, *Science*, 1994; 264: 1415–21.
34. Darnell, J.E., STATs and gene regulation, *Science*, 1997; 277: 1630–35.
35. Yokosawa, N., Kubota, T., and Fujii, N., Poor induction of interferon-induced 2',5'-oligoadenylate synthetase (2-5 AS) in cells persistently infected with mumps virus is caused by decrease of STAT-1α, *Arch. Virol.*, 1998; 143: 1985–92.
36. Hartmann, R. et al., Activation of 2'-5' oligoadenylate synthetase by single-stranded and double-stranded RNA aptamers, *J. Biol. Chem.*, 1998; 273: 3236–46.
37. Rebouillat, D. and Hovanessian, A.G., The human 2',5'-oligoadenylate synthetase family: interferon-induced proteins with unique enzymatic properties, *J. Interferon Cytokine Res.*, 1999; 19: 295–308.
38. Ghosh, A. et al., Enzymatic activity of 2'-5'-oligoadenylate synthetase is impaired by specific mutations that affect oligomerization of the protein, *J. Biol. Chem.*, 1997; 272: 33220–6.
39. Sarkar, S.N. et al., The nature of the catalytic domain of 2'-5'-oligoadenylate synthetases, *J. Biol. Chem.*, 1999; 274: 25535–42.
40. Marie, I. et al., 69-kDa and 100-kDa isoforms of interferon-induced (2',5') oligoadenylate synthetase exhibit differential catalytic parameters, *Eur. J. Biochem.*, 1997; 248: 558–66.
41. Sarkar, S.N. et al., Enzymatic characteristics of recombinant medium isozyme of 2'-5' oligoadenylate synthetase, *J. Biol. Chem.*, 1999; 274: 1848–55.
42. Hovnanian, A. et al., The human 2',5'-oligoadenylate synthetase-like gene (OASL) encoding the interferon-induced 56kDa protein maps to chromosome 12q24.2 in the proximity of the 2',5'-OAS locus, *Genomics*, 1999; 56: 362–3.
43. Lee, J.W. et al., Two classes of proteins dependent on either the presence or absence of thyroid hormone for interaction with the thyroid hormone receptor, *Mol. Endocrinol.*, 1995; 9: 243–54.
44. Hartmann, R. et al., p59OASL, a 2'-5' oligoadenylate synthetase-like protein: a novel human gene related to the 2'-5' oligoadenylate synthetase family, *Nucleic Acids Res.*, 1998; 28: 4121–7.
45. Englebienne, P. et al., Interactions between RNase L ankyrin-like domain and ABC transporters as a possible origin for pain, ion transport, CNS, and immune disorders of chronic fatigue immune dysfunction syndrome, *J. Chronic Fatigue Syndrome*, 2001; 8: 83–102.

46. Diaz-Guerra, M., Rivas, C., and Esteban, M., Full activation of RNase L in animal cells requires binding of 2-5A within ankyrin repeats 6 to 9 of this interferon-inducible enzyme, *J. Interferon Cytokine Res.,* 1999; 19: 113–9.
47. Kobe, B. and Kemp, B.E., Active site–site directed protein regulation, *Nature*, 1999; 402: 373–6.
48. Torrence, P.F. et al., Oligonucleotide structural parameters that influence binding of 5'-O-triphosphoadenylyl-(2'-5')-adenylyl-(2'-5')-adenosine to the 5'-O-triphosphoadenylyl-(2'-5')-adenylyl-(2'-5')-adenosine dependent endoribonuclease: chain length, phosphorylation state, and heterocyclic base, *J. Med. Chem.,* 1984; 27: 726–33.
49. Krause, D. et al., Activation of 2-5A-dependent RNase by analogs of 2-5A (5'-O-triphosphoryladenylyl(2'-5')adenylyl(2'-5')adenosine) using 2',5'-tetraadenylate (core)-cellulose, *J. Biol. Chem.,* 1986; 261: 6836–9.
50. Caroll, S.S. et al., Activation of RNase L by 2',5'-oligoadenylates. Kinetic characterization, *J. Biol. Chem.,* 1997; 272: 19193–8.
51. Cole, J.L., Carrol, S.S., and Kuo, L.C., Stoichiometry of 2',5'-oligoadenylate-induced dimerization of ribonuclease L, *J. Biol. Chem.,* 1996; 271: 3979–81.
52. Cole, J.L. et al., Activation of RNase L by 2',5'-oligoadenylates. Biophysical characterization, *J. Biol. Chem.,* 1997; 272: 19187–92.
53. Dong, B. et al., Intrinsic molecular activities of the interferon-induced 2-5A-dependent RNase, *J. Biol. Chem.,* 1994; 269: 14153–8.
54. Carrol, S.S. et al., Cleavage of oligoribonucleotides by the 2',5'-oligoadenylate-dependent ribonuclease L, *J. Biol. Chem.,* 1996; 271: 4988–92.
55. Lesiak, K., Khamnei, S., and Torrence, P.F., 2',5'-Oligoadenylate:antisense chimeras. Synthesis and properties, *Bioconj. Chem.,* 1993; 4: 467–72.
56. Li, G., Xiao, W., and Torrence, P.F., Synthesis and properties of second generation 2-5A-antisense chimeras with enhanced resistance to exonucleases, *J. Med. Chem.,* 1997; 40: 2959–66.
57. Squire, J. et al., Localization of the interferon-induced, 2-5A-dependent RNase gene (RNS4) to human chromosome 1q25, *Genomics*, 1994; 19: 174–5.
58. Diaz-Guerra, M., Rivas, C., and Esteban, M., Activation of the IFN-inducible RNase L causes apoptosis in animal cells, *Virology*, 1997; 236: 354–63.
59. Nilsen, T.W., Weissma, S.G., and Baglioni, C., Role of 2',5'-oligo(adenylic acid) polymerase in the degradation of ribonucleic acid linked to double-stranded ribonucleic acid by extract of interferon-treated cells, *Biochemistry*, 1980; 19: 5574–9.
60. Stark, G.R. et al., 2-5A synthetase: assay, distribution and variation with growth or hormone status, *Nature*, 1979; 278: 471–3.
61. Cohrs, R.J., Goswami, B.B., and Sharma, O.K., Occurrence of 2-5A and RNA degradation in the chick oviduct during rapid estrogen withdrawal, *Biochemistry*, 1988; 27: 3246–52.
62. Castelli, J., Wood, K.A., and Youle, R.J., The 2-5A system in viral infection and apoptosis, *Biomed. Pharmacother.,* 1998; 52: 386–90.
63. Zhou, A. et al., Interferon action and apoptosis are defective in mice devoid of 2',5'-oligoadenylate-dependent RNase L, *EMBO J.,* 1997; 16: 6355–63.
64. Castelli, J.C. et al., A study of the interferon antiviral mechanism: apoptosis activation by the 2-5A system, *J. Exp. Med.,* 1997; 186: 967–72.
65. Bisbal, C. et al., The 2'-5' oligoadenylate/RNase L/RNase L inhibitor pathway regulates both MyoD mRNA stability and muscle cell differentiation, *Mol. Cell. Biol.,* 2000; 20: 4959–69.

66. Li, X.-L., Blackford, J.A., Judge, C.S., et al., RNase-L-dependent destabilization of interferon-induced mRNAs. A role for the 2-5A system in attenuation of the interferon response, *J. Biol. Chem.,* 2000; 275: 8880–8.
67. Rusch, L., Zhou, A., and Silverman, R.H., Caspase-dependent apoptosis by 2',5'-oligoadenylate activation of RNase L is enhanced by IFN-beta, *J. Interferon Cytokine Res.,* 2000; 20: 1091–100.
68. Schubert, U. et al., Rapid degradation of a large fraction of newly synthesized protein by proteasomes, *Nature*, 2000; 404: 770–4.
69. Salehzada, T. et al., 2',5'-Oligoadenylate-dependent RNase L is a dimer of regulatory and catalytic subunits, *J. Biol. Chem.,* 1993; 268: 7733–40.
70. Diriong, S. et al., Localization of the ribonuclease L inhibitor gene (RNS4I, a new member of the interferon-regulated 2-5A pathway), to 4q31 by fluorescence *in situ* hybridization, *Genomics*, 1996; 82: 488–90.
71. Aubry, F. et al., Chromosomal localization and expression pattern of the RNase L inhibitor gene, *FEBS Lett.,* 1996; 381: 135–9.
72. Jordan, C. et al., Identification of a binding motif for ankyrin on the α-subunit of Na^+, K^+-ATPase, *J. Biol. Chem.,* 1995; 270: 29971–75.
73. Ding, Y., Kobayashi, S., and Kopito, R., Mapping of ankyrin binding determinants on the erythroid anion exchanger AE1, *J. Biol. Chem.,* 1996; 271: 22494–8.
74. Martinand, C. et al., RNase L inhibitor (RLI) antisense constructions block partially the down regulation of the 2-5A/RNase L pathway in encephalomyocarditis virus (EMCV)-infected cells, *Eur. J. Biochem.,* 1998; 254: 238–47.
75. Martinand, C. et al., RNase L inhibitor is induced during human immunodeficiency virus type 1 infection and down regulates the 2-5A/RNase L pathway in human T cells, *J. Virol.,* 1999; 73: 290–6.
76. Suhadolnik, J. et al., Upregulation of the 2-5A synthetase/RNase L antiviral pathway associated with chronic fatigue syndrome, *Clin. Infect. Dis.,* 1994; 18: S96–104.
77. Roelens., S. et al., G-actin cleavage parallels 2-5A-dependent RNase L cleavage in peripheral blood mononuclear cells. Relevance to a possible serum-based screening test for dysregulations in the 2-5A pathway, *J. Chronic Fatigue Syndrome,* 2001, in press.
78. Lebleu, B. et al., A truncated form of RNase L accumulates in PBMC of chronic fatigue syndrome patients, Abstract, Third Joint Meeting ICS/ISICR, November 5–9, 2000, Amsterdam.
79. Saido, T.C., Sorimachi, H., and Suzuki, K., Calpain: new perspectives in molecular diversity and physiological-pathological involvement, *FASEB J.,* 1994; 8: 814–22.
80. Vojdani, A. et al., Elevated apoptotic cell population in patients with chronic fatigue syndrome: the pivotal role of protein kinase RNA, *J. Intern. Med.,* 1997; 242: 465–78.
81. Hengartner, M.O., The biochemistry of apoptosis, *Nature*, 2000; 407: 770–83.
82. Camire, R.M., Kalafatis, M., and Tracy, P.B., Proteolysis of factor V by cathepsin G and elastase indicates that cleavage at Arg 1545 optimizes cofactor function by facilitating factor Xa binding, *Biochemistry,* 1998; 37: 11896–906.
83. Earnshaw, W.C., Martins, M.L., and Kaufmann, S.H., Mammalian caspases: structure, activation, substrates, and functions during apoptosis, *Annu. Rev. Biochem.,* 1999; 68: 383–424.
84. Babine, R.E. and Bender, S.L., Molecular recognition of protein-ligand complexes: applications to drug design, *Chem. Rev.,* 1997; 97: 1359–1472.
85. Leung, D., Abbenante, G., and Fairlie, D.P., Protease inhibitors: current status and future prospects, *J. Med. Chem.,* 2000; 43: 305–41.

86. Wedemeyer, W.J. et al., Disulfide bonds and protein folding, *Biochemistry*, 2000; 39: 4207–16.
87. Roitt, I., *Essential Immunology,* 7th ed., Blackwell Scientific Publications, Oxford, 1991, p.116.
88. Chomarat, P. et al., IL-6 switches the differentiation of monocytes from dendritic cells to macrophages, *Nature Immunol.,* 2000; 1: 510–4.
89. Langenkamp, A. et al., Kinetics of dendritic cell activation: impact on priming of T_H1, T_H2 and nonpolarized T cells, *Nature Immunol.,* 2000; 1: 311–6.
90. Baker, D., A surprising simplicity to protein folding, *Nature*, 2000; 407: 39–42.
91. Fuentes-Prior, P., Iwanaga, Y., Huber, R., et al., Structural basis for the anticoagulant activity of the thrombin–thrombomodulin complex, *Nature*, 2000; 404: 518–25.
92. Komaroff, A.L. and Buchwald, D.S., Chronic fatigue syndrome: an update, *Annu. Rev. Med.,* 1998; 49: 1–13.
93. Ogawa, M. et al., Decreased nitric oxide-mediated natural killer cell activation in chronic fatigue syndrome, *Eur. J. Clin. Invest.,* 1998; 28: 937–43.
94. Chao, C.C. et al., Altered cytokine release in peripheral blood mononuclear cell cultures from patients with the chronic fatigue syndrome, *Cytokine*, 1991; 3: 292–8.
95. Patarca, R. et al., Interindividual immune status variation patterns in patients with chronic fatigue syndrome. Association with gender and the tumor necrosis factor system, *J. Chronic Fatigue Syndrome,* 1996; 2: 13–39.
96. Patarca, R. et al., Dysregulated expression of soluble immune mediator receptors in a subset of patients with chronic fatigue syndrome: cross-sectional categorization of patients by immune status, *J. Chronic Fatigue Syndrome,* 1995; 1: 81–96.
97. Bonnevie-Nielsen, V., Husum, G., and Kristiansen, K., Lymphocytic 2',5'-oligoadenylate synthetase is insensitive to dsRNA and interferon stimulation in autoimmune BB rats, *J. Interferon Res.,* 1991; 11: 351–6.
98. Podevin, P. et al., Bile acids modulate the interferon signalling pathway, *Hepatology*, 1999; 29: 1840–7.
99. Kestinen, P. et al., Impaired antiviral response in human hepatoma cells, *Virology*, 1999; 263: 364–75.
100. Bourgeade, M.F. and Besancon, F., Induction of 2',5'-oligoadenylate synthetase by retinoic acid in two transformed human cell lines, *Cancer Res.,* 1984; 44: 5355–60.
101. Ho, C.K. et al., Enhancement of interferon-induced 2-5 oligoadenylate synthetase activity by retinoic acid in human histiocytic lymphoma U937 cells and WISH cells, *Differentiation*, 1989; 40: 70–5.
102. Schilbach, K. et al., Reduction of N-myc expression by antisense RNA is amplified by interferon: possible involvement of the 2-5A system, *Biochem. Biophys. Res. Commun.,* 1990; 170: 1242–8.
103. Russo, N. and Shapiro, R., Potent inhibition of mammalian ribonucleases by 3', 5'-pyrophosphate-linked nucleotides, *J. Biol. Chem.,* 1999; 274: 14902–8.
104. Leonidas, D.D. et al., Toward rational design of ribonuclease inhibitors: high resolution crystal structure of a ribonuclease A complex with a potent 3', 5'-pyrophosphate-linked dinucleotide inhibitor, *Biochemistry*, 1999; 38: 10287–97.
105. Arni, R.K. et al., Three-dimensional structure of ribonuclease T1 complexed with an isosteric phosphonate substrate analogue of GpU: alternate substrate binding modes and catalysis, *Biochemistry*, 1999; 38: 2452–61.
106. Juminaga, D. et al., Tyrosyl interactions in the folding and unfolding of bovine pancreatic ribonuclease A: a study of tyrosine-to-phenylalanine mutants, *Biochemistry*, 1997; 36: 10131–45.

3 A 37-kDa RNase L: A Novel Form of RNase L Associated with Chronic Fatigue Syndrome

Robert J. Suhadolnik, Susan E. Shetzline, Camille Martinand-Mari, and Nancy L. Reichenbach

CONTENTS

3.1 INTRODUCTION

The state of chronic immune activation observed in many individuals with chronic fatigue syndrome (CFS) led to the first investigations of the status of the 2-5A/RNase L pathway in this syndrome. CFS is an illness of unknown etiology, associated with sudden onset, flu-like symptoms, debilitating fatigue, low-grade fever, myalgia, and neurocognitive dysfunction.[1,2] Diagnosis of CFS remains one of exclusion. An accumulating body of evidence suggests that CFS is associated with dysregulation of humoral and cellular immunity, including reactivation of viruses, abnormal cytokine production, diminished natural killer cell function, and changes in intermediary metabolites.[3-14]

A number of viruses have been associated with CFS. IgG antibody to human herpesvirus 6 (HHV-6) and detection of HHV-6 antigen in PBMC with monoclonal antibodies to HHV-6 DNA are elevated in some individuals with CFS.[15-18] Epstein–Barr virus reactivation in CFS has been inferred on the basis of relatively

0-8493-1046-6/02/$0.00+$1.50

high serum titers of this virus by immunofluorescence assay.[19,20] The possible role of other viruses in CFS, including enteroviruses and a human T-lymphotrophic virus, has been explored.[21-25] The chronic immune activation often observed in CFS is also associated with changes in serum levels of cytokines which, along with depressed function of natural killer cell function and increased numbers of $CD8^+$ cytotoxic T cells that bear antigenic markers of activation, may contribute to immune activation, persistent illness, and fatigue.[10,26-28] On the basis of these indicators of an activated immune state, we reasoned that the clinical and immunological abnormalities observed in CFS might include defects in the dsRNA-dependent, interferon-inducible 2-5A/RNase L antiviral defense pathways.

3.2 STATUS OF THE 2-5A/RNase L PATHWAY IN CFS

The first evidence that the 2-5A/RNase L pathway in CFS was abnormal was given in a pilot study of 15 individuals with CFS who were severely disabled by their illness.[29,30] Levels of 2-5A synthetase (2-5OAS), bioactive 2-5A, and RNase L activity were measured in extracts of PBMC from these 15 individuals who met the CDC criteria for CFS. Patients differed significantly from controls in having a lower mean basal level of latent 2-5OAS ($p < .0001$), a higher level of bioactive 2-5A ($p = .002$), and a higher level of RNase L activity ($p < .0001$) (Figure 3.1).[29] The

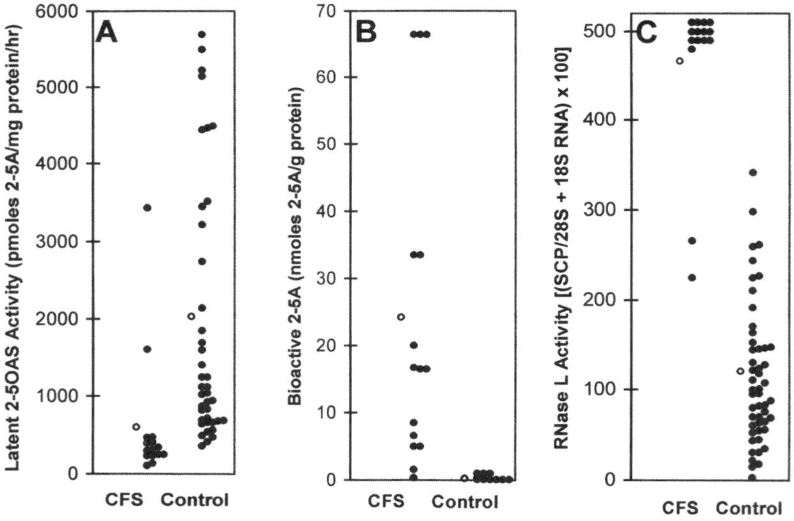

FIGURE 3.1 Latent 2-5OAS activity (panel A), intracellular concentration of bioactive 2-5A (panel B), and 2-5A-dependent RNase L activity (panel C) in extracts of PBMC from individuals with CFS and matched healthy controls.[29] Closed circles represent activities in individual subjects; open circles represent the group mean for CFS patients and controls. Latent 2-5A synthetase activity is expressed as picomoles of 2-5A per milligram of protein per hour; bioactive 2-5A as nanomoles of 2-5A per gram of protein; 2-5A-dependent RNase L activity as the ratio of specific cleavage product formation to 28S and 18S rRNA multiplied by 100. (With permission from Suhadolnik, R. J. et al., *Clin. Infect. Dis.*, 18, S96, 1994.)

decreased level of latent 2-5OAS in PBMC extracts from CFS patients suggested that most of the 2-5OAS present was in the activated form.

To test this hypothesis, bioactive 2-5A was isolated from extracts of PBMC and quantitated by its ability to compete for binding to RNase L in 2-5A core-cellulose affinity binding assays.[29] Concentrations of bioactive 2-5A were elevated up 220-fold in extracts of PBMC from all 15 individuals with CFS compared to controls ($p = .002$). Thirteen of the 15 individuals with CFS exhibited a dramatically upregulated level of RNase L enzyme activity — up to 1500-fold above control levels.[29] In 13 of the 15 individuals with CFS, RNase L activity resulted in complete hydrolysis of 28S and 18S ribosomal RNA (rRNA) and the specific cleavage products. Time studies and protein dilution assays verified that the increased cleavage of 28S and 18S rRNA observed with CFS PBMC extracts was due to RNase L and not other nonspecific RNases.[29,31]

In an open-label study of these 15 individuals severely disabled with CFS, therapy with the biological response modifier, poly(I)-poly($C_{12}U$) (Ampligen) resulted in downregulation of the 2-5A/RNase L pathway and a significant decrease in HHV-6 activity ($p < .01$) in temporal association with clinical and neuropsychological improvement.[29,30] Poly(I)-poly($C_{12}U$) is a specifically configured dsRNA with both antiviral and immunomodulatory activities.[32-34] In extracts of PBMC from CFS patients, levels of bioactive 2-5A were significantly decreased at 12 weeks ($p = .002$) and had declined to normal levels by 24 weeks of poly(I)-poly($C_{12}U$) therapy ($p = .003$).[31] RNase L activity also decreased significantly in PBMC extracts from all 15 individuals with CFS ($p < .0001$ at 12 and 24 weeks of therapy). The profile of RNase L activity for a representative PBMC extract from a CFS patient before and during treatment with poly(I)-poly($C_{12}U$) is shown in Figure 3.2. Before therapy, no

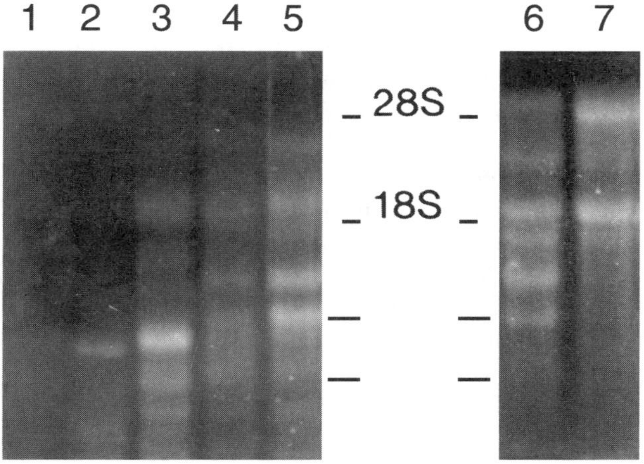

FIGURE 3.2 RNase L activity profile for an extract of peripheral blood mononuclear cells (PBMC) from an individual with CFS before and during therapy with poly(I)-poly($C_{12}U$).[29] RNase L activity was measured by specific cleavage product formation in ribosomal RNA cleavage assays. Lane 1, pretherapy; lanes 2 to 5, after 4, 8, 12, and 16 weeks of therapy; lane 6, healthy control; lane 7, 28S and 18S ribosomal RNA controls. (With permission from Suhadolnik, R. J. et al., *Clin. Infect. Dis.,* 18, S96, 1994.)

28S rRNA, 18S rRNA, or specific cleavage products were visible (Figure 3.2, lane 1), indicating elevated RNase L activity. RNase L activity decreased toward normal during therapy, as evidenced by the appearance of specific cleavage products after 12 and 16 weeks (Figure 3.2, lanes 4 and 5, respectively).

In order to determine whether the phenotype noted in studies of this subset of severely disabled CFS patients could be generalized to a larger, more heterogeneous CFS population, components of the 2-5A/RNase L pathway were measured as part of a randomized, multicenter placebo-controlled, double-blind study of poly(I)-poly(C_{12}U).[31,32] This expanded study of 97 individuals confirmed that the mean values for bioactive 2-5A and RNase L activity were significantly elevated in individuals with CFS compared to controls ($p < .0001$ and $p = .001$, respectively).[31] Clinical response in CFS patients to 24 weeks of poly(I)-poly(C_{12}U) therapy included improvement in Karnofsky performance score ($p < .03$), enhanced capacity to perform activities of daily living ($p < .04$), greater ability to complete exercise treadmill testing ($p = .01$), and reduced cognitive problems ($p = .05$).[32]

3.3 THE 37-kDa 2-5A BINDING PROTEIN: A NOVEL FORM OF RNase L IN CFS

Further characterization of the upregulated RNase L in CFS PBMC was accomplished with azido photolabeling and immunoprecipitation methodologies that specifically identify 2-5A binding, RNase L immunoreactive proteins in extracts of PBMC. Photolabeling of 2-5A binding proteins was accomplished by incubation of PBMC extracts with the 2-5A photoprobe, [^{32}P]pApAp(8-azidoA), and ultraviolet irradiation.[35] Following photolabeling, immunoprecipitation of PBMC extracts was accomplished with an affinity-purified RNase L polyclonal antibody. These methodologies eliminate proteins which immunoreact with the polyclonal antibody to recombinant human 80-kDa RNase L but are not 2-5A binding proteins, as well as 2-5A binding proteins which are not immunoreactive to the polyclonal antibody to RNase L.

Using these methodologies, marked differences have been observed in the molecular mass and RNase L enzyme activity of 2-5A binding proteins in extracts of PBMC from individuals with CFS compared to healthy controls.[35] Under denaturing conditions, three 2-5A binding proteins with molecular masses of 80-, 42-, and 37-kDa were observed in healthy control PBMC (Figure 3.3, lanes 2, 3) and in a subset of CFS PBMC (Figure 3.3, lanes 1, 4, 5). However, photoaffinity labeling and immunoprecipitation studies revealed a second subset of CFS PBMC in which only one 2-5A binding protein with an estimated molecular mass of 37-kDa was observed; no 80- or 42-kDa immunoreactive 2-5A binding proteins were observed (Figure 3.3, lanes 6, 7, 8, 9). Formation of the 37-kDa form of RNase L was not due to the effects of proteolytic degradation of the native 80-kDa RNase L during PBMC extract preparation and processing.[35]

Characterization of the 2-5A binding proteins detected in CFS PBMC extracts by photoaffinity labeling or immunoprecipitation by SDS-PAGE (denaturing conditions) was continued by analytical gel permeation HPLC and assay of 2-5A binding, and 2-5A dependent RNase L enzyme activity as determined by the hydrolysis of poly(U)-3'-[^{32}P]pCp under native (nondenaturing) conditions.[35] In healthy control

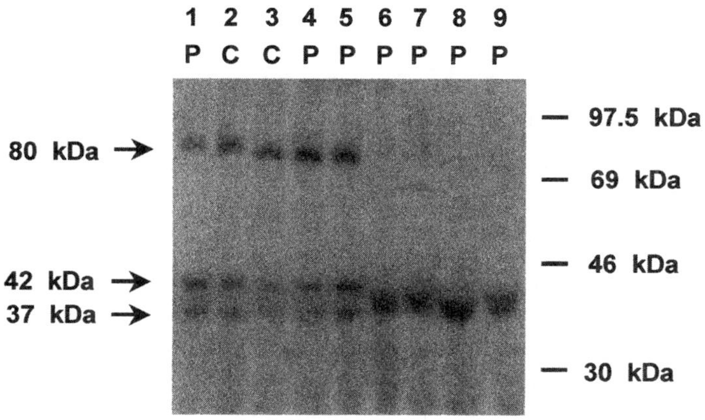

FIGURE 3.3 Azido photoaffinity labeling and immunoprecipitation of 2-5A binding proteins in extracts of PBMC from CFS patients and healthy controls.[35] PBMC extracts were incubated with a radiolabeled 2-5A photoprobe and UV-irradiated. The photolabeling mixture was incubated with polyclonal antibody to recombinant human 80-kDa RNase L and Protein A Sepharose. The Sepharose was collected and the bound proteins were fractionated by SDS-PAGE and subjected to autoradiography. P, CFS patient; C, control. Photoaffinity labeled or immunoreactive 2-5A binding proteins are indicated by arrows. Molecular weight markers are indicated at the right. (With permission from Suhadolnik, R. J. et al., *J. Interferon and Cytokine Res.*, 17, 377, 1997.)

PBMC extracts prepared in the absence of protease inhibitors and analyzed under native conditions, 2-5A binding and 2-5A-dependent RNase L enzyme activities were observed at 80- and 42-kDa.[35] In one subset of CFS PBMC extracts, three 2-5A binding proteins with 2-5A-dependent RNase L enzyme activity were observed at 80-, 42-, and 30-kDa. In a second subset of CFS PBMC extracts (as shown in Figure 3.3, lanes 6, 7, 8, 9), no 2-5A binding or 2-5A-dependent RNase L enzyme activity was observed at 80- or 42-kDa; however, 2-5A binding and 2-5A-dependent RNase L enzyme activity was observed at 30-kDa. The 2-5A binding protein observed at 37-kDa under denaturing conditions is in reasonable agreement with the 30-kDa protein observed under native conditions, based on precedents in the literature accounting for differences in molecular mass observed under denaturing vs. native conditions.[36] Control assays with poly(C)-3'-[^{32}P]pCp verified that the 2-5A-dependent RNase L enzyme activity observed in PBMC extracts was not the result of nonspecific RNase activity.[35]

A subsequent study was designed to determine the extent of the biochemical dysregulation in the 2-5A/RNase L pathway in a larger cohort of CFS patients and to identify clinical correlates to these biochemical findings. CFS patients who met the CDC criteria for CFS and matched healthy controls were assessed with respect to their general health, depression, and pain. Concomitant biochemical assays were completed for three blood draws over a period of 1 year. Analysis of the mean values for bioactive 2-5A, RNase L activity, 37-kDa RNase L in CFS PBMC extracts confirmed the statistically significant upregulation of the 2-5A/RNase L pathway compared to control PBMC extracts ($p < .001$, .002, and .007, respectively).[37] In

TABLE 3.1
Relationship between Components of the 2-5A/RNase L Pathway and Clinical Findings in CFS[37]

Correlation	r-Coefficient[a]	P Value[b]
RNase L activity vs. Bioactive 2-5A	r = .546	< .001
Bioactive 2-5A vs. KPS	r = –.154	.025
RNase L activity vs. KPS	r = -.210	.002
RNase L activity vs. MSQ (total score)	r = .200	.010
IFN-α vs. 37 kDa RNase L	r = .133	.05
80 kDa RNase L vs. 37 kDa RNase L	r = –.538	< .001

[a] Comparison by the Pearson correlation coefficient.
[b] p-values were extrapolated from r coefficients by binomial distribution.

Modified from Suhadolnik, R. J. et al., *J. Chronic Fatigue Syndrome*, 5, 223, 1999. With permission.

addition, several significant correlations were also observed between the biochemical findings and clinical parameters. A summary of the significant positive and negative correlations observed between the biochemical findings and the clinical findings is presented in Table 3.1.

Consistent with two earlier CFS clinical studies,[29,31] a significant positive correlation was observed between RNase L activity and bioactive 2-5A concentration ($p < .001$). The same two biochemical parameters also demonstrated significant negative correlations ($p < .002$ and $p < .025$, respectively) to Karnofsky performance score (KPS), a scale used by physicians to describe how well patients can complete their daily activities. RNase L activity also positively correlated with a second clinical measure, the total score on the metabolic screening questionnaire (MSQ) ($p < .01$), indicating that the upregulation of the 2-5A synthetase/RNase L pathway is an indication of a lower state of general health. A positive correlation was also observed between IFN-α level and the 37-kDa RNase L ($p < .05$). An even stronger correlation was observed between IFN-α level and the 37-kDa RNase L in a subset of the highest 100 IFN-α values from all study subjects, independent of clinical diagnosis ($p < .001$). These positive correlations between IFN-α level and presence of the 37-kDa RNase L are consistent with a viral etiology for CFS.

A recent independent study demonstrated that a reduction in maximal oxygen consumption (VO_2 max) and lower exercise duration during cardiopulmonary exercise testing are associated with elevated expression of the 37-kDa RNase L in CFS patients.[38] Both abnormal RNase L activity and low oxygen consumption are observed in most CFS patients, suggesting that their extremely low tolerance for physical activity may be linked to abnormal oxidative metabolism. It is noteworthy that those CFS patients exhibiting elevated levels of the 37-kDa RNase L performed significantly lower on an exercise test than those with normal levels.[38] Researchers from the University of New Castle and University of Sydney have confirmed that elevated RNase L activity is a good predictor of development of fatigue, muscle pain, and reduced mood in individuals with CFS.[39]

A highly significant negative correlation was also observed between the 80-kDa RNase L and the 37-kDa RNase L ($p < .0001$) (Table 3.1), suggesting that the 37-kDa RNase L may be derived from the 80-kDa RNase L. We have previously established that the 37-kDa RNase L does not originate as the result of protease-mediated degradation of the native 80-kDa RNase L at the time of PBMC extract preparation and processing.[35] However, several lines of biochemical evidence are consistent with the possibility that a cellular or virus-encoded protease may be involved in the origin of the 37-kDa RNase L. Numerous proteases have been demonstrated to have functional impact in normal and virus-infected cells.[40]

It is important to note that the 37-kDa RNase L in extracts of CFS PBMC is not recognized by a monoclonal antibody to human 80-kDa RNase L, even though it is detected by photolabeling and immunoprecipitation with a polyclonal antibody to human 80-kDa RNase L.[41] This is consistent with results described for truncation mutants prepared from the cloned sequence to the human 80-kDa RNase L.[42] One deletion mutant (CΔ399), consisting of the first 342 amino acids from the N-terminus of the cloned 80-kDa RNase L, has a calculated molecular weight of 37.4 kDa.[42] The CΔ399 mutant binds 2-5A, but does not have 2-5A-dependent RNase activity, nor is the mutant recognized by a monoclonal antibody to 80-kDa RNase L. Unlike the CΔ399 deletion mutant, the 37-kDa RNase L has 2-5A-dependent RNase activity as measured by the specific hydrolysis of poly(U), but not poly(C).[35]

Further support for the importance of the 37-kDa RNase L in CFS was the blinded study conducted by De Meirleir et al. in which they detected differences in the distribution of 2-5A binding proteins in CFS PBMC extracts compared with controls.[43] Using a 3'-oxidized 2-5A[^{32}P]pC probe and a different labeling technique in a larger group of CFS patients and controls, they found the 37-kDa RNase L in patients with CFS along with 80- and 40-kDa 2-5A binding proteins. The 37-kDa RNase L was more frequently found in CFS PBMC extracts. The ratio of the novel 37-kDa form of RNase L to the native 80-kDa form of RNase L was high (72%) in CFS patients, low (1%) in healthy controls and, most significantly, absent in depression or fibromyalgia controls (Figure 3.4).[43] The 80-, 40-, and 37-kDa proteins detected by their technique were proven to be authentic 2-5A binding proteins in competition experiments. Binding of the radioactive probe, 3'-oxidized 2-5A[^{32}P]pC, was prevented in incubations of PBMC extracts with unlabeled 2-5A, but not with ATP (Figure 3.5).[43] Incubations with recombinant ribonuclease L inhibitor (RLI) also prevented binding of the probe. Incubations of PBMC extracts at 37°C revealed that the 37-kDa 2-5A binding protein does not arise from protease-mediated degradation of the 80-kDa protein in the laboratory. However, viral or cellular-derived proteases *in vivo* could explain the conversion of the native 80-kDa to smaller molecular weight forms of RNase L.

3.4 MOLECULAR CHARACTERIZATION OF THE 2-5A-DEPENDENT 37-kDa RNase L

In continued studies designed to characterize the molecular and biochemical properties of this newly discovered 2-5A-dependent 37-kDa RNase L, the molecular mass of RNase L in PBMC extracts from control and individuals with CFS has been

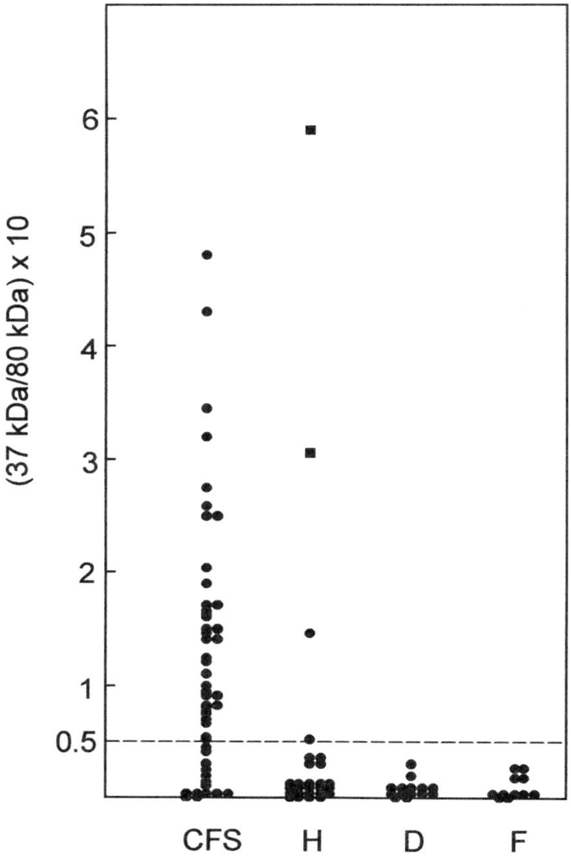

FIGURE 3.4 2-5A binding proteins in extracts of PBMC from CFS patients and controls. The amounts of 2-5A binding proteins in extracts of PBMC from CFS patients, healthy controls, depressed controls, and fibromyalgia controls were determined in assays with a radiolabeled 2-5A photoprobe.[43] Proteins were fractionated by polyacrylamide gel electrophoresis under denaturing conditions. The size distribution of 2-5A binding proteins was visualized by autoradiography. A ratio of 37-kDa/80-kDa RNase L × 10 was calculated and plotted for each sample. The dotted line corresponds to a 0.5 threshold value for the ratio. Each point represents a single study subject. Two contact controls in the healthy control group are identified by squares. (With permission from De Meirleir, K. et al., *Am. J. Med.*, 108, 99, 2000.)

estimated by analytic gel permeation FPLC and two-dimensional gel electrophoresis. Fractionation of control and CFS PBMC extracts by analytic gel permeation FPLC under native conditions revealed the presence of catalytically active and inactive forms of the native 80-kDa RNase L and the 37-kDa RNase L.[44] The 80-kDa RNase L observed in control and CFS PBMC extracts was present as an 80-kDa catalytically active monomer and as a 148-kDa catalytically inactive complex with another protein. Western blot analysis revealed that RLI co-eluted with the 148-kDa catalytically inactive protein, strongly suggesting that the 148-kDa protein is a heterodimer

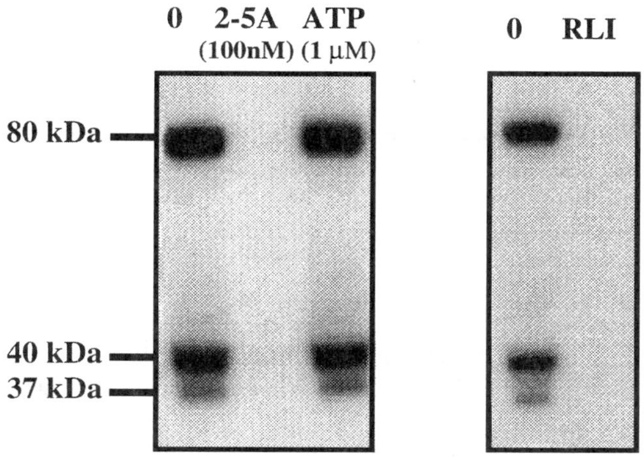

FIGURE 3.5 2-5A binding specificity in PBMC extracts. Peripheral blood mononuclear cell extracts were incubated with the radiolabeled photoprobe, 2-5A [^{32}P]pCp, and either unlabeled 2-5A trimer 5'-triphosphate (100 nM), ATP (1 μM), or recombinant ribonuclease L inhibitor (RLI) (54 pg) as indicated.[43] (With permission from De Meirleir, K. et al., *Am. J. Med.*, 108, 99, 2000.)

complex of 80-kDa RNase L and RLI. RLI is a negative regulator of RNase L that functions by inhibiting the binding of the allosteric activator, 2-5A, to RNase L.[45]

Two forms of the 37-kDa RNase L are observed in CFS PBMC extracts following fractionation and Western blot analysis.[44] One form is the 37-kDa RNase L that was determined to have 2-5A-dependent RNase L enzyme activity as determined by hydrolysis of poly(U)-3'-[^{32}P]pCp. The second form of the 37-kDa RNase L in CFS PBMC extracts has an apparent molecular mass of 105-kDa and is catalytically inactive. Western blot analysis revealed that RLI co-elutes with the 105-kDa form of the 37-kDa RNase L, indicating that the 37-kDa RNase L in CFS PBMC extracts forms a catalytically inactive heterodimer complex with RLI. The expression of RLI has been reported to be decreased in CFS compared to control PBMC.[46] We have also demonstrated that 90% of the native 80-kDa RNase L and 90% of the 37-kDa RNase L exist as a heterodimer complex with RLI. Only 10% of the native 80-kDa RNase L and 10% of the 37-kDa RNase L exists in a free, enzymatically active form. Taken together, these data indicate that RLI regulates the activity of the 37-kDa RNase L in CFS PBMC extracts through protein–protein interactions.

The ability of RLI to form a heterodimer complex with the 80- and the 37-kDa forms of RNase L indicates that the 80- and the 37-kDa forms of RNase L are structurally similar at the N-terminus, which contains the ankyrin repeats and the 2-5A binding site. No homodimer form of the 37-kDa RNase L is observed when CFS PBMC extracts were fractionated in the presence of 2-5A by analytic gel permeation FPLC,[35] which suggests the 37-kDa RNase L lacks the protein kinase homology domain required for homodimerization of the native 80-kDa RNase L. Thus, the 37-kDa is distinct from the 80-kDa RNase L, which requires positive cooperativity for homodimerization and enzyme activity.

To determine if disulfide bonds are required for structural stability of the native, enzymatically active 37-kDa form of RNase L, CFS and control PBMC extracts were covalently photoaffinity labeled using [^{32}P]pApAp(8-azidoA), immunoprecipitated with a polyclonal antibody to RNase L and incubated under reducing and nonreducing conditions prior to SDS-PAGE analysis.[44] Azido photolabeled or immunoprecipitated control and CFS PBMC extracts were treated with iodoacetic acid to produce S-carboxymethyl derivatives of cysteine residues.[47] Following alkylation of the free sulfhydryl groups, the disulfide bonds in the photolabeled and immunoprecipitated PBMC extracts were reduced with DTT and subsequently treated with iodoacetic acid to ensure that the newly reduced disulfide bonds did not re-oxidize during electrophoresis.[47] Control PBMC extracts contained an 80- and a 42-kDa RNase L under nonreducing conditions; the electrophoretic migration of the 80- and 42-kDa forms of RNase L did not change significantly under reducing conditions. In CFS PBMC extracts, three 2-5A photolabeled or immunoprecipitated proteins were observed at 80-, 42-, and 37-kDa under nonreducing conditions; there was no difference in the electrophoretic migration of the 80- and 37-kDa, forms of RNase L under reducing conditions, suggesting that neither form of RNase L contains reducible dimers stabilized by disulfide bonds.[44] An observed difference in migration of the 42-kDa form of RNase L in CFS PBMC extracts under nonreducing and reducing conditions suggests that intradisulfide bonds stabilize the 42-kDa form of RNase L in CFS PBMC extracts and that this protein is structurally different from the 42-kDa RNase L observed in control PBMC extracts.

The electrophoretic migration of the 37-kDa RNase L in CFS PBMC extracts has been further characterized by two-dimensional (2-D) gel electrophoresis.[44] Extracts of CFS PBMC were pooled and fractionated by analytic gel permeation FPLC. Fractions containing the native, catalytically active 37-kDa RNase L were combined with fractions containing the 37-kDa RNase L:RLI heterodimer complex, then pooled and incubated in the presence or absence of the photoprobe, [^{32}P]pApAp(8-azidoA). Coomassie-blue staining and subsequent PhosphorImager analysis visualized the photolabeled proteins on the 2-D gels. The fractions that contained catalytically active 37-kDa RNase L were also analyzed by immunoblotting with a polyclonal antibody to RNase L following 2-D gel electrophoresis. Identification of the 37-kDa RNase L was determined by superimposing the pattern of radiolabeled proteins with the pattern of immunoreactive proteins observed in the 2-D gels. Only one protein with an apparent molecular weight of 37-kDa and a pI of 6.1 was determined to be present in the 2-D gels. The pI of 6.1 for the 37-kDa RNase L compares favorably with a pI of 6.2 for the native 80-kDa RNase L.[48]

Initial peptide mapping studies of the 37-kDa RNase L have been completed using the highly sensitive technique of matrix-assisted laser desorption or ionization mass spectrometry (MALDI-MS) to determine whether peptide sequence similarities exist between the 37- and the 80-kDa RNase L.[44] The 37-kDa RNase L purified by 2-D gel electrophoresis was excised from the 2-D gel, in-gel digested with trypsin, and the peptide mixture was subjected to MALDI-MS. Computer-assisted searches of the NCBI and EMBL protein databases have identified three peptide masses in the 37-kDa RNase L with amino acid sequences identical to the known amino acid sequence of human 80-kDa RNase L. Two of the three peptides in the 37-kDa RNase

L have amino acid sequences identical to sequences located in the ankyrin repeat region of the 80-kDa RNase L. A third peptide is identical to an amino acid sequence in the catalytic domain of human 80-kDa RNase L. No peptides in the MALDI-MS analyses of the 37-kDa RNase L are detected with amino acid sequences equivalent to the protein kinase homology region of the 80-kDa RNase L. These MALDI-MS data indicate that the 37-kDa RNase L retains essential structural and functional features of the native 80-kDa RNase L, in particular, the 2-5A binding domain and the catalytic domain.[44] We are currently completing experiments to obtain the total amino acid sequence of the 37-kDa RNase L.

Binding and enzyme kinetic studies have been completed to delineate the specific interactions involved in the allosteric 2-5A binding and activation of the 80- and the 37-kDa forms of RNase L in PBMC extracts from healthy controls and individuals diagnosed with CFS, respectively. To determine the specificity of the azido photoprobe, [^{32}P]pApAp(8-azidoA), for the 2-5A binding site on the 80- and the 37-kDa forms of RNase L, competition experiments were completed with unlabeled photoprobe.[49] Concentration-dependent protection against photoaffinity labeling of the 80- and 37-kDa RNase L was observed when control or CFS PBMC extracts were incubated with a mixture of [^{32}P]pApAp(8-azidoA) and unlabeled pApAp(8-azidoA) (Figure 3.6).[49] Binding of the photoprobe to 80-kDa RNase L was inhibited by 50% with 1.7×10^{-8} M unlabeled pApAp(8-azidoA) (Figure 3.6A, B), compared to 50% inhibition of photoinsertion of 37-kDa RNase L with 3.9×10^{-9} M unlabeled pApAp(8-azidoA) (Figure 3.6C, D). An increase in covalent photoincorporation of [^{32}P]pApAp(8-azidoA) into the 2-5A binding site of 80-kDa RNase L was observed when photolabeling mixtures contained unlabeled photoprobe at concentrations less than 1×10^{-8} M (Figure 3.6B, inset).

Simultaneous incubation of 1×10^{-8} M [^{32}P]pApAp(8-azidoA) with unlabeled pApAp(8-azidoA) at concentrations below 1×10^{-8} M did not result in increased covalent photolabeling of [^{32}P]pApAp(8-azidoA) to the 37-kDa RNase L (Figure 3.6D, inset). Complete protection against photolabeling of the 80- and 37-kDa RNase L could be achieved with 100-fold excess unlabeled pApAp(8-azidoA) (Figure 3.6B, D, and insets), demonstrating that the [^{32}P]pApAp(8-azidoA) binds specifically to the 2-5A binding site of the 37- and the 80-kDa RNase L.[49]

In view of the significantly elevated RNase L activity previously observed in CFS PBMC using the ribosomal RNA cleavage assay,[29,31] we hypothesized that the 37-kDa RNase L would be more active than the 80-kDa RNase L in the presence of 2-5A and poly(U)-3'-[^{32}P]pCp. Time-dependent cleavage of poly(U)-3'-[^{32}P]pCp was observed with both 80- and 37-kDa RNase L (Figure 3.7).[49] Maximum hydrolysis of poly(U)-3'-[^{32}P]pCp by the 80-kDa RNase L in control PBMC extracts is attained at 60 min with half-maximal hydrolysis at 24 min (Figure 3.7A). In control PBMC extracts, maximum hydrolysis (65%) of poly(U)-3'-[^{32}P]pCp is observed at 120 min (Figure 3.7A). A delay in activation was observed with 80-kDa RNase L in control PBMC extracts, possibly attributed to the time required for the 80-kDa RNase L to dissociate from the core 2-5A-cellulose and form its activated homodimer complex with poly(U)-3'-[^{32}P]pCp and 2-5A. A similar lag time for activation of 80-kDa RNase L immobilized on core 2-5A-cellulose was previously observed with murine cell extracts.[50-52]

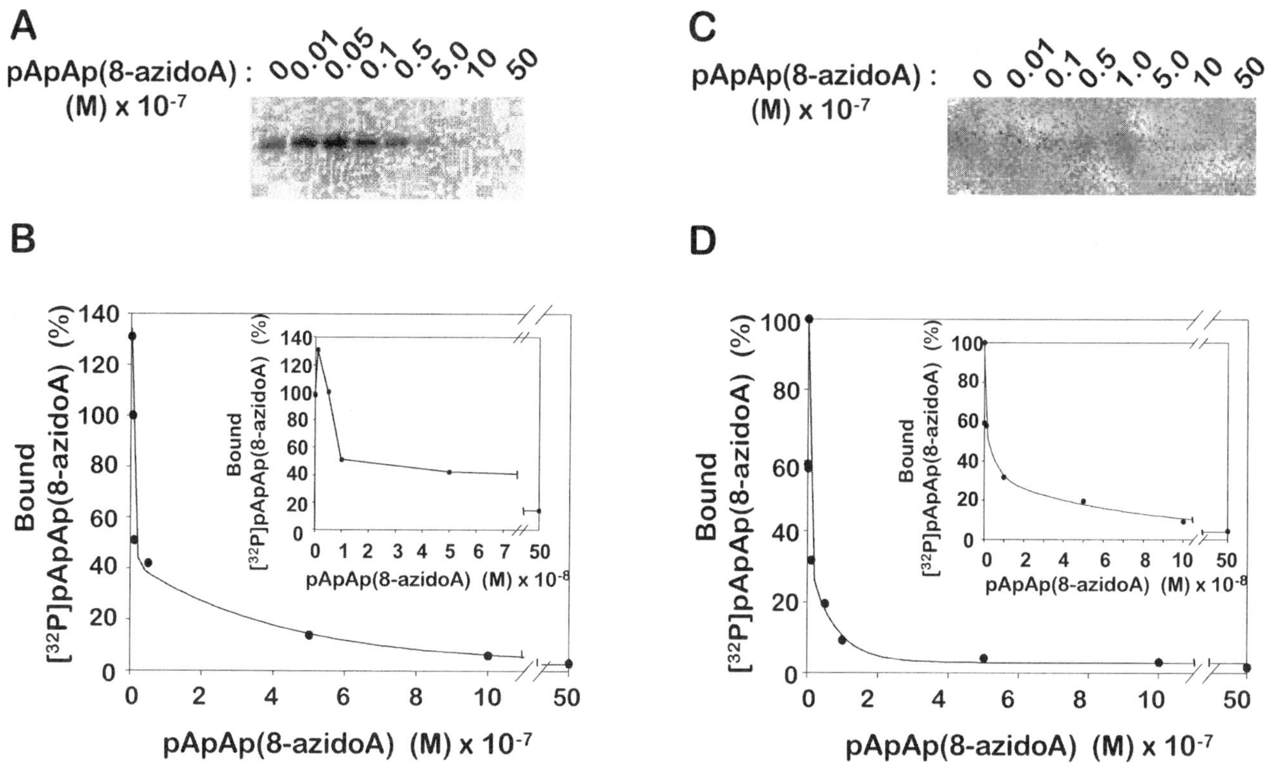

FIGURE 3.6 Displacement of the radiolabeled [^{32}P]pApAp(8-azidoA) photoprobe from RNase L by pApAp(8-azidoA) in PBMC extracts.[49] PBMC extracts were incubated simultaneously with [^{32}P]pApAp(8-azidoA) (1×10^{-8} M; 6×10^{6} Ci/mol; 10,000 dpm) and unlabeled pApAp(8-azidoA) (0 to 5×10^{-6} M), UV-irradiated, immunoprecipitated, and analyzed. Photolabeling of the 80- and 37-kDa RNase L by [^{32}P]pApAp(8-azidoA) in the absence of unlabeled pApAp(8-azidoA) represents 100% photoinsertion. Competition experiments were completed for three healthy control PBMC extracts containing only the 80-kDa RNase L and for three CFS PBMC extracts containing only the 37-kDa RNase L. Results from a representative healthy control (panels A and B) and from a representative CFS patient (panels C and D) are shown. (With permission from Shetzline, S. E. and Suhadolnik, R. J., *J. Biol. Chem.*, 276, 23707, 2001.)

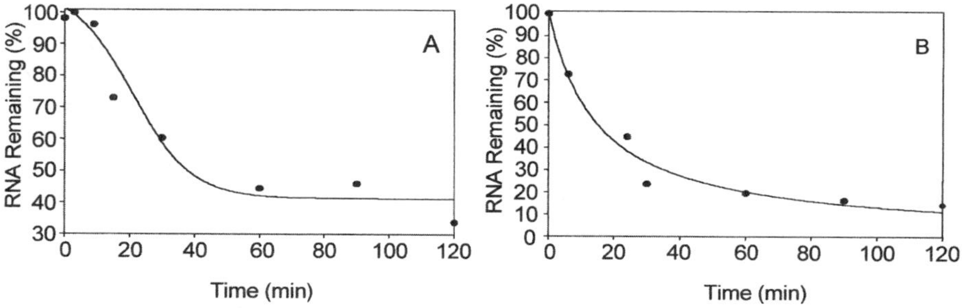

FIGURE 3.7 Time-dependent hydrolysis of poly(U)-3'-[^{32}P]Cp by RNase L in PBMC extracts.[49] RNase L in PBMC extracts from healthy controls (panel A) and from CFS patients (panel B) were immobilized on core(2-5A)-cellulose as described.[49] Activation of 2-5A-dependent RNaseL was measured at different time intervals following the addition of poly(U)-3'-[^{32}P]Cp and 2-5A to the affinity-purified RNase L. The data shown represent the mean of three separate determinations completed for healthy control and CFS PBMC extracts. (With permission from Shetzline, S. E. and Suhadolnik, R. J., *J. Biol. Chem.*, 276, 23707, 2001.)

In contrast to the 80-kDa RNase L, maximum hydrolysis of poly(U)-3'-[^{32}P]pCp by the 37-kDa RNase L was attained at 30 min with half-maximal hydrolysis at 8 min (Figure 3.7B). Unlike the 80-kDa RNase L, the 37-kDa RNase L did not have a lag time for activation. Taken together, these results demonstrate that the 37-kDa RNase L hydrolyzes poly(U)-3'-[^{32}P]pCp at a rate three times faster than the 80-kDa RNase L, suggesting that the elevated RNase L activity observed in CFS PBMC extracts may be attributed to the increased affinity of the RNA substrate and to the increased efficiency of RNA cleavage by the 37-kDa RNase L.

3.5 RELEVANCE OF THE 37-kDa RNase L TO THE PATHOPHYSIOLOGY OF CFS

The evidence described here for the upregulation of the 2-5A/RNase L pathway and the 37-kDa RNase L in particular has contributed to the understanding of the underlying pathophysiology of CFS. As currently defined, CFS may well represent a heterogeneous group of disorders. The working hypothesis of the research presented here is that the characteristic signs and symptoms of CFS are associated with a dysregulation of the 2-5A/RNase L pathway. This IFN-inducible RNA degradation pathway is involved in hydrolysis of cellular and viral RNA and is tightly regulated by dsRNA, 2-5A, and RLI.[45,53] The upregulation of the 2-5A/RNase L pathway has been reported from four independent laboratories with CFS patients on three continents; North America, Europe, and Australia.[29,31,35,39,43,46] An upregulated 2-5A/RNase L pathway in CFS is consistent with an activated immune state and a role for persistent viral infection in the pathogenesis of CFS.

The increased enzyme activity of the 37-kDa RNase L in CFS PBMC presents several pathophysiological questions. What is the biosynthetic origin of the 37-kDa RNase L? What are the physiological/biochemical effects of the elevated RNase L activity in CFS? What is the impact of upregulated RNase L activity on RNA metabolism? What are the therapeutic implications of the 37-kDa RNase L in CFS? Whatever the biochemical mechanism responsible for the 37-kDa RNase L, it exhibits the biochemical properties of an authentic 2-5A dependent RNase L. Based on the differences in 2-5A binding and 2-5A-dependent RNase L activity observed in CFS PBMC, it is tempting to speculate that the presence of the 37-kDaRNase L is related to the severity of CFS symptoms. Two studies have indicated that the absence of the 80- and 42-kDa forms of RNase L and presence of the 37-kDa form of RNase L correlate with the severity of CFS clinical presentation.[37,38] A potential subgroup of individuals with CFS has been identified with an RNase L enzyme dysfunction that is profound, i.e., individuals with only the 37-kDa form of RNase L.[29] It is noteworthy that extracts of PBMC from the most severely disabled individuals with CFS in this study (KPS = 40) contained only the 37-kDa RNase L.

The first evidence has been presented that the presence of the 37-kDa RNase L may distinguish individuals with CFS from those with clinically similar illnesses, i.e., depression and fibromyalgia.[43] It will be most interesting to determine the extent to which the 37-kDa RNase L may be used in the diagnosis of CFS and if control of the RNase L enzyme dysfunction could provide therapeutic benefit. For example, we have reported on the inhibition of 2-5A-dependent RNase L activation by several stereochemically modified phosphorothioate derivatives of 2-5A.[54-56] Such 2-5A derivatives may function as antagonists of the 2-5A-mediated activation of RNase L in CFS.

Based on previous reports on the posttranslational modification of proteins, several biochemical mechanisms can be proposed to explain the biochemical origin of the 37-kDa RNase L and its relationship to the native 80-kDa RNase L. First, the 37-kDa RNase L may be the result of proteolytic degradation of the 80-kDa RNase L via a limited N- or C-truncation followed by protein splicing in which the protein kinase homology region is proteolytically excised.[42,57,58] These posttranslational peptidase reactions would be required to retain 2-5A allosteric activation and RNA binding by the 37-kDa RNase L. There is some evidence that incubation of CFS PBMC extracts without protease inhibitors leads to the disappearance of the 80-kDa RNase L and accumulation of a 37-kDa form (B. Lebleu, private communication). If proteolytic degradation is involved in formation of the 37-kDa RNase L, protease inhibitors may find therapeutic application in CFS.[59] Other possible mechanisms for formation of the 37-kDa RNase L would involve proteolytic degradation by a virus-encoded protease, by the ubiquitin proteasome pathway, or by proinflammatory cytokines such as IFN-α and TNF-α, both of which are elevated in CFS.[40,60-62] Alternatively, it is possible that the 37-kDa RNase L is formed by mRNA splicing or from a novel gene.

The results summarized here strongly suggest that the 80- and 37-kDa forms of RNase L in PBMC extracts share structural and functional features; in particular, the 2-5A binding and catalytic domains. Even though the 80- and the 37-kDa RNase L bind 2-5A and hydrolyze single-stranded RNA in a 2-5A-dependent manner, the two forms of RNase L have different parameters for activation. Future characterization of the 37-kDa RNase L will lead to a better understanding of the role that alterations in the 2-5A/RNase L pathway play in the pathophysiology of CFS.

REFERENCES

1. Holmes, G.P. et al., Chronic fatigue syndrome: a working case definition, *Annu. Intern. Med.*, 108, 387, 1998.
2. Fukuda, K. et al., International chronic fatigue syndrome study group. The chronic fatigue syndrome: a comprehensive approach to its definition and study, *Annu. Intern. Med.*, 121, 953, 1994.
3. Komaroff, A. L. and Buchwald, D. S., Chronic fatigue syndrome: an update, *Annu. Rev. Med.*, 49, 1, 1998.
4. Bates, D. et al.,Clinical laboratory test findings in patients with chronic fatigue syndrome, *Arch. Intern. Med.,* 155, 97, 1995.
5. Chao, C. C. et al., Altered cytokine release in peripheral blood mononuclear cell cultures from patients with chronic fatigue syndrome, *Cytokine*, 3, 292, 1991.
6. McGregor, N. R. et al., Preliminary determination of a molecular basis to chronic fatigue syndrome, *Biochem. Mol. Med.*, 57, 73, 1996.
7. Levy, J. A., Viral studies of chronic fatigue syndrome, *Clin. Infect. Dis*., 18, S117, 1994.
8. Straus, S. et al., Lymphocyte phenotype and function in the chronic fatigue syndrome, *J. Clin. Immunol.*, 13, 30, 1993.
9. Demitrack, M. et al., Evidence for impaired activation of the hypothalamic–pituitary–adrenal axis in patients with chronic fatigue syndrome, *J. Clin. Endocrinol. Metab.*, 73, 1224, 1991.
10. Caligiuri, M. et al., Phenotypic and functional deficiency of natural killer cells in patients with chronic fatigue syndrome, *J. Immunol.,* 139, 3306, 1987.
11. Buchwald, D. and Komaroff, A. L., Review of laboratory findings for patients with chronic fatigue syndrome, *Rev. Infect. Dis*., 13 (Suppl. 1), S12, 1991.
12. Jones, J. et al., Evidence of active Epstein–Barr virus infection in patients with persistent, unexplained illnesses: elevated anti-early antigen antibodies, *Annu. Intern. Med.*, 102, 1, 1985.
13. Klimus, N. G. et al., Immunologic abnormalities in chronic fatigue syndrome, *J. Clin. Microbiol.,* 28, 1403, 1990.
14. See, D. M. and Tilles, J. G., Alpha interferon treatment of patients with chronic fatigue syndrome, *Immunol. Invest*., 25, 153, 1996.
15. Josephs, S. F. et al., HHV-6 reactivation in chronic fatigue syndrome, *Lancet*, 337, 1346, 1991.
16. Balachandran, N. et al., Identification of proteins specific for human herpesvirus 6-infected human T cells, *J. Virol.*, 63, 2835, 1989.
17. Buchwald, D. et al., A chronic illness characterized by fatigue, neurologic, and immunologic disorders, and active human herpesvirus type 6 infection, *Annu. Intern. Med.*, 116, 103, 1992.
18. Peterson, D. L. et al., Clinical improvements obtained with Ampligen in patients with severe chronic fatigue syndrome and associated encephalopathy, in *The Clinical and Scientific Basis of Myalgic Encaphalomyeletis/Chronic Fatigue Syndrome*, Hyde, B., ed., Nightingale Research Foundation, Ottawa, 1992, 634.
19. Jones, J. F. et al., Evidence for active Epstein–Barr virus infection in patients with persistent unexplained illnesses: elevated anti-early antigen antibodies, *Annu. Intern. Med.*, 102, 1, 1985.
20. Straus, S. E. et al., Persisting illness and fatigue in adults with evidence of Epstein–Barr virus infection, *Annu. Intern. Med.*, 102, 7, 1985.
21. Archard, L. C. et al., Postviral fatigue syndrome: persistence of enterovirus RNA in muscle and elevated creatine kinase, *J. R. Soc. Med.,* 81, 326, 1988.

22. Cunningham, L. et al., Persistence of enteroviral RNA in chronic fatigue syndrome is associated with the abnormal production of equal amounts of positive and negative strands of enteroviral RNA, *J. Gen. Virol.*, 71, 1399, 1990.
23. Gow, J. W. et al., Enteroviral RNA sequences detected by polymerase chain reaction in muscle of patients with postviral fatigue syndrome, *Brit. Med. J.*, 302, 692, 1991.
24. DeFreitas, E. et al., Retroviral sequences related to human T-lymphotropic virus type II in patients with chronic fatigue immune dysfunction syndrome, *Proc. Natl. Acad. Sci. U.S.A.*, 88, 2922, 1991.
25. Khan, A. S. et al., Assessment of a retrovirus sequence and other possible risk factors of the chronic fatigue syndrome in adults, *Annu. Intern. Med.,* 118, 241, 1993.
26. Landay, A. L. et al., Chronic fatigue syndrome: clinical conditions associated with immune activation, *Lancet,* 338, 707, 1991.
27. Aoki, T. et al., Low NK and its relationship to chronic fatigue syndrome, *Clin. Immunol. Immunopathol.,* 69, 253, 1993.
28. Morrison, L. J. A., Behan, W. M. H., and Behan, P. O., Changes in natural killer cell phenotype in patients with post-viral fatigue syndrome, *Clin. Exp. Immunol.*, 83, 441, 1991.
29. Suhadolnik, R. J. et al., Upregulation of the 2-5A synthetase/RNase L antiviral pathway associated with chronic fatigue syndrome, *Clin. Infect. Dis.*, 18, S96, 1994.
30. Suhadolnik, R. J. et al., Biochemical defects in the 2-5A synthetase/RNase L pathway associated with chronic fatigue syndrome with encephalopathy, in *The Clinical and Scientific Basis of Myalgic Encephalomyelitis/Chronic Fatigue Syndrome,* Hyde, B., ed., The Nightingale Foundation Press, Ottawa, 1992, 634.
31. Suhadolnik, R. J. et al., Changes in the 2-5A synthetase/RNase L antiviral pathway in a controlled clinical trial with poly(I)-poly($C_{12}U$) in chronic fatigue syndrome, *In Vivo*, 8, 599, 1994.
32. Strayer, D. R. et al., A controlled clinical trial with a specifically configured RNA drug, poly(I)-poly($C_{12}U$), in chronic fatigue syndrome, *Clin. Infect. Dis.,* 18 (suppl 1), S88, 1994.
33. Strayer, D. R. et al., The antitumor activity of Ampligen, a mismatched double-stranded RNA that modulates the 2-5A synthetase/RNase L pathway in cancer and AIDS, in *Advances in Chemotherapy of AIDS,* Diassio, R. B. and Sommadossi, J.-P., eds., Pergamon Press, New York, 1990, 23.
34. Carter, W. A. et al., Clinical, immunological, and virological effects of Ampligen, A mismatched double-stranded RNA, in patients with AIDS or AIDS-related complex, *Lancet*, 1, 1286, 1987.
35. Suhadolnik, R. J. et al., Evidence for a novel low molecular weight RNase L in chronic fatigue syndrome, *J. Interferon Cytokine Res.*, 17, 377, 1997.
36. Somerville, C., Nishino, S. F., and Spain, J. H., Purification and characterization of nitrobenzene nitroreductase from *Pseudomonas pseudoalcaligenes* JS45, *J. Bacteriol.,* 177, 3837, 1995.
37. Suhadolnik, R. J. et al., Biochemical dysregulation of the 2-5A synthetase/RNase L antiviral defense pathway in chronic fatigue syndrome, *J. Chronic Fatigue Syndrome*, 5, 223, 1999.
38. Snell, C. R. et al., Comparison of maximal oxygen consumption and RNase L enzyme in patients with chronic fatigue syndrome, abstract 27, in Proc. 5th Intl. Res., Clinical and Patient Conference, Am. Assoc. Chronic Fatigue Syndrome, Seattle, WA, 2001.
39. McGregor, N. R. et al., The biochemistry of chronic pain and fatigue, Proc. 2nd World Congress on Chronic Fatigue Syndrome and Related Disorders, Brussels, 1999.

40. Sen, G. C. and Ransohoff, R. M., Interferon-induced antiviral actions and their regulation, *Adv. Virus Res.*, 42, 57, 1993.
41. Shetzline, S. E. et al., Characterization of RNase L dysfunction in chronic fatigue syndrome, abstract 95, Proc. 5th Intl. Res., Clinical and Patient Conference, Am. Assoc. Chronic Fatigue Syndrome, Seattle, WA, 2001.
42. Dong, D. and Silverman, R. H., A bipartite model of 2-5A-dependent RNase L, *J. Biol. Chem.*, 272, 22236, 1997.
43. De Meirleir, K. et al., A 37 kDa 2-5A binding protein as a potential biochemical marker for chronic fatigue syndrome, *Am. J. Med.*, 108, 99, 2000.
44. Shetzline, S. E. et al., Characterization of a 2-5A dependent 37-kDa RNase L. Two-dimensional gel electrophoresis and matrix-assisted laser desorption/ionization mass spectrometry, manuscript submitted.
45. Bisbal, C. et al., Cloning and characterization of a RNase L inhibitor, a new component of the interferon-regulated 2-5A pathway, *J. Biol. Chem.*, 270, 13308, 1995.
46. Vojdani, A., Choppa, P.C., and Lapp, C.V., Downregulation of RNase L inhibitor correlates with upregulation of interferon-induced proteins (2-5A synthetase and RNase L) in patients with chronic fatigue immune dysfunction syndrome, *J. Clin. Lab. Immunol.*, 50, 1, 1998.
47. Aitken, A. and Learmonth, M., Quantitation of cysteine residues and disulfide bonds by electrophoresis, (1996) in *The Protein Protocols Handbook,* Walker, J. M., ed., Humana Press, NJ, 1996, 489.
48. Silverman, R. H., 2-5A-Dependent RNase L: a regulated endoribonuclease in the interferon system, in *Ribonucleases. Structures and Functions*, D'Alessio, G. and Riordan, J. F., eds., Academic Presss, New York, 1997, chap. 16.
49. Shetzline, S. E. and Suhadolnik, R. J., Characterization of a 2-5A dependent 37-kDa RNase L. 2. Azido photoaffinity labeling and 2-5A dependent activation, *J. Biol. Chem.*, 276, 23707, 2001.
50. Silverman, R. H., Functional analysis of 2-5A-dependent RNase and 2-5A using 2',5'-oligoadenylate-cellulose, *Anal. Biochem.*, 144, 450. 1985.
51. Suhadolnik, R.J. et al., 2- and 8-Azido photoaffinity probes. I. Enzymatic synthesis, characterization, and biological properties of 2- and 8-azido photoprobes of 2-5A and photolabeling of 2-5A binding proteins, *Biochemistry,* 27, 8840, 1988.
52. Nolan-Sorden, N. L. et al., Photochemical cross-linking in oligonucleotide-protein complexes between a bromine-substituted 2-5A analog and 2-5A-dependent RNase by ultraviolet lamp or laser, *Anal. Biochem.*, 184, 298, 1990.
53. Stark, G. R. et al., How cells respond to interferons, *Annu. Rev. Biochem.*, 67, 227, 1998.
54. Kariko, K. et al., Phosphorothioate analogues of 2',5-oligoadenylate. Activation of 2',5'-oligoadenylate-dependent endoribonuclease by 2',5'-phosphorothioate cores and 5'-monophosphates, *Biochemistry,* 26, 7136, 1987.
55. Charachon, G. et al., Phosphorothioate analogues of (2'-5')(A_4): Agonist and antagonist activities in intact cells, *Biochemistry,* 29, 2550, 1990.
56. Sobol, R. W. et al., Inhibition of HIV-1 replication and activation of RNase L by phosphorothioate/phosphodiester 2',5'-Oligoadenylate derivatives, *J. Biol. Chem.*, 270, 5963, 1995.
57. Diaz-Guerra, M., Rivas, C., and Esteban, M., Full activation of RNase L in animal cells requires binding of 2-5A within ankyrin repeats 6 to 9 of this interferon-inducible enzyme, *J. Interferon Cytokine Res.*, 19, 113, 1999.
58. Cooper, A. A. and Stevens, T. H., Protein splicing: excision of intervening sequences at the protein level, *BioEssays,* 15, 667, 1993.

59. Herst, C. V. et al., The low molecular weight ribonuclease L present in peripheral blood mononuclear cells of CFS patients is formed by proteolytic cleavage of the native enzyme, abstract 65, Proc. 5th Intl. Res., Clinical and Patient Conference, Am. Assoc. Chronic Fatigue Syndrome, Seattle, WA, 2001.
60. Voges, D., Zwickle, P., and Baumeister, W., The 26S proteasome: a molecular machine designed for controlled proteolysis, *Annu. Rev. Biochem.,* 68, 1015, 1999.
61. Borish, L. et al., Chronic fatigue syndrome: identification of distinct subgroups on the basis of allergy and psychologic variables, *J. Allergy Clin. Immunol.,* 102, 222, 1998.
62. Patarca, R. et al., Interindividual immune status variation patterns in patients with chronic fatigue syndrome: association with gender and tumor necrosis factor system, *J. Chronic Fatigue Syndrome,* 2, 13, 1996.

4 Ribonuclease L Inhibitor: A Member of the ATP-Binding Cassette Superfamily

Patrick Englebienne, C. Vincent Herst, Anne D'Haese, Kenny De Meirleir, Lionel Bastide, Edith Demettre, and Bernard Lebleu

CONTENTS

4.1 INTRODUCTION

The activity of ribonuclease L (RNase L) is regulated by its natural inhibitor, the RNase L inhibitor (RLI). This protein was isolated from an expression library by binding 2',5'-oligoadenylates (2-5A) and subsequently cloned and characterized.[1] This polypeptide is likely to regulate the activity of RNase L by inhibiting the binding of 2-5A by the enzyme, and hence its dimerization and activation.[2] Counter to the other proteins involved in the 2-5A pathway, the expression of RLI is not regulated by type I interferons.[1] Recently, RLI has been classified as a member of the ATP-binding cassette (ABC) superfamily,[3] and we have suggested[4] that its downregulation observed in chronic fatigue syndrome (CFS)[5] might contribute to the development

0-8493-1046-6/02/$0.00+$1.50

of the chanelopathies observed in this illness.[6] These data are likely to suggest that the interactions between RNase L and RLI play an important role in CFS so far unsuspected. This aspect will be reviewed and extended in the present chapter.

4.2 CHARACTERISTICS OF THE PRIMARY STRUCTURE OF RNase L INHIBITOR

Human RLI is a polypeptide comprising 599 amino acids.[1] The physical characteristics of the protein are summarized in Table 4.1. The protein is composed to one-fourth (26.2%) by charged amino acid residues, of which 72 are negatively and 85 positively charged, respectively. The protein is particularly rich in proline and phenylalanine residues (respectively 5 and 4.2%) but poor in histidine (1.5%) and tryptophane (0.3%) residues. A further statistical analysis of the protein sequence[7] reveals neither specific charge clusters nor significant charge patterns, but reveals

TABLE 4.1
Physical Characteristics of Human RLI

Number of amino acids	599
Molecular weight	67,558.7 Da
pI	8.86
Extinction coefficient ($\varepsilon_{280\ nm}$)	37,460 ± 678 $mol^{-1}\ cm^{-1}$
Number of negatively charged residues (Asp + Glu)	72
Number of positively charged residues (Arg + Lys)	85

Amino Acid Composition:	**n**	**%**
Ala (A)	36	6.0
Arg (R)	33	5.5
Asn (N)	23	3.8
Asp (D)	35	5.8
Cys (C)	17	2.8
Gln (Q)	23	3.8
Glu (E)	37	6.2
Gly (G)	36	6.0
His (H)	9	1.5
Ile (I)	43	7.2
Leu (L)	59	9.8
Lys (K)	52	8.7
Met (M)	15	2.5
Phe (F)	25	4.2
Pro (P)	30	5.0
Ser (S)	30	5.0
Thr (T)	29	4.8
Trp (W)	2	0.3
Tyr (Y)	20	3.3
Val (V)	45	7.5

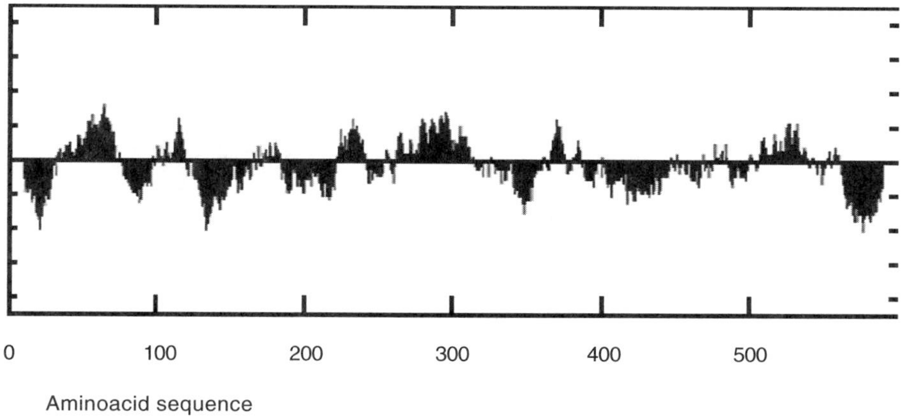

FIGURE 4.1 Hydrophobicity (hydropathy) profile of human RLI.

two tandem and periodic motif repeats: LSGGELQR (residues 217 to 224 and 462 to 469) and VVEHD (residues 271 to 275 and 517 to 521), respectively. The hydrophobicity profile of the protein displayed in Figure 4.1 does not show any significant hydrophobic segment which could be indicative of the presence of a transmembrane region (spanning helix). RLI is a globular soluble protein. The protein sequence contains a four-amino acid KPKK pattern (AA 16 to 19) indicative of nuclear targeting. The protein is also likely to be present in mitochondria and lysosomes. The RLI sequence contains two N-glycosylation sites and six N-myristoylation sites identified in Table 4.2. The protein sequence also contains several possible phosphorylation sites, summarized in Table 4.3.

The analysis of the amino acid sequence of RLI allows us to identify several significant features. First, starting from the N-terminal end, the sequence contains a ferredoxin iron-binding motif.[1,8] This is a 4Fe-4S signature from residue 55 to 66 (CIGCGICIKKCP). The possible role for this homology is presently unknown. It has been suggested that these motifs could act either by promoting RLI–DNA

TABLE 4.2
N-glycosylation and N-myristoylation Sites in the Human RLI Sequence

	Motif	AA Sequence
N-glycosylation	NGTG	381–384
	NVSY	408–411
N-myristoylation	GCGICI	57–62
	GLVGTN	107–112
	GIGKSA	113–118
	GTVGSI	182–187
	GVPSAY	291–296
	GTGKTT	382–387

TABLE 4.3
Possible Phosphorylation Sites Present in the Amino Acid Sequence of Human RLI

Kinase	Motif	RNase L AA Sequence
Protein kinase C	THR	84–86
	TIR	258–260
	SVR	304–306
	TGK	383–385
	SYK	410–412
	SPK	417–419
	SVR	423–425
	TFR	565–567
	SIK	582–584
Tyrosine kinase	RGSELQNY	148–155
Caseine kinase II	SNLE	77–80
	SILD	186–189
	SGGE	218–221
	SYLD	245–248
	SVLD	277–280
	SVRE	304–307
	TANE	336–339
	TDSE	370–373
	SGGE	463–466
	SQLE	560–563
	SIKD	582–585

interactions,[1] or by impairing RNase L dimerization,[4] respectively. The sequence of RLI contains two conserved ATP/GTP-binding site motifs (phosphate-binding loop motifs, P-loops) between residues 110 to 117 (GTNGIGKS) and 379 to 386 (GENGTGKT), respectively. Finally, the human RLI sequence contains two conserved ABC transporters family motifs, respectively LSGGELQRFACAVV (residues 217 to 230) and LSGGELQRVRLRLCL (residues 462 to 476). The first of these motifs is situated between the two ATP-binding sites, which is the signature pattern for this class of proteins.[9,10] These characteristic features of the RLI sequence are summarized in Table 4.4.

4.3 RNase L INHIBITOR IS A MEMBER OF THE ABC SUPERFAMILY

Every cell of our organism is surrounded by a plasma membrane which protects it from the external world. However, proper cellular function requires a cross-talk with the external world, the uptake of nutrients, and the excretion of metabolic waste and toxic substances. With the aim at communication, cells have developed membrane receptors which send specific internal signals once triggered by their

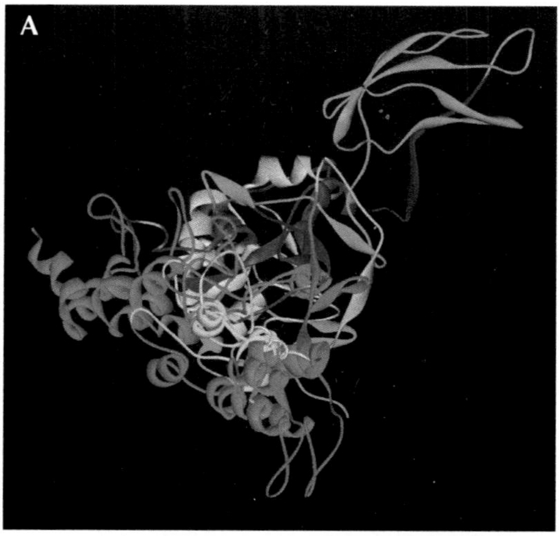

CHAPTER 2 FIGURES 2.9A and 2.9B Part A. 3D-model of human RNase L. The protein is presented in ribbon colored according to the parts presented in Figure 2.3. The ankyrin domain is in red and includes the 2-5A binding site (yellow) and the two P-loops (GKT, magenta). The protein kinase-like domain is in blue and includes the cysteine-rich region (blue-green). The catalytic domain is in green. Part B. Structure of the 2-5A-binding site filled with the 2-5A trimer (ball and stick). The interacting residues are numbered and represented by their surface atoms.

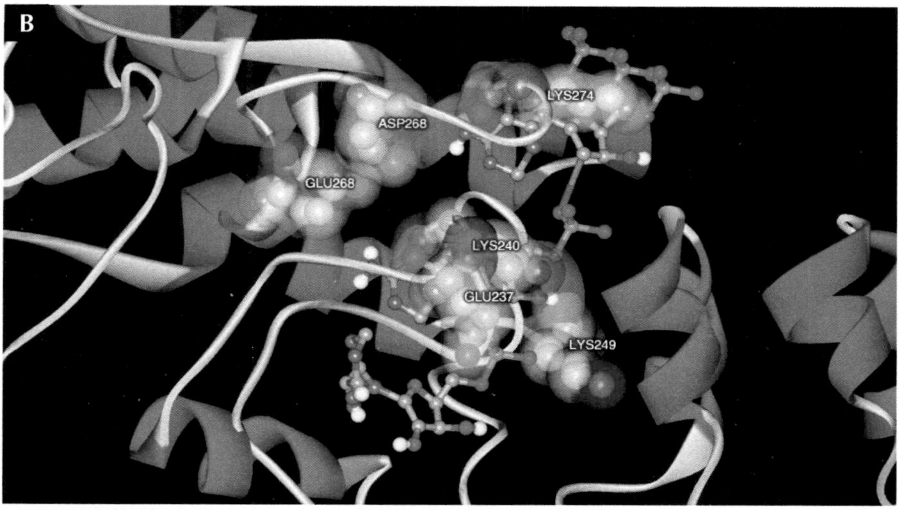

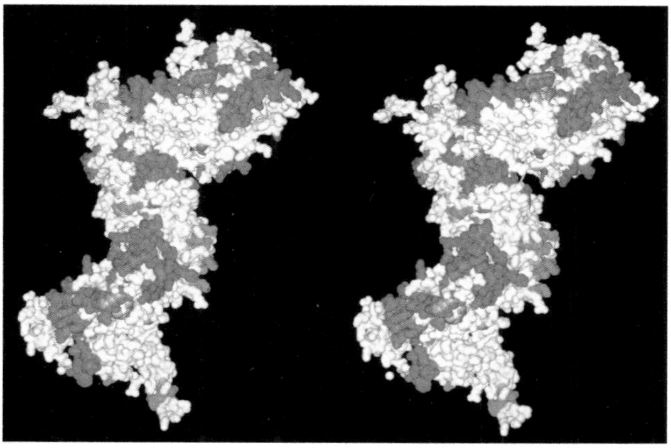

CHAPTER 2 FIGURE 2.10 Stereo view of the 3D-model of RNase L homodimer. The proteins are shown as surfaces colored according to the secondary structure. The atoms of the 2-5A trimers docked in the respective binding sites are presented as balls.

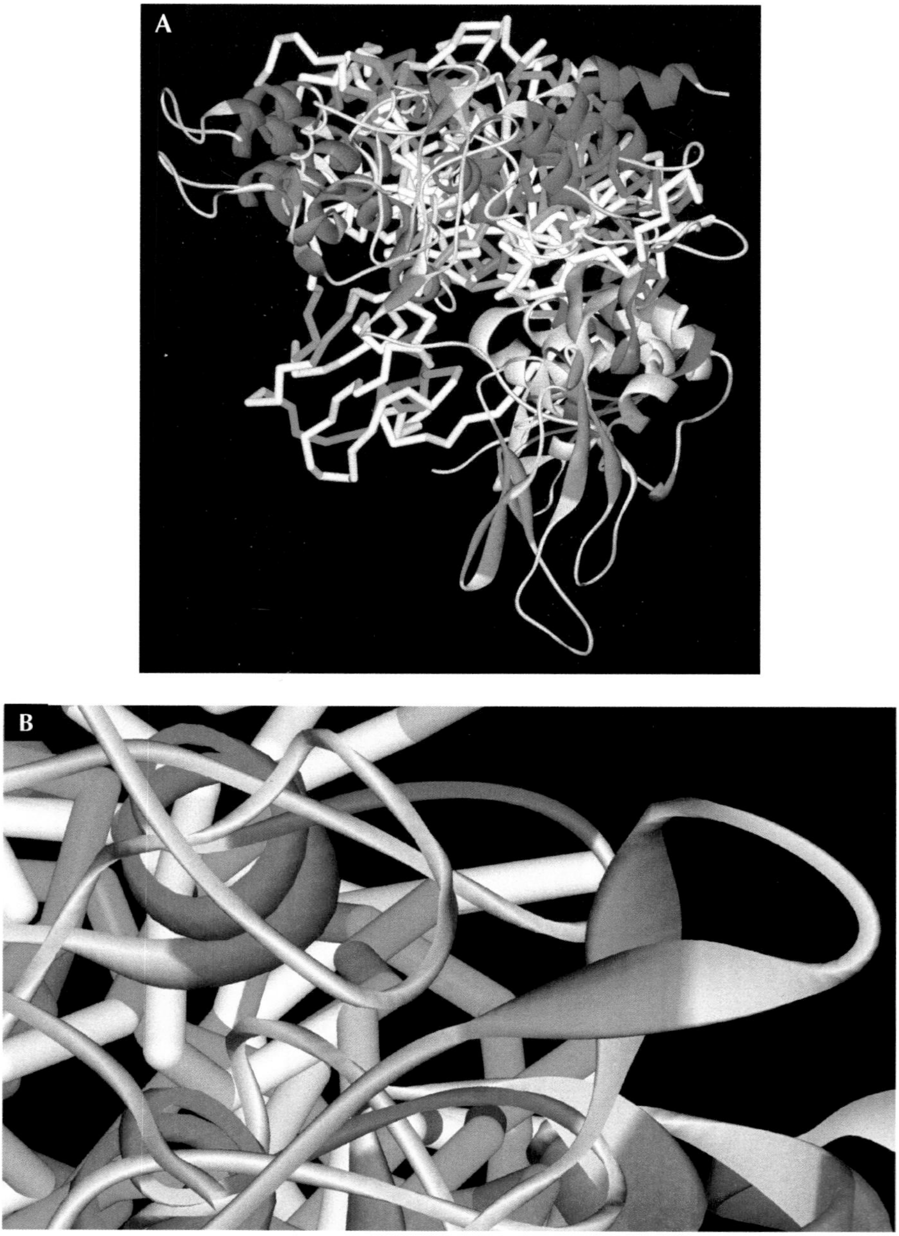

CHAPTER 4 FIGURES 4.6A and 4.6B Model showing the interaction between the AIIK motif of RLI (tube) with the ankyrin domain of RNase L (ribbon). Part A. General overview of the complex. The two P-loops of RNase L (magenta) are imparied to bind 2-5A by steric hindrance and the ferredoxin domain of RLI (green) lines up with the finger-like region of RNase L. Part B. Close-up of the 2-5A binding site.

TABLE 4.4
Summary of the Characteristic Features of Human RLI Sequence. Alternative possibilities in the consensus are in brackets and X represents any amino acid. Amino acids in curly brackets are those not acceptable at the given position.

Signature	Consensus	Motif in RLI	Residues
4Fe-4S ferredoxins, iron-sulfur binding	$CX_2CX_2CX_3C$[PEG]	CIGCGICIKKCP	55–66
ATP/GTP-binding site	[AG]X_4GK[ST]	GTNGIGKS	110–117
		GENGTGKT	379–386
ABC transporters family	[LIVMFYC][SA][SAPGLVFYKQH]G [DENQMW][KRQASPCLIMFW] [KRNQSTAVM][KRACLVM][LIVMF YPAN]{PHY}[LIVMFW][SAGCLI VP]{FYWHP}{KRHP}[LIVMFYW STA]	LSGGELQRFACAVV	219–230
		LSGGELQRVRLLCL	462–476

ligand (Chapter 5). For the exchange of chemicals and ions, cells have developed transporters and channels. These membrane proteins function in different ways: transporters bind the substrate that they transfer across the membrane, and channels undergo conformational changes which allow them to open or close for the passage of ions across the membrane.

A common feature of these transport systems is their sensitivity to ATP hydrolysis.[11] This sensitivity to ATP can be dependent or independent to protein kinases and phosphatases.[12] Protein kinase-independent mechanisms involve lipid kinases, metal chelation, hydrolysis by depolymerized actin, and direct modulation by ATP binding.[11] The ATP-binding cassette (ABC) superfamily is made of membrane proteins containing two P-loops (ATP/GTP-binding sites); these two P-loops function as ATPases. These ATP-binding units are either equivalent and work alternatively to hydrolyze ATP or nonequivalent; i.e., one binding unit only is associated with catalysis.[13-15] These differences in ATP binding and capacity for hydrolysis probably explain why the proteins of this superfamily are capable of acting as active transporters (like the multidrug resistance proteins), channels (like the cystic fibrosis transmembrane conductance regulator), or switches (i.e., receptors) taking part in more complex systems (like the sulfonylurea receptor which regulates the Kir6.2 subunit of the ATP-sensitive potassium channel).[13,16]

Besides the two conserved P-loops (Walker A and B motifs) and the ABC signature sequence located between these two ATP-binding sites, most members of the ABC proteins superfamily are also characterized by transmembrane domains. These comprise six spanning helices. Consequently, the structure of these proteins is characterized by various alternating arrangements of the transmembrane domains (TMB) and ABC signatures (ABC) in the complete sequence. The different arrangements so far observed in the superfamily are summarized in Figure 4.2. The minimal requirement for an active membrane protein transporter seems to be two TMB and

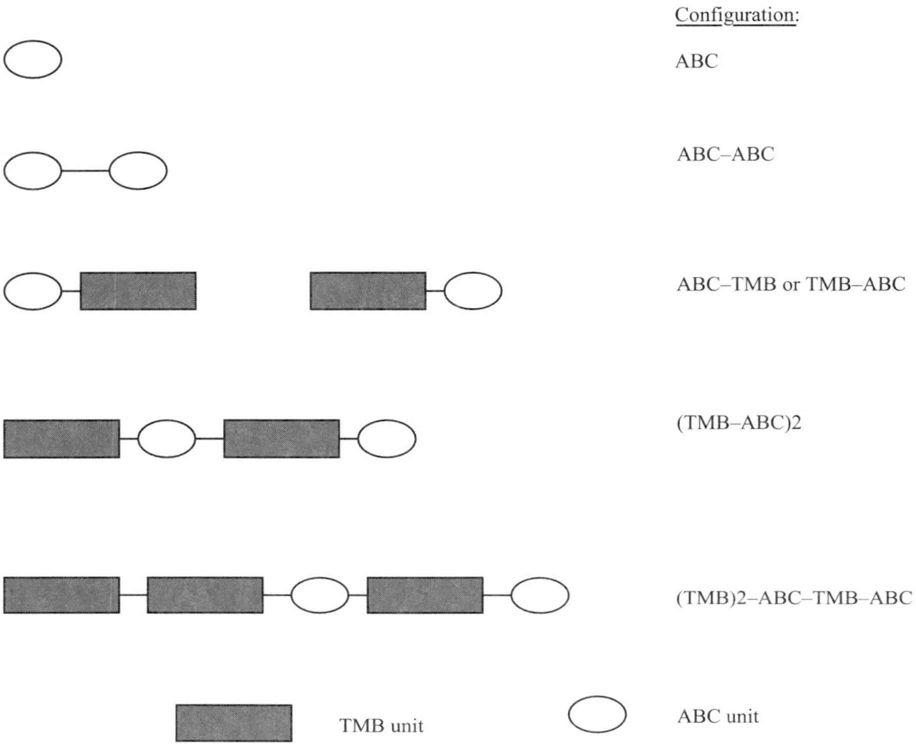

FIGURE 4.2 Various arrangements of the ABC and transmembrane (TMB) domains in the ABC superfamily.

two ABC units. (The proteins of the White family [ABCG] that have the TMB–ABC configuration dimerize to become active membrane transporters.) According to the sequence homologies within the members of the superfamily, eight subfamilies have been identified, of which RLI (also known as the organic anion-binding protein, OABP) currently makes a subfamily by itself. Table 4.5 summarizes the characteristics of the known human members of these different subfamilies.[3,17,18] The complexity of the protein structure decreases progressively in the subfamilies. RLI is classified in the ABC subfamily E as member one with a structural ABC–ABC arrangement.

4.4 PHYLOGENETIC ORIGIN OF THE RNase L INHIBITOR

A search into the protein domains database (ProDom)[19] allows us to relate the amino acid sequence of RLI to many ABC transporters which include the iron-binding signature in other species. Such comparative analysis permits us to draw tentative phylogenetic trees among the species, which are summarized in Figure 4.3. This tree is likely to indicate that mammalian, and more specifically human, RLI did not

TABLE 4.5
Characteristics of the Members of the Human ABC Superfamily Classified According to Their Subfamilies

Subfamily	Name	Synonyms	Length of Sequence	Configuration
ABCA	ABCA1	ABC1	2259	$(TMB–ABC)_2$
	ABCA2	ABC2	>2174	$(TMB–ABC)_2$
	ABCA3	ABC3, ABC-C	1704	$(TMB–ABC)_2$
	ABCA4	ABCR, Rim protein	2273	$(TMB–ABC)_2$
	ABCA5	EST90625	?	?
	ABCA6	EST155051	?	?
	ABCA7	ABC4	>1840	$(TMB–ABC)_2$
	ABCA8	KIAA0822	1581	?
	ABCA9	EST640918	?	?
	ABCA10	EST698739	?	?
	ABCA11	EST1133530	?	?
ABCB	ABCB1	P-glycoprotein 1 (Pgp), MDR1	1280	$(TMB–ABC)_2$
(MDR-multidrug	ABCB2	TAP1	686	TMB–ABC
resistance/TAP-	ABCB3	TAP2	748	TMB–ABC
antigen transport)	ABCB4	Pgp3, MDR3	1279	$(TMB–ABC)_2$
	ABCB7	ABC7	752	TMB–ABC
	ABCB8	M-ABC1	718	TMB–ABC
	ABCB11	BSEP, SPGP	1321	$(TMB–ABC)_2$
ABCC	ABCC1	MRP1	1531	$(TMB)_2ABC–TMB–ABC$
(MRP, multidrug	ABCC2	MRP2, cMOAT, cMRP	1545	$(TMB)_2ABC–TMB–ABC$
resistance	ABCC3	MRP3, MOAT-D, cMOAT-2	1527	$(TMB)_2ABC–TMB–ABC$
proteins/CFTR)	ABCC4	MRP4, MOAT-B	1325	$(TMB–ABC)_2$
	ABCC5	MRP5, MOAT-C, pABC11	1437	$(TMB–ABC)_2$
	ABCC6	MRP6, MLP-1	1503	$(TMB)_2ABC–TMB–ABC$
	ABCC7	CFTR	1480	$(TMB–ABC)_2$
	ABCC8	Sulfonylurea receptor (SUR)1	1581	$(TMB)_2ABC–TMB–ABC$
	ABCC9	SUR2	1549	$(TMB)_2ABC–TMB–ABC$
	ABCC10	MRP7		
ABCD	ABCD1	ALD, ALDP	745	TMB–ABC
	ABCD2	ALDL1, ALDR	740	TMB–ABC
	ABCD3	PXMP1, PXMP70	659	TMB–ABC
	ABCD4	PXMP1L, P70R	606	TMB–ABC
ABCE	ABCE1	RNase L inhibitor, OABP	599	ABC–ABC
ABCF	ABCF1	ABC50	807	ABC–ABC
ABCG	ABCG1	ABC8, White	638	ABC–TMB
(White)	ABCG2	MXR1 (mitoxanthrone resistance), BRCP	655	ABC–TMB
ANSA	ANSA1		332	ABC
	ANSA2		348	ABC

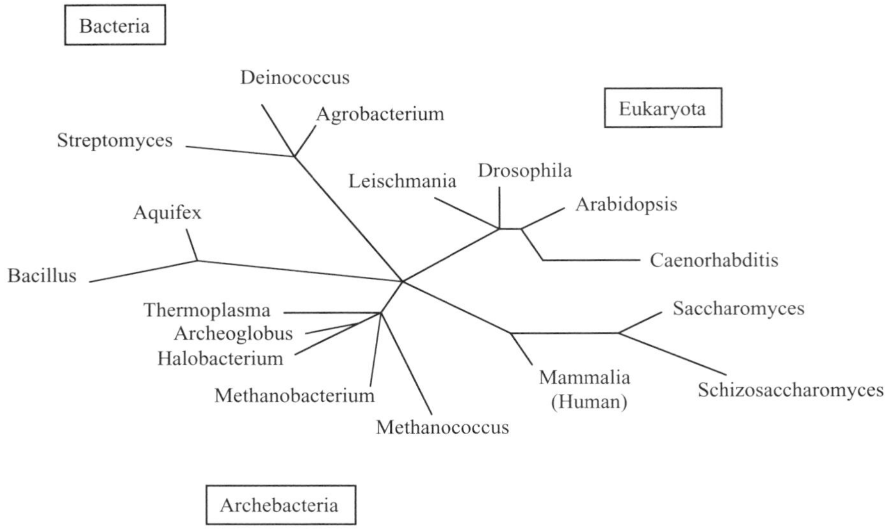

FIGURE 4.3 Tentative phylogenetic relationships between RLI-like (ABC transporter) proteins in various species.

differentiate very much from the root and is related to similar proteins in yeasts, worms, plants, and even Archebacteria. To illustrate this point, it is certainly worth mentioning that the sequence of human RLI shares, respectively, 93% identity with mouse RLI (599 aa, accession GenBank GI 3273417), 70% identity with an RNase L inhibitor-like protein from *Arabidopsis thaliana* (600 aa, accession emb CAA16710.1), 62% identity with an unidentified protein from *Caenorhabditis elegans* (610 aa, accession emb CAB54424.1), 60% identity with a putative RNase L inhibitor from *Schizosaccharomyces pombe* (593 aa, accession emb CAA19324.1), 45% identity with a hypothetical ABC transporter from *Methanococcus jannaschii* (600 aa, accession SwsissProt Q58129), and 43% identity with the RNase L inhibitor from *Pyrococcus abyssi* (593 aa, accesssion emb CAB50155.1).

Among the human ABC transporter superfamily, RLI is currently the single known member of subfamily E (Table 4.5). However, RLI shares important similarities to other members of this superfamily, which include ABC3 (multidrug resistance, ABCA3[20]), the Rim protein (photoreceptor-associated transporter, ABCA4[21]), TAP1 (antigen peptide transporter, ABCB2[22]), MDR3 (multidrug resistance, ABCB4[23]), ABC7 (iron transport, ABCB7[24]), multidrug resistance proteins MRP3 (ABCC3[25]), MRP6 (ABCC6[26]), the cystic fibrosis transmembrane conductance regulator (CFTR, ABCC7[27]), the sulfonylurea receptor (SUR1, ABCC8[28]), and the white protein (ABCG1[29]). As shown in Figure 4.4, the identities are particularly strinking in the P-loops and the ABC domains (residues 20 to 60 and 140 to 180, respectively). The similarities between the complete amino acid sequences of these proteins, including RLI, are significant and allow us to draw a tree showing the relationship between these proteins and common ancestor genes, illustrated in Figure 4.5.

```
                        10         20         30         40         50         60
                ....|....| ....|....| ....|....| ....|....| ....|....| ....|....|
RLI 81-298      ---KETTHRY CANAFKLHRL PIPRPGEVLG LVGTNGIGKS AALKILAGKQ KP-NLGKYDD
RLI 348-599     YKYPGMKKKM GEFELAIVAG EFT-DSEIMV MLGENGTGKT TFIRMLAGRL KPDEGGEVPV
SUR1 818-905    ---------- ---------- ---------- ---------- ---------- ----------
MRP6 646-785    ---------- ----LHRINL TVP-QGCLLA VVGPVGAGKS SLLSALLGEL SK-VEGFVSI
MRP3 644-860    ---------- ----LHSLDI QVP-KGALVA VVGPVGCGKS SLVSALLGEM EK-LEGKVHM
ABC8 86-275     ---PWWRKKG YKTLLKGISG KFN-SGELVA IMGPSGAGKS TLMNILAGYR ETGMKGAVLI
CFTR 446-610    ---------- ---------F KIE-RGQLLA VAGSTGAGKT SLLMMIMGEL EP-SEGKIKH
CFTR 1236-1415  ---------- ---------- ----PGQRVG LLGRTGSGKS TLLSAFLRLL NT--EGEIQI
TAP1 529-720    ---------- ---------- --R-PGEVTA LVGPNGSGKS TVAALLQNLY QP-TGGQLLL
MDR3 422-675    ---------- ---------- -----GQTVA LVGSSGCGKS TTVQLIQRLY DP-DEGTINI
MDR3 1058-1275  ---------- ---------- EVK-KGQTLA LVGSSGCGKS TVVQLLERFY DP-LAGTVLL
ABC3 558-790    ---------- ---------- ----EGQITV LLGHNGAGKT TTLSMLTGLF PP-TSGRAYI
ABC3 1409-1601  ---------- ---------- -----GECFG LLGFNGAGKT TTFKMLTGEE SL-TSGDAFV
Rim 953-1135    ---------- ---------- -FY-ENQITA FLGHNGAGKT TTLSILTGLL PP-TSGTVLV
Rim 1963-2175   ---------- ---------- --R-PGECFG LLGVNGAGKT TTFKMLTGDT TV-TSGDATV
ABC7 484-735    ---------- GQKVLSGISF EVP-AGKKVA IVGGSGSGKS TIVRLLFRFY EP-QKGSIYL

                        70         80         90        100        110        120
                ....|....| ....|....| ....|....| ....|....| ....|....| ....|....|
RLI 81-298      PPD---WQEI LTYFRGSELQ NYFTKILEDD LKAIIKPQYV ARFLRLAKGT VGSILDRKDE
RLI 348-599     LN-------- ------VSYK PQK--ISPKS TGSVRQLLHE K-------IR DAYTHP----
SUR1 818-905    ---------- ---------- ---------- ---------- ---------- ----------
MRP6 646-785    EG-------- -----AVAYV PQEAWVQNTS VVENVCFGQE LD---PPWLE RVLEAC---A
MRP3 644-860    KG-------- -----SVAYV PQQAWIQNCT LQENVLFGKA LN---PKRYQ QTLEAC---A
ABC8 86-275     NGLP---RDL RCFRKVSCYI MQDDMLLPHL TVQEAMMVSA H-----LKLQ EKDEGR---R
CFTR 446-610    SG-------- -----RISFC SQFSWIMPGT IKENIIFGVS YD---EYRYR SVIKAC---Q
CFTR 1236-1415  DGVSWDSITL QQWRKAFGVI PQKVFIFSGT FRKNLDPYEQ WS---DQEIW KVADEV---G
TAP1 529-720    DGKPLPQYEH RYLHRQVAAV GQEPQVFGRS LQENIAYGLT QK-PTMEEIT AAAVKS---G
MDR3 422-675    DGQDIRNFNV NYLREIIGVV SQEPVLFSTT IAENICYGRG N--VTMDEIK KAVKEA---N
MDR3 1058-1275  DGQEAKKLNV QWLRAQLGIV SQEPILFDCS IAENIAYGDN SRVVSQDEIV SAAKAA---N
ABC3 558-790    SGYE-ISQDM VQIRKSLGLC PQHDILFDNL TVAEHLYFYA Q----LKGLS RQKCPE---E
ABC3 1409-1601  GGHR-ISSDV GKVRQRIGYC PQFDALLDHM TGREMLVMYA R----LRGIP ERHIGA---C
Rim 953-1135    GGRD-IETSL DAVRQSLGMC PQHNILFHHL TVAEHMLFYA Q----LKGKS QEEAQL---E
Rim 1963-2175   AGKS-ILTNI SEVHQNMGYC PQFDAIDELL TGREHLYLYA R----LRGVP AEEIEK---V
ABC7 484-735    AGQNIQDVSL ESLRRAVGVV PQDAVLFHNT IYYNLLYGNI SAS--PEEVY AVAKLA---G

                       130        140        150        160        170        180
                ....|....| ....|....| ....|....| ....|....| ....|....| ....|....|
RLI 81-298      TKTQAIVCQQ LDLTHLKERN VEDLSGGELQ RFACAVVCIQ KADIFMFDEP SSYLDVKQRL
RLI 348-599     Q-FVTDVMKP LQIENIIDQE VQTLSGGELQ RVRLRLCLGK PADVYLIDEP SAYLDSEQR-
SUR1 818-905    ---------- -DQTQIGER- GINLSGGQRQ RISVARALYQ HANVVFLDDP FSALDIHLSD
MRP6 646-785    LQPDVDSFPE GIHTSIGEQ- GMNLSGGQKQ RLSLARAVYR KAAVYLLDDP LAALDA----
MRP3 644-860    LLADLEMLPG GDQTEIGEK- GINLSGGQRQ RVSLARAVYS DADIFLLDDP LSAVDSHVAK
ABC8 86-275     EMVKEILTAL GLLSCANTR- TGSLSGGQRK RLAIALELVN NPPVMFFDEP TSGLDSASCF
CFTR 446-610    LEEDISKFAE KDNIVLGEG- GITLSGGQRA RISLARAVYK DADLYLLDSP FGYLDVLTEK
CFTR 1236-1415  LRSVIEQFPG KLDFVLVDG- GCVLSHGHKQ LMCLARSVLS KAKILLLDEP SAHLDPVTYQ
TAP1 529-720    AHSFISGLPQ GYDTEVDEA- GSQLSGGQRQ AVALARALIR KPCVLILDDA TSALDANSQ-
MDR3 422-675    AYEFIMKLPQ KFDTLVGER- GAQLSGGQKQ RIAIARALVR NPKILLLDEA TSALDTESE-
MDR3 1058-1275  IHPFIETLPH KYETRVGDK- GTQLSGGQKQ RIAIARALIR QPQILLLDEA TSALDTESE-
ABC3 558-790    VKQMLHII-- GLEDKWNSR- SRFLSGGMRR KLSIGIALIA GSKVLILDEP TSGMDAISRR
ABC3 1409-1601  VENTLRGL-- LLEPHANKL- VRTYSGGNKR KLSTGIALIG EPAVIFLDEP STGMDPVARR
Rim 953-1135    MEAMLEDT-- GLHHKRNEE- AQDLSGGMQR KLSVAIAFVG DAKVVILDEP TSGVDPYSRR
Rim 1963-2175   ANWSIKSL-- GLTVYADCL- AGTYSGGNKR KLSTAIALIG CPPLVLLDEP TTGMDPQARR
ABC7 484-735    LHDAILRMPH GYDTQVGER- GLKLSGGEKQ RVAIARAILK DPPVILYDEA TSSLDSITEE
```

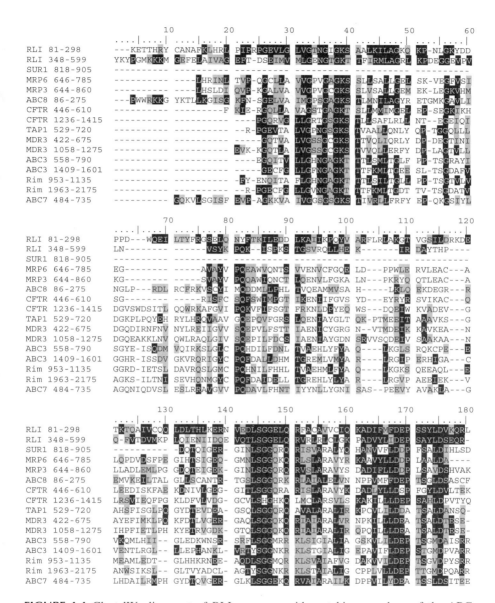

FIGURE 4.4 ClustalW alignment of RLI sequence with matching members of the ABC superfamily of proteins. For proteins containing two ABC domains, these latter were aligned separately. Identities and similarities are boxed black and gray, respectively.

```
                 ....|....| ....|....| ....|....| ....|....| ....|....| ....|....|
RLI 81-298       KAAITIRSLI N--PDRYIIV VEHDLSVLDY LSDFICCLY- GVP--SAYG- ----------
RLI 348-599      LMAARVVKRF ILHAKKTAFV VEHDFIMATY LADRVIVFD- GVPSKNTVAN SPQTLLAGMN
SUR1 818-905     HLMQAGILEL LRDDKRTVVL VTHKLQYLPH -ADWIIAMKD G--------- ----------
MRP6 646-785     ---------- ---------- ---------- ---------- ---------- ----------
MRP3 644-860     HIFDHVIGPE GVLAGKTRVL VTHGISFLPQ T-DFIIVLAD GQ---VSEMGP YPALLQRNG
ABC8 86-275      QVVSLMKGLA Q--GGRSIIC TIHQPSAK-- ---------- ---------- ----------
CFTR 446-610     EIFESCVCK- -LMANKTRIL VTSKMEHL-- ---------- ---------- ----------
CFTR 1236-1415   IIRRTLKQAF ---ADCTVIL CEHRIEAMLE CQQFLV---- ---------- ----------
TAP1 529-720     LQVEQLLYES PERYSRSVLL ITQHLSLVEQ -ADHILFLEG GA---I---- ----------
MDR3 422-675     AEVQAALDKA RE--GRTTIV IAHRLSTVRN -ADVIAGFED GV---IVEQGS HSELMKKEG
MDR3 1058-1275   KVVQEALDKA RE--GRTCIV IAHRLSTIQN -ADLIVVFQN GR---VKEHGT HQQLLAQKG
ABC3 558-790     AIWDLLQRQ- -- SDRTIVL TTHFMDEADL LGDRIAIMAK GE---LQCCGS SLFLKQKYG
ABC3 1409-1601   LLWDTVARAR --ESGKAIII TSHSMEECEA LCTRLAIMVQ GQ---FKCLGS PQHL------
Rim 953-1135     SIWDLLLKY- --RSGRTIIM STHHMDEADL LGDRIAIIA- ---------- ----------
Rim 1963-2175    MLWNVIVSII --REGRAVVL TSHSMEECEA LCTRLAIMVK GA---FRCMGT IQHLKSKFG
ABC7 484-735     TILGAMKDVV K---HRTSIF IAHRLSTVVD -ADEIIVLDQ GK---VAERGT HHGLLANPH

                          250        260        270        280        290
                 ....|....| ....|....| ....|....| ....|....| ....|....| ...
RLI 81-298       ---------- ---------- ---------- ---------- ---------- ---
RLI 348-599      KFLSQLEITFR RDPNNYRPRI NKLNSIKDVE QKKSGNYFFL DD-------- --
SUR1 818-905     ---------- ---------- ---------- ---------- ---------- ---
MRP6 646-785     ---------- ---------- ---------- ---------- ---------- ---
MRP3 644-860     SFANFLCNYAP DEDQGH---- ---------- ---------- ---------- --
ABC8 86-275      ---------- ---------- ---------- ---------- ---------- ---
CFTR 446-610     ---------- ---------- ---------- ---------- ---------- ---
CFTR 1236-1415   ---------- ---------- ---------- ---------- ---------- ---
TAP1 529-720     ---------- ---------- ---------- ---------- ---------- ---
MDR3 422-675     VYFKLVNMQTS GSQIQSEEFE LNDEKAATRM APNGWKSRLF RHSTQKNLKN SQ
MDR3 1058-1275   IYFSMVSVQAG ---------- ---------- ---------- ---------- --
ABC3 558-790     AGYHMTLVKEP HCNPEDISQL VHHHVPNATL ESSA------ ---------- --
ABC3 1409-1601   ---------- ---------- ---------- ---------- ---------- ---
Rim 953-1135     ---------- ---------- ---------- ---------- ---------- ---
Rim 1963-2175    DGYIVTMKIKS PK-------- ---------- ---------- ---------- --
ABC7 484-735     SIYSEMWHTQS SRVQNHDNPK WEAKKENISK EEERKK---- ---------- --
```

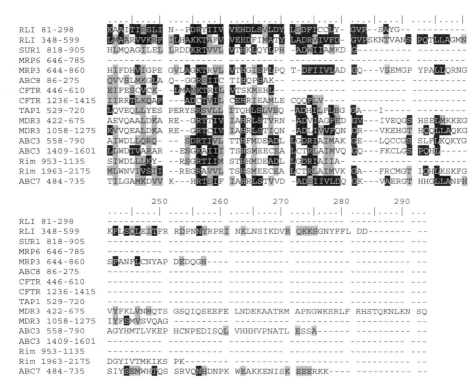

FIGURE 4.4 *Continued.*

```
       +-------------------------Rim  (ABCA4)
       !
       !                    +-----SUR1 (ABCC8)
    +--7   +----------------8
    !  !   !                !  +--MRP6 (ABCC6)
    !  !   !                +--6
    !  +--5                    +--MRP3 (ABCC3)
    !     !
    !     !        +-------------CFTR (ABCC7)
    !     !        !
    !     +--------3  +-----------MDR3 (ABCB4)
  --1              !  !
    !              +--4  +--------TAP1 (ABCB2)
    !                 !  !
    !                 +--9      +--RLI  (ABCE1)
    !                    !  +-10
    !                    +--2  +--ABC8 (ABCG1, white)
    !                       !
    !                       +-----ABC7 (ABCB7)
    !
    +----------------------------ABC3 (ABCA3)
```

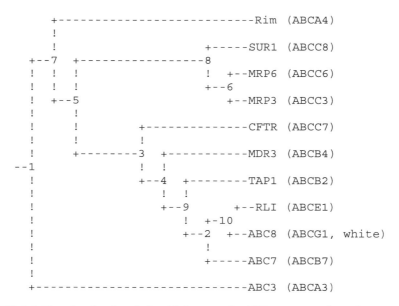

FIGURE 4.5 Tree showing the relationship between the ABC transporters homologous to RLI.

4.5 THE RNS4I GENE, LOCALIZATION, AND BIOLOGICAL ROLES OF THE RNase L INHIBITOR

The gene coding for the members of the ABC superfamily is not chromosome-clustered. The gene coding for RLI has been mapped to chromosome locus 4q31.[8,30] Among the currently known members of the superfamily, only the ABCG2 gene has been localized on the same chromosome.[3] As mentioned above, RLI is not induced by interferons.[1] However, RLI is induced by viruses such as encephalomyocarditis virus (EMCV), and evidence has been provided for the induction of RLI, at mRNA and protein levels, under the influence of synthetic dsRNA such as poly(I):poly(C).[31] Consequently, RLI is likely to be induced by a process very similar to that inducing type I interferons.

As implied by its name, the primary role of RLI is the regulation by inhibition of RNase L.[1,8] However, its membership in the ABC superfamily suggests that it might exert other biological duties. Data available on the tissue distribution of RLI[8] are likely to indicate a wide distribution, like the protein which it regulates, i.e., RNase L. RLI expression at the mRNA level, however, is likely to depend on the proliferative capacity of the tissue considered, the mRNA being expressed at a lesser extent in non- or poorly proliferative tissues, than in tissues with a high proliferative capacity.[8] No data are available on RLI subcellular distribution, except that, according to its amino acid sequence (see above), the protein is soluble and thus probably present in the cytoplasm, and the presence of a nuclear localization signal (KPKK) might indicate nuclear targeting. Interestingly, similar nuclear localization signals are also present in the amino acid sequence of ABC50, the only member of ABC subfamily F identified so far in the human genome.[3] More striking is the fact that ABC50 is the only member of the ABC superfamily known to share an ABC–ABC structure similar to that of RLI (Table 4.5). Moreover, ABC50 is likely to be induced by the tumor necrosis factor α (TNF-α),[32] which might suggest important roles for the members of subfamilies E and F in infection and inflammation.

RLI is particularly effective in the downregulation of RNase L activity. The transfection of HeLa cells with RLI antisense cDNA leads to the reversal of RNase L inhibition observed in EMCV-infected cells.[33] Similarly, RLI increases during the time-course of T-cell infection by immunodeficiency virus type 1, which results in a downregulation of RNase L.[34] But surprisingly, RLI mRNA has recently been reported to be suppressed in chronic hepatitis C,[35] despite an overexpression of the dsRNA-dependent protein kinase. Because RNase L is capable of degrading not only viral, but also cellular mRNA (Chapter 2), a possible role for RLI in mRNA protection has been suggested for some time; this role progressively receives the attention it deserves. Mouse models have shown that, in some tissues such as lungs, RLI mRNA increases by 40% in young animals when compared to older animals.[36] More recently,[37] the protective role of RLI on mouse MyoD mRNA expression and hence its regulating role in muscle cell differentiation has been demonstrated. However, our understanding of the role of RLI and its relationship to RNase L during infection and in cellular development and regulation remains severely limited.

4.6 INTERACTIONS OF RNase L INHIBITOR WITH RNase L

RLI inhibits both the binding of 2-5A to RNase L and its enzymatic activation in a dose-dependent manner, but this effect can be reversed.[1] It has been demonstrated that RNase L inhibition occurs neither through a competition by RLI for 2-5A-binding, nor from a degradation by RLI of the 2-5A molecules. Rather, the inhibition occurs from the interaction of both proteins, which exerts a blockade by RLI of the 2-5A-binding site of RNase L.[1,38] Therefore, the interaction between RLI and RNase L takes place on the ankyrin domain of the enzyme (Chapter 2).

So far, the ankyrin-binding motif of RLI remains speculative. It was suggested earlier[1,8] that the ferredoxin iron-binding signature of RLI could be involved in the interaction, due to its rich cysteine content. However, repetitive cysteine residues are barely involved in the interaction of proteins with ankyrins. Rather, arginine-rich domains are involved such as in the interaction of the tumor suppressor p53 with the ankyrin domain of 53BP2,[39] or tyrosine-containing motifs such as the L[IL]GY domain in the proteins of the ETS family[40] and the FIGQY domain of the L1CAM family of proteins, which all interact with ankyrins.[41] We have analyzed RLI sequence for possible ankyrin-interacting peptides[4] and have identified a tetrapeptide motif (AIIK, residues 166 to 169), which is strongly analogous to the tetra- or pentapeptide clusters (respectively, ALLK or ALLLK) involved in the interaction between Na^+, K^+-ATPase, or the erythroid anion exchanger with ankyrins.[42-44] The only difference between these motifs is the conservative replacement of leucines by isoleucines in RLI. However, it has been noted[45] in the case of thyroid receptors' interaction with the co-regulators that such isoleucine could substitute to leucine in the LXXLL motif without any impairment in the protein-receptor interactions. Consequently, we can argue that this motif in RLI sequence serves for the interaction of the protein with the ankyrin repeat domain of RNase L.

In order to better appreciate this type of interaction, we made a three dimensional- (3D) model of RLI and tethered the AIIK motif on the ankyrin β-hairpin tips of the 3D-model of RNase L. The results are displayed in Figure 4.6.* The binding of RLI to the ankyrin domain of RNase L results in RLI completely surrounding the RNase L ankyrin structure, including the two P-loops (magenta in Figure 4.6), which is likely to induce a steric hindrance that could account for the loss of 2-5A-binding activity in the heterodimeric complex. Moreover, the interaction results in relaxing the cysteine-rich finger-like domain of RNase L, which has been shown to be involved in homodimerization and activation.[46] This effect is accompanied by a positioning of the cysteine-rich ferredoxin-like domain of RLI (colored green in Figure 4.6A) very close to the latter RNase L site, which might act as a mock target site for dimerization.[4]

The interaction between RLI and the ankyrin domain of RNase L can take place through hydrophobic contacts by the alanine–isoleucine–isoleucine triad (residues 166-168) and an ionic contact by lysine (residue 169). The target residues on the RNase L ankyrin domain might be situated either in the first and second hairpin

* Color insert figures follow page 76.

A

B

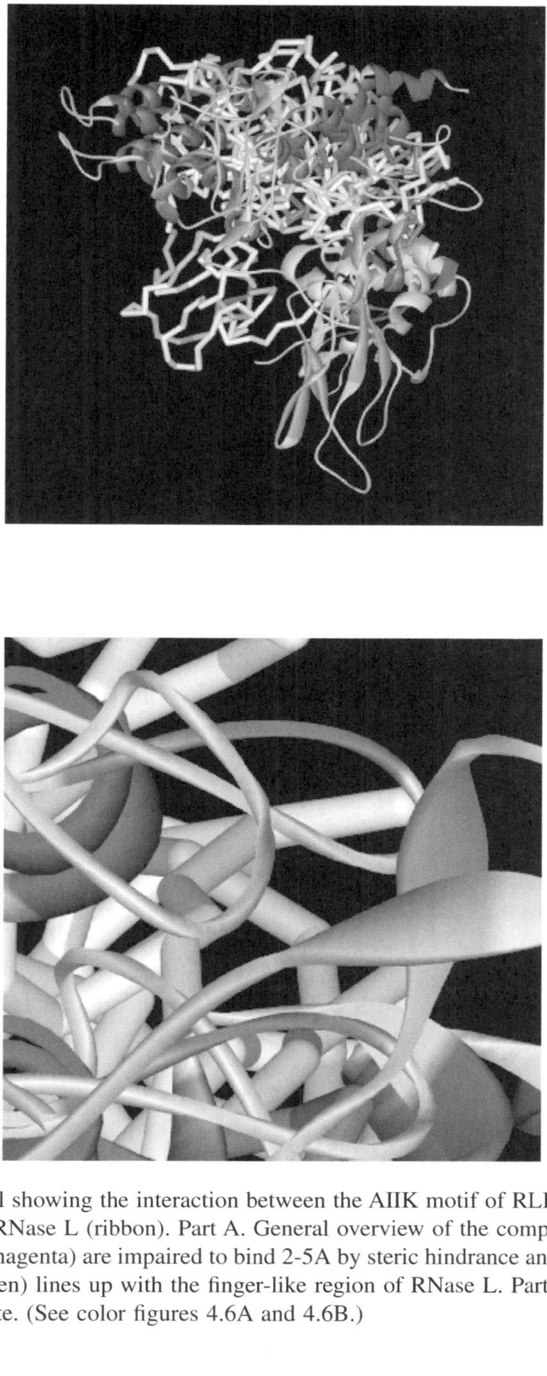

FIGURE 4.6 Model showing the interaction between the AIIK motif of RLI (tube) with the ankyrin domain of RNase L (ribbon). Part A. General overview of the complex. The two P-loops of RNase L (magenta) are impaired to bind 2-5A by steric hindrance and the ferredoxin domain of RLI (green) lines up with the finger-like region of RNase L. Part B. Close-up of the 2-5A-binding site. (See color figures 4.6A and 4.6B.)

loops (hydrophobic contacts with phenylalanine-53, tryptophane-60, first loop, and alanine-93, second loop; ionic contact with threonine-94, second loop), or in the third and fourth hairpin loops (hydrophobic contacts with phenylalanines-123 and 130, third loop, and alanine-169, fourth loop; ionic contact with threonine-170, fourth loop), respectively. The third and fourth loops of the ankyrin-repeat domain of RNase L are located just before the two P-loops that make the 2-5A-binding site (hairpin loops six and seven). These two possibilities for interaction might explain why the RLI molecule is capable of interaction with both the 80-kDa and the truncated 37-kDa RNase L.[38] The latter form is likely to lose the first ankyrin-repeats motifs upon proteolytic cleavage of the former, native protein.[4]

4.7 RNase L INHIBITOR AND CHRONIC FATIGUE SYNDROME

Besides chronic debilitating fatigue, chronic fatigue syndrome (CFS) is characterized by a series of other symptoms reminiscent of voltage-gated ion channelopathies,[47,48] such as hypokaliemic periodic paralysis, skeletal muscle pain, ventricular hypercontractility, drenching night sweats, and cognitive defects.[49] The involvement of improper ion channel function in CFS has been proposed[6,50] as an explanation to these symptoms.

Besides the study by Vojdani's group[5] which mentions a downregulation of RLI at both mRNA and protein expression levels in the peripheral blood mononuclear cells (PBMC) of CFS patients as compared to controls, little is known about RLI in CFS. Recently, we have cloned and expressed recombinant RLI, which allowed us to raise polyclonal antibodies. We used these antibodies to study the molecular forms of RLI in the PBMC of patients classified according to the ratio of 37- over 80-kDa RNase L.[38] This ratio reflects the presence of proteases in these cells, including calpain, which we suspect to be partly responsible for the cleavage of the native RNase L into its truncated form (Chapters 2 and 3, and Reference 51). As shown in Figure 4.7, the detection of RLI by immunoblotting in PBMC extracts indicates the progressive disappearance of the 68-kDa RLI form with increasing RNase L ratios, which confirms the previous observations.[5] In the samples with the highest RNase L ratios, some low molecular weight fragments are visible, which are not present in the lower ratio samples. This suggests that RLI might not only be downregulated in CFS, but could also experience some cleavage by proteases.

The progressive downregulation of RLI in CFS PBMC undoubtedly plays a role in the upregulation of RNase L already described some years ago.[52,53] This downregulation of RLI decreases the interaction of the protein with the ankyrin domain of RNase L. Ankyrins constitute a family of proteins containing a consensus repeat sequence which links integral membrane components (by N-terminal interaction) with cytoskeletal elements (by C-terminal interaction).[54] These interactions play fundamental roles in diverse biological activities by controlling membrane topology, elasticity, and protein composition.[55] The ankyrin domain of native RNase L (Chapter 2) is unable to fulfill these roles because its C-terminal portion is linked to the kinase homology region of the protein and is consequently unavailable for interaction.

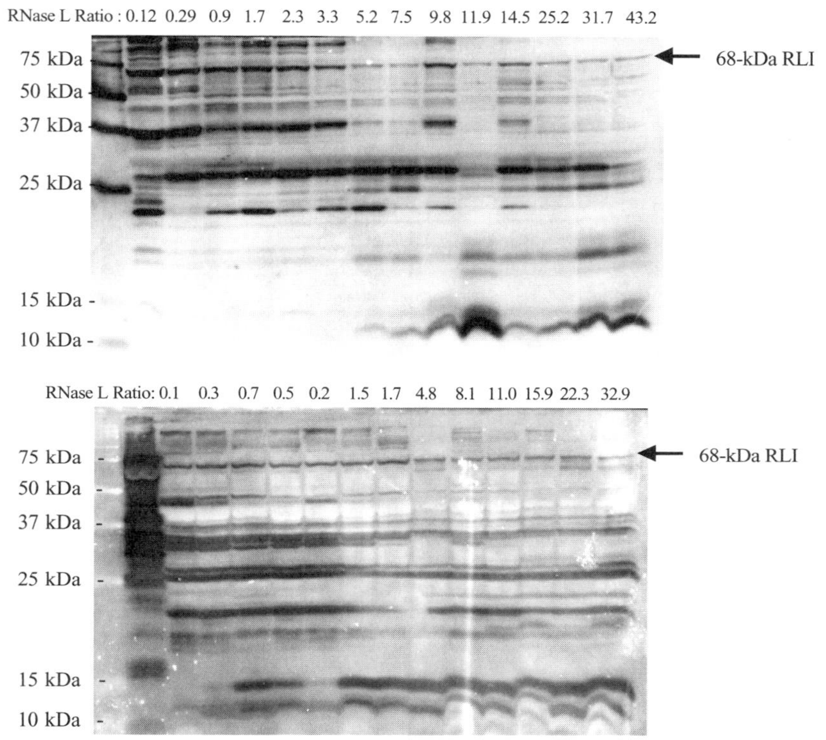

FIGURE 4.7 Molecular forms of RLI in PBMC extracts as a function of RNase L ratio analyzed by immunoblotting with two different antibodies. The 68-kDa RLI molecule is less expressed in samples with high RNase L ratio and some cleavage is likely to occur.

However, cleavage of RNase L by proteases such as calpain, which is likely to occur in CFS PBMC,[4] releases the ankyrin domain of the protein which then comprises both N- and C-terminal ends. This suggests that this ankyrin protein is capable of interaction with integral membrane components. The removal of terminal domains of erythrocyte ankyrin by calpain has similarly been suggested to play a role in the modulation of its activity.[56] The ankyrin domain of RNase L interacts with a tetrapeptide motif (AIIK) homologous to the motif by which membrane ion exchangers interact with their cognate ankyrin proteins.[42-44]

Surprisingly, all the ABC transporters that share strong sequence similarities with RLI (Figures 4.4 and 4.5) contain variants of the same motif. The 15 amino acid stretches of the ABC transporters AE1 and the Na^+, K^+-ATPase containing this motif are aligned in Figure 4.8. The transporters are aligned according to their degree of relationship to RLI as displayed on Figure 4.5. The striking similarities and identities within the ALLK/AIIK motif suggest that all these transporters would be capable of interacting with the ankyrin fragment of RNase L upon its release by proteolytic cleavage. As depicted schematically in Figure 4.9, such a competition with the cognate ankyrin protein exerted by the RNase L ankyrin fragment for

```
                    1         10
                    ....|....|  ....|
RLI         161     EDDLKAII-K  PQYVAR
ABC8        627     FQKSEAILRE  LD-VEN
TAP1        651     VALARALIRK  PC-VLI
SUR1        359     VLLFLALLLQ  RTFLQ
Rim        1351     LQHVQALLVK  R-FQHT
MRP6       1410     LCLARALLRK  TQ-ILI
MRP3       1434     VCLARALLRK  SR-ILV
MDR3        151     QKFFHAILRQ  EIGWF
CFTR        321     SVLPYALI-K  GIILRK
ABC7        617     VAIARAIL-K  DPPVIL
ABC3        881     SDGIGALI-E  EERTAV
AE1         151     EELLRALLLK  HSHAG
NaKATPase   461     DASETALL-K  FSELTL
```

FIGURE 4.8 ClustalW alignment of the 15 amino acid stretch of the ABC transporter similar to RLI containing the motif homologous to the motif by which the anion exchanger AE1 and the Na, K-ATPase interact with ankyrins.

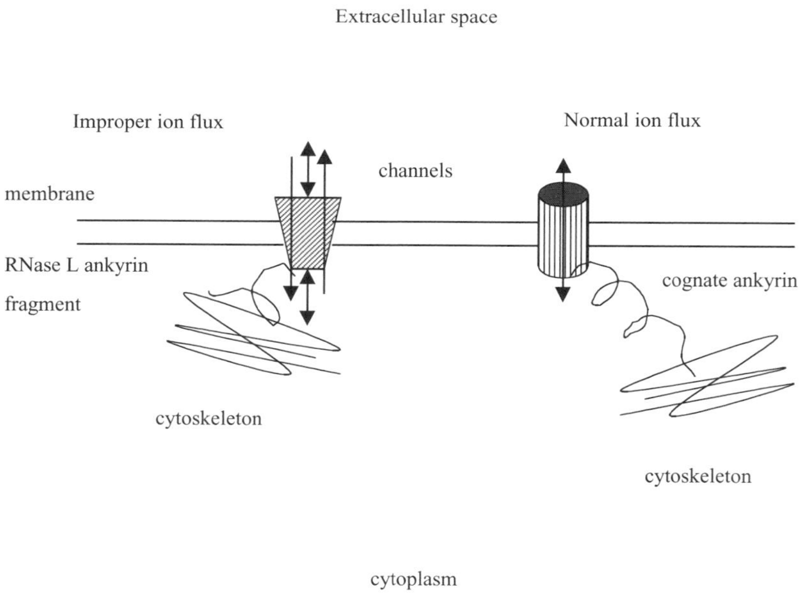

FIGURE 4.9 Schematic representation of the dysfunctions possibly occurring upon competition with a cognate ankyrin protein by the RNase L ankyrin fragment for binding ABC transporters on the cell membranes. The dysfunction can lead to a blockade of ion transport or to excess transport in one or another direction.

binding the ion channel can lead to its improper functioning. Interestingly, all the ABC transporters which can be related to RLI by sequence similarity exert important cellular channel functions that can be associated with various symptoms or abnormalities observed in CFS. Many of the ABC channels which share sequence similarities with RLI are members of the multidrug resistance transporters group (MDR and MRP),[20,57] including MRP3 (ABCC3), MRP6 (ABCC6), and MDR3 (ABCB3). These multidrug transporters bind chemically dissimilar cytotoxic compounds and pump them out of the cells.[58]

Hypersensitivity to chemical exposure is frequently reported by CFS patients[59,60] and has even been proposed as the etiology of some cases.[61] Consequently, hypersensitivity to chemical exposure might be explained by a dysfuncton in these channels. Besides the MDR and MRP channels, the amino acid sequence of RLI also shares striking similarities (Figure 4.4) with CFTR (ABCC7), SUR1 (ABCC8), ABC3 (ABCA3), ABC7 (ABCB7), ABC8 (ABCG1), TAP1 (ABCB2), and Rim (ABCA4). CFTR is a chloride channel that regulates anion movement across a transmembrane pore. The mutated *cftr* gene leads to cystic fibrosis.[62] CFTR regulates the exocrine function of many epithelial tissues including the pancreas, intestinal glands, biliary tree, bronchial, and sweat glands.[3] The normal function of CFTR also involves the regulation of Na^+ channels,[63] which play a primary role in the generation of pain and inflammatory hyperalgesia in peripheral neurones.[64-66] Moreover, in normal conditions, the MDR3 channel and CFTR share a similar physiological role.[67] Any dysfunction of these channels may therefore lead not only to the drenching night sweats reported by CFS patients, but also to the muscular and cardiac symptoms, the irritable bowel, and the shift of pain sensitivity threshold experienced by subsets of these patients, including those suffering from secondary fibromyalgia.[49,68-72]

The sulfonylurea receptor (SUR 1) takes part of the ATP-sensitive potassium channel (K_{ATP}) of which it regulates the opening function in the presence of magnesium ions.[73] SUR1 switches K_{ATP} on or off, depending on the concommitant binding of ATP and ADP.[13,73] Any increase in the ATP/ADP ratio closes the channel, which leads to the opening of the voltage-dependent calcium channel. In the β-cell of the pancreas, this mechanism plays a key role in the regulation of glucose-induced insulin secretion, and any dysfunction in this highly regulated system might explain the transient abnormalities in glucose metabolism observed in subsets of CFS sufferers. The human ABC3 protein, which is homologous to ced-7 of *Caenorhabditis elegans*, plays a key role in the engulfment of apoptotic cell bodies. ABC3 has been proposed to function in both dying and engulfing (macrophages) cells by translocating to the membrane the molecules that mediate homotypic recognition and adhesion between the respective surfaces of these cells.[3,17] ABC8, which is the mammalian homolog of the drosophila white protein involved in eye pigmentation,[74] is a regulator of macrophage cholesterol and phospholipid transport in humans.[75] Similarly intriguing, TAP1 plays a critical role in antigen processing and presentation by MHC class I molecules.[76] Consequently, any dysfunction in these three channels can be at the origin of the altered immune function and reactivity associated with CFS.[49,68,77-79]

Human ABC8 has also been implied in the cellular uptake of tryptophan, a precursor of the neurotransmitter serotonin,[80] and mutations in the gene coding for

ABC7 have been shown to lead to ataxia.[81] Their improper function, along with that of MDR, which is also normally involved in monoamine neurotransmitter transport in the brain,[82] could consequently lead to or at least reinforce the many central nervous system abnormalities associated with CFS.[49,68,83-89] Finally, the absence of retinal ABC transporter Rim in retinal pigment epithelium results in all-trans retinaldehyde accumulation, which leads to delayed dark adaptation with secondary photoreceptor degeneration.[90,91] The physiological roles of these various ABC transporters and their associated genetic disorders, along with the CFS symptoms that their dysfunctions might explain, are summarized in Table 4.6.

4.8 CONCLUSIONS AND PROSPECTS

RLI downregulation in CFS PBMC undoubtedly plays a major role in the RNase L upregulation previously observed[52,53] in these same cells. However, the upregulation in RNase L catalytic activity is not likely to result only from RLI downregulation, but also from a difference in dependence to 2-5A between the 37-kDa cleaved form found in CFS PBMC and the native 80-kDa enzyme.[92]

In Chapter 2, as well as in a previous publication,[4] we show that this pathological cleavage of the native RNase L is accompanied by the release of small ankyrin fragments which contain the site of interaction with RLI. Here we show that RLI is a member of the ABC transporters family and that its amino acid sequence is strongly similar to other members of that family. Taken together with the simultaneous downregulation of RLI and the release of ankyrin fragments from RNase L, these observations consequently suggest that the ankyrin fragments are capable of interacting with the ABC transporters similar to RLI. This competitive interaction with their natural cognate ankyrin protein can lead to a dysfunction of these transporters as schematically displayed in Figure 4.9. Such discussion is based on a hypothesis that requires further biological support. Currently, we are transfecting RNase L-deficient cells with DNA sequences coding for the ankyrin fragment of the RNase L molecule. An analysis of the ion channeling functions of these cells will allow us to gather further insight into the proposed pathological mechanism.

This hypothesis is far from gratuitous, however. Interestingly enough, all the ABC transporters identified as analogous to RLI play physiological roles of which a dysfunction can be related to various CFS symptoms. The occurence of channelopathies in CFS had already been suspected on grounds of clinical observations with cardiac muscle thallium uptake.[6,50] We extend this suspicion to other channels and to cells other than muscle, including neuronal and immune cells. Such extension is backed up by the linear correlation observed between the extent of RNase L cleavage and the increase in intracellular calcium concentration in PBMC (Chapter 6). The regulation of different ion channels at the cytoplasmic level is such a fine-tuning system that any dysfunction may have tremendous consequences.[93] In CFS, the consequences can span from autonomic nervous system and neuroimmunomodulation dysfunction up to oxidative alterations in muscle,[94-98] and through apoptotic dysfunctions (Chapter 6), in which ion channel dysregulations are known to play critical roles in immune cells as well as neurones.[99-101] We hypothesize that the effect of ion channel dysregulation might further be modulated by the opportunistic viral

TABLE 4.6
Summary of the Physiological Roles of the ABC Transporters Similar to RLI Including Their Associated Genetic Diseases with the CFS Symptoms That Their Dysfunctions Might Explain

ABC Transporter	Synonym	Physiological Role/Defect	CFS-Related Symptoms
ABCC8	Sulfonylurea receptor; SUR1	Switches K_{ATP} on/off; in pancreas, regulation of glucose-induced insulin secretion	Transient hypoglycemia
ABCC7	Cystic fibrosis transmembrane conductance regulator; CFTR	Regulates exocrine function of many epithelial tissues; interacts with Na^+ channel; mutation leads to cystic fibrosis	Drenching night sweats; sarcoidosis; shift in pain sensitivity threshold
ABCC6	MRP6	Terminal excretion of cytotoxic substances	Hypersensitivity to toxic chemicals
ABCC3	MRP3; MOAT-D	Terminal excretion of cytotoxic substances	Hypersensitivity to toxic chemicals
ABCG1	White homolog; ABC8	Regulator of macrophage cholesterol and phospholipid homeostasis; tryptophan uptake	Immunodeficiency/macrophage dysfunction; depression
ABCB4	Pgp3; MDR3	Transport of monoamine neurotransmitters; interacts with Na^+ channel	Dysfunction in transport of monoamine neurotransmitter; pain sensitivity threshold reduced; muscle K^+ loss
ABCB2	TAP1	Processing and presentation of antigens by MHC class I	Immunodeficiency/Th2 switch
ABCB7	ABC7	Transport of heme from mitochondria to cytosol; mutation leads to ataxia	Anemia; CNS abnormalities
ABCA3	ABC3; ABC-C	Recognition of apoptotic cells by macrophages for engulfment	Immunodeficiency/Th2 switch
ABCA4	ABCR; Rim	Dysfunction leads to all-trans retinaldehyde accumulation	Visual problems

infections or reactivations currently observed in the course of CFS evolution.[48,69,102] Retroviral sequences related to human T-lymphotropic virus type II (HTLV II)[103] as well as mycoplasmas[104] have been retrieved by PCR in CFS patients. Viruses such as human immunodeficiency virus type-1 have been recently demonstrated to activate ion channel functions in macrophages,[105] and some viruses have developed strategies to escape immune surveillance by affecting the function of other ABC transporters such as TAP1.[106] Mycoplasma infection, in turn, has been implied in the sensitization of immune cells to apoptosis by various mechanisms.[107]

Taken together, all the considerations laid in the present chapter undoubtedly shed a new light on the development of CFS. The only intent of the authors is to pave the way to new research avenues.

REFERENCES

1. Bisbal, C. et al., Cloning and characterization of a RNase L inhibitor. A new component of the interferon-regulated 2-5A pathway, *J. Biol. Chem.,* 1995; 270: 13308–17.
2. Dong, B. and Silverman, R. H., 2-5A-dependent RNase molecules dimerize during activation by 2-5A, *J. Biol. Chem.,* 1995; 270: 4133–7.
3. Klein, I., Sarkadi, B., and Varadi, A. An inventory of the human ABC proteins, *Biochim. Biophys. Acta,* 1999; 1461: 237–62.
4. Englebienne, P. et al., Interactions between RNase L ankyrin-like domain and ABC transporters as a possible origin for pain, ion transport, CNS, and immune disorders of chronic fatigue immune dysfunction syndrome, *J. Chronic Fatigue Syndrome,* 2001; 8: 83–102.
5. Vojdani, A., Choppa, P. C., and Lapp, C. W., Down-regulation of RNase L inhibitor correlates with up-regulation of interferon-induced proteins (2-5A synthetase and RNase L) in patients with chronic fatigue immune dysfunction syndrome, *J. Clin. Lab. Immunol.,* 1998; 50: 1–16.
6. Chaudhuri, A. et al., The symptoms of chronic fatigue syndrome are related to abnormal ion channel function, *Med. Hypotheses,* 2000; 54: 59–63.
7. Brendel, V. et al., Methods and algorithms for statistical analysis of protein sequences, *Proc. Nat. Acad. Sci. U.S.A.,* 1992; 89: 2002–6.
8. Aubry, F. et al., Chromosomal localization and expression pattern of the RNase L inhibitor, *FEBS Lett.,* 1996; 381: 135–9.
9. Higgins, C. F. et al., A family of related ATP-binding subunits coupled to distinct biological processes in bacteria, *Nature*, 1986; 323: 448–50.
10. Doolittle, R. F. et al., Domainal evolution of a prokaryotic DNA repair protein and its relationship to active-transport proteins, *Nature*, 1986; 323: 451–3.
11. Hilgemann, D. W., Cytoplasmic ATP-dependent regulation of ion transporters and channels: mechanisms and messengers, *Annu. Rev. Physiol.,* 1997; 59: 193–220.
12. Herzig, S. and Neumann, J., Effects of serine/threonine protein phosphatases on ion channels in excitable membranes, *Physiol. Rev.,* 2000; 80: 173–210.
13. Ueda, K. et al., Comparative aspects of the function and mechanisms of SUR1 and MDR1 proteins, *Biochim. Biophys. Acta,* 1999; 1461: 305–13.
14. Nagel, G., Differential function of the two nucleotide binding domains on cystic fibrosis transmembrane conductance regulator, *Biochim. Biophys. Acta,* 1999; 1461: 263–74.
15. Seibert, F. S. et al., Influence of phosphorylation by protein kinase A on CFTR at the cell surface and endoplasmic reticulum, *Biochim. Biophys. Acta,* 1999; 1461: 275–283.
16. Bryan, J. and Aguilar-Bryan, L., Sulfonylurea receptors: ABC transporters that regulate ATP-sensitive K^+ channels, *Biochim. Biophys. Acta,* 1999; 1461: 285–303.

17. Broccardo, C., Luciani, M.-F., and Chimini, G., The ABCA subclass of mammalian transporters, *Biochim. Biophys. Acta,* 1999; 1461: 395–404.
18. Borst, P. et al., The multidrug resistance superfamily, *Biochim. Biophys. Acta,* 1999; 1461: 347–57.
19. Corpet, F. et al., ProDom and ProDOmCG: tools for protein domain analysis and whole genome comparisons, *Nucleic Acids Res.,* 2000; 28: 267–9.
20. Klugbauer, N. and Hofmann, F., Primary structure of a novel ABC transporter with a chromosomal localization on the band encoding the multidrug resistance-associated protein, *FEBS Lett.,* 1996; 391: 61–5.
21. Allikmets, R. et al., A photoreceptor cell-specific ATP-binding transporter gene (ABCR) is mutated in recessive Stargardt macular dystrophy, *Nature Genet.,* 1997; 15: 236–46.
22. Beck, S. et al., DNA sequence analysis of 66kb of the human MHC class II region encoding a cluster of genes for antigen processing, *J. Mol. Biol.,* 1992; 228: 433–41.
23. Van der Bliek, A. M. et al., Sequence of mdr3 cDNA encoding a human P-glycoprotein, *Gene*, 1988; 71: 401–11.
24. Shimada, Y. et al., Cloning and chromosomal mapping of a novel ABC transporter gene (hABC7), a candidate for X-linked sideroblastic anemia with spinocerebellar ataxia, *J. Hum. Genet.,* 1998; 43: 115–22.
25. Kiuchi, Y. et al., cDNA cloning and inducible expression of human multidrug resistance associated protein 3 (MRP3), *FEBS Lett.,* 1998; 433: 149–52.
26. Kool, M. et al., Expression of human MRP6, a homologue of the multidrug resistance protein gene MRP1, in tissues and cancer cells, *Cancer Res.,* 1999; 59: 175–82.
27. Riordan, J. R. et al., Identification of the cystic fibrosis gene: cloning and characterization of complementary DNA, *Science*, 1989; 245: 1066–73.
28. Thomas, P. M. et al., Mutations in the sulfonylurea receptor gene in familial persistent hyperinsulinemic hypoglycemia of infancy, *Science*, 1995; 268: 426–9.
29. Chen, H. et al., Cloning of the cDNA for a human homologue of the *Drosophila* white gene and mapping to chromosome 21q22.3, *Am. J. Hum. Genet.,* 1996; 59: 66–75.
30. Diriong, S. et al., Localization of the ribonuclease L inhibitor gene (RNS4I), a new member of the interferon-regulated 2-5A pathway, to 4q31 by fluorescence *in situ* hybridization, *Genomics*, 1996; 82: 486–90.
31. Martinand, C. et al., The RNase L inhibitor (RLi) is induced by double-stranded RNA, *J. Interferon Cytokine Res.,* 1998; 18: 1031–8.
32. Richard, M., Drouin, R., and Beaulieu, A. D., ABC50, a novel human ATP-binding cassette protein found in tumor necrosis factor-alpha-stimulated synoviocytes, *Genomics*, 1998; 53: 137–45.
33. Martinand, C. et al., RNase L inhibitor (RLI) antisense constructions block partially the down regulation of the 2-5A/RNase L pathway in encephalomyocarditis-virus-(EMCV)-infected cells, *Eur. J. Biochem.,* 1998; 254: 248–55.
34. Martinand, C. et al., RNase L inhibitor is induced during human imunodeficiency virus type 1 infection and down-regulates the 2-5A/RNase L pathway in human T cells, *J. Virol.,* 1999; 73: 290–6.
35. Yu, S. H. et al., Intrahepatic mRNA expression of interferon-inducible antiviral genes in liver diseases: dsRNA-dependent protein kinase overexpression and RNase L inhibitor suppression in chronic hepatitis C, *Hepatology*, 2000; 32: 1089–95.
36. Semsei, I. and Goto, S., Expression of mRNAs of pancreatic and L type RNase inhibitors as a function of age in different tissues of SAMP8 and BDF1 mice, *Mech. Ageing Dev.,* 1997; 97: 249–61.

37. Bisbal, C. et al., The 2'-5' oligoadenylate/RNase L/RnaseL inhibitor pathway regulates both MyoD mRNA stability and muscle cell differentiation, *Mol. Cell Biol.*, 2000; 20: 4959–69.
38. De Meirleir, K. et al., A 37kDa 2-5A binding protein as a potential biochemical marker for chronic fatigue syndrome, *Am. J. Med.*, 2000; 108: 99–105.
39. Gorina, S. and Pavletich, N. P., Structure of the p53 tumor suppressor bound to the ankyrin and SH3 domains of 53BP2, *Science*, 1996; 274: 1001–5.
40. Batchelor, A. H. et al., The structure of GABPα/β: an ETS domain-ankyrin-repeat heterodimer bound to DNA, *Science*, 1998; 279: 1037–41.
41. Zhang, X. et al., Structural requirements for association of neurofascin with ankyrin, *J. Biol. Chem.*, 1998; 273: 30785–94.
42. Jordan, C. et al., Identification of a binding motif for ankyrin on the α-subunit of NA^+, K^+-ATPase, *J. Biol. Chem.*, 1995; 270: 29971–5.
43. Ding, Y., Kobayashi, S., and Kopito, R., Mapping of ankyrin-binding determinants on the erythroid anion exchanger, AE1, *J. Biol. Chem.*, 1996; 271: 22494–8.
44. Zhang, Z. et al., Structure of the ankyrin-binding domain of α-Na, K-ATPase, *J. Biol. Chem.*, 1998; 273: 18681–4.
45. Mahajan, M. and Samuels, H. H., A new family of nuclear receptor coregulators that integrate nuclear receptor signaling through CREB-binding protein, *Mol. Cell. Biol.*, 2000; 20: 5048–63.
46. Dong, B. and Silverman, R. H., Alternative function of a protein kinase homology domain in 2',5'-oligoadenylate-dependent RNase L, *Nucleic Acids Res.*, 1999; 27: 439–45.
47. Lehmann-Horn, F. and Jurkat-Rott, K., Voltage-gated ion channels and hereditary disease, *Physiological Rev.*, 1999; 70: 1317–72.
48. Hoffman, E. P., Voltage-gated ion channelopathies: inherited disorders caused by abnormal sodium, chloride, and calcium regulation in skeletal muscle, *Annu. Rev. Med.*, 1995; 46: 431–41.
49. Komaroff, A. L. and Buchwald, D. S., Chronic fatigue syndrome: an update, *Annu. Rev. Med.*, 1998; 49: 1–13.
50. Watson, W. S. et al., Possible cell membrane transport defect in chronic fatigue syndrome? *J. Chronic Fatigue Syndrome,* 1997; 3: 1–13.
51. Roelens, S. et al., G-actin cleavage parallels 2-5A-dependent RNase L cleavage in peripheral blood mononuclear cells. Relevance to a possible serum-based screening test for dysregulations in the 2-5A pathway, *J. Chronic Fatigue Syndrome,* 2001; 8: 63–82.
52. Suhadolnik, R. J. et al., Up-regulation of the 2-5A synthetase/RNase L antiviral pathway associated with chronic fatigue syndrome, *Clin. Infect. Dis.*, 1994; 18: S96–104.
53. Suhadolnik, R. J. et al., Biochemical dysregulation of the 2-5A synthetase/RNase L antiviral defense pathway in chronic fatigue syndrome, *J. Chronic Fatigue Syndrome,* 1999; 5: 223–42.
54. Bennett, V. et al., Diversity of ankyrins in the brain, *Biochem. Soc. Trans.*, 1991; 19: 1034–9.
55. Michaely, P. et al., A requirement for ankyrin binding to clathrin during coated pit budding, *J. Biol. Chem.*, 1999; 274: 35908–13.
56. Hall, T. G. and Bennett, V., Regulatory domains of erythrocyte ankyrin, *J. Biol. Chem.*, 1987; 262: 10537–45.
57. König, J. et al., Conjugate export pumps of the multidrug resistance protein (MRP) family: localization, substrate specificity, and MRP2-mediated drug resistance, *Biochim. Biophys. Acta,* 1999; 1461: 377–94.

58. Zheleznova, E. E. et al., A structure-based mechanism for drug binding by multidrug transporters, *Trends. Biochem. Sci.,* 2000; 25: 39–43.
59. Nawab, S. S. et al., Self-reported sensitivity to chemical exposures in five clinical populations and healthy controls, *Psychiatry Res.,* 2000; 95: 67–74.
60. Bell, I. R., Baldwin, C. M., and Schwartz, G. E., Illness from low levels of environmental chemicals: relevance to chronic fatigue syndrome and fibromyalgia, *Am. J. Med.,* 1998; 105: 74S–82S.
61. Racciatti, D. et al., Chronic fatigue syndrome following a toxic exposure, *Sci. Total. Environ.,* 2001; 270: 27–31.
62. Seibert, F. S. et al., Disease-associated mutations in the cytoplasmic loops 1 and 2 of cystic fibrosis transmembrane conductance regulator impede processing or opening of the channel, *Biochemistry,* 1997; 36: 11966–74.
63. Kunzelmann, K. and Schreiber, R., CFTR, a regulator of channels, *J. Membr. Biol.,* 1999; 168: 1–8.
64. Dubner, R. and Gold, M., The neurobiology of pain, *Proc. Natl. Acad. Sci. U.S.A.,* 1999; 96: 7627–30.
65. Gold, M. S., Tetrodoxin-resistant Na^+ currents and inflammatory hyperalgesia, *Proc. Natl. Acad. Sci. U.S.A.,* 1999; 96: 7645–9.
66. Waxman, S. G. et al., Sodium channels and pain, *Proc. Natl. Acad. Sci. U.S.A.,* 1999; 96: 7635–9.
67. Valverde, M. A. et al., Volume-regulated chloride channels associated with the human multidrug-resistance P-glycoprotein, *Nature,* 1992; 355: 830–3.
68. Evengard, B., Schacterle, R. S., and Komaroff, A. L., Chronic fatigue syndrome: new insights and old ignorance, *J. Intern. Med.,* 1999; 246: 455–69.
69. Azpiroz, F. et al., Nongastrointestinal disorders in the irritable bowel syndrome, *Digestion,* 2000; 62: 66–72.
70. Bennett, R., Fibromyalgia, chronic fatigue syndrome, and myofascial pain, *Curr. Opin. Rheumatol.,* 1998; 10: 95–103.
71. Aaron, L. A., Burke, M. M., and Buchwald, D., Overlapping conditions among patients with chronic fatigue syndrome, fibromyalgia, and temporomandibular disorder, *Arch. Intern. Med.,* 2000; 160: 221–7.
72. White, K. P. et al., Co-existence of chronic fatigue syndrome with fibromyalgia syndrome in the general population. A controlled study, *Scand. J. Rheumatol.,* 2000; 29: 44–51.
73. Schwanstecher, M. et al., Potassium channel openers require ATP to bind to and act through sulfonylurea receptors, *EMBO J.,* 1998; 17: 5529–35.
74. Savary, S. et al., Molecular cloning of a mammalian ABC transporter homologous to *Drosophila* white gene, *Mamm. Genome,* 1996; 7: 673–6.
75. Klucken, J. et al., ABCG1 (ABC8), the human homolog of the *Drosophila* white gene, is a regulator of macrophage cholesterol and phospholipid transport, *Proc. Natl. Acad. Sci. U.S.A.,* 2000; 97: 817–22.
76. Cresswell, P. et al., The nature of the MHC class I peptide loading complex, *Immunol. Rev.,* 1999; 172: 21–8.
77. Konstantinov, K. et al., Autoantibodies to nuclear envelope antigens in chronic fatigue syndrome, *J. Clin. Invest.,* 1996; 98: 1888–96.
78. Vedhara, K. et al., Consequences of live poliovirus vaccine administration in chronic fatigue syndrome, *J. Neuroimmunol.,* 1997; 75: 183–95.
79. Sirois, D. A. and Natelson, B., Clinicopathological findings consistent with primary Sjogren's syndrome in a subset of patients diagnosed with chronic fatigue syndrome: preliminary observations, *J. Rheumatol.,* 2001; 28: 126–31.

80. Croop, J. M. et al., Isolation and characterization of a mammalian homolog of the *Drosophila* white gene, *Gene*, 1997; 185: 77–85.
81. Allikmets, R. et al., Mutation of a putative mitochondrial iron transporter gene (ABC7) in X-linked sideroblastic anemia and ataxia (XLSA/A), *Hum. Mol. Genetics,* 1999; 8: 743–9.
82. Koepsell, H., Organic cation transporters in intestine, kidney, liver and brain, *Annu. Rev. Physiol.,* 1998; 60: 243–66.
83. Kavelaars, A. et al., Disturbed neuroendocrine-immune interactions in chronic fatigue syndrome, *J. Clin. Endocrinol. Metab.,* 2000; 85: 692–6.
84. Brooks, J. C. et al., Proton magnetic resonance spectroscopy and morphometry of the hippocampus in chronic fatigue syndrome, *Br. J. Radiol.,* 2000; 73: 1206–8.
85. Moorkens, G. et al., Characterization of pituitary function with emphasis on GH secretion in the chronic fatigue syndrome, *J. Clin. Endocrinol.* (Oxf.), 2000; 53: 99–106.
86. Lawrie, S. M. et al., The difference in patterns of motor and cognitive function in chronic fatigue syndrome and severe depressive illness, *Psychol. Med.,* 2000; 30: 433–42.
87. Lange, G. et al., Neuroimaging in chronic fatigue syndrome, *Am. J. Med.,* 1998; 105: 50S–3S.
88. Rowe, P. C. and Calkins, H., Neurally mediated hypotension and chronic fatigue syndrome, *Am. J. Med.,* 1998; 105: 15S–21S.
89. Cook, D. B. et al., Relationship of brain MRI abnormalities and physical functional status in chronic fatigue syndrome, *Int. J. Neurosci.,* 2001; 107: 1–6.
90. Weng, J. et al., Insights into the function of Rim protein in photoreceptors and etiology of Stargardt's disease from the phenotype in ABCR knockout mice, *Cell*, 1999; 98: 13–23.
91. Sun, H., Molday, R. S., and Nathans, J., Retinal stimulates ATP hydrolysis by purified and reconstituted ABCR, the photoreceptor-specific ATP-binding cassette transporter responsible for Stargardt disease, *J. Biol. Chem.,* 1999; 274: 8269–81.
92. Shetzline, S. E. and Suhadolnik, R. J., Characterization of a 2-5A dependent 37-kDa RNase L. 2. Azido photoaffinity labeling and 2-5A dependent activation, *J. Biol. Chem.,* 2001;276: in press.
93. Trimmer, J. S., Regulation of ion channel expression by cytoplasmic subunits, *Curr. Opin. Neurobiol.,* 1998; 8: 370–4.
94. Pagani, M. and Lucini, D., Chronic fatigue syndrome: a hypothesis focusing on the autonomic nervous system, *Clin. Sci.,* 1999; 96: 117–25.
95. Soderlund, A., Skoge, A. M., and Malterud, K., "I could not lift my arm holding the fork…" Living with chronic fatigue syndrome, *Scand. J. Prim. Health Care,* 2000; 18: 165–9.
96. Fulle, S. et al., Specific oxidative alterations in vastus lateralis muscle of patients with the diagnosis of chronic fatigue syndrome, *Free Rad. Biol. Med.,* 2000; 29: 1252–9.
97. Visser, J. T. J., De Kloet, E. R., and Nagelkerken, L., Altered glucocorticoid regulation of the immune response in the chronic fatigue syndrome, *Annu. N.Y. Acad. Sci.,* 2000; 917: 868–75.
98. Altemus, M. et al., Abnormalities in response to vasopressin infusion in chronic fatigue syndrome, *Psychoneuroendocrinology,* 2001; 26: 175–88.
99. Yu, S. P. and Choi, D. W., Ions, cell volume, and apoptosis, *Proc. Natl. Acad. Sci. U.S.A.,* 2000; 97: 9360–2.

100. Maeno, E. et al., Normotonic cell shrinkage because of disordered volume regulation is an early prerequisite to apoptosis, *Proc. Natl. Acad. Sci. U.S.A.,* 2000; 97: 9487–92.
101. Shimizu, S. et al., BH4 domain of antiapoptotic Bcl-2 family members closes voltage-dependent anion channel and inhibits apoptotic mitochondrial changes and cell death, *Proc. Natl. Acad. Sci. U.S.A.,* 2000; 97: 3100–5.
102. Ablashi, D. V. et al., Frequent HHV-6 reactivation in multiple sclerosis (MS) and chronic fatigue syndrome (CFS) patients, *J. Clin. Virol.,* 2000; 16: 179–91.
103. De Freitas, E. et al., Retroviral sequences related to human T-lymphotropic virus type II in patients with chronic fatigue immune dysfunction syndrome, *Proc. Natl. Acad. Sci. U.S.A.,* 1991; 88: 2922–6.
104. Buskila, D., Fibromyalgia, chronic fatigue syndrome, and myofascial pain syndrome, *Curr. Opin. Rheumatol.,* 2000; 12: 113–23.
105. Liu, Q.-H. et al., HIV-1 gp120 and chemokines activate ion channels in primary macrophages through CCR5 and CXCR4 stimulation, *Proc. Natl. Acad. Sci. U.S.A.,* 2000; 97: 4832–7.
106. Abele, R. and Tampe, R., Function of the transport complex TAP in cellular immune recognition, *Biochim. Biophys. Acta,* 1999; 1461: 405–19.
107. Sokolova, I. A., Vaughan, A. T. M., and Khodarev, N. N., Mycoplasma infection can sensitize host cells to apoptosis through contribution of apoptotic-like endonuclease(s), *Immunol. Cell Biol.,* 1998; 76: 526–34.

5 The 2-5A Pathway and Signal Transduction: A Possible Link to Immune Dysregulation and Fatigue

Patrick Englebienne, C. Vincent Herst, Marc Frémont, Thierry Verbinnen, Michel Verhas, and Kenny De Meirleir

CONTENTS

5.1 INTRODUCTION

Signal transduction consists of a set of cascades of biochemical events that occur intracellularly upon receptor triggering by external signals. The process transmits the external signal into and within the cell to elicit either a stimulating or an inhibiting response by the cell.[1] The response occurs in the nucleus where gene transcription is adapted to the external signal. The mechanisms of signal transduction make use of different enzymes (protein kinases and phosphatases) that have relatively broad substrate specificities. These enzymes are used in different combinations to achieve distinct biological responses.[1] The ways by which these different players are

0-8493-1046-6/02/$0.00+$1.50

TABLE 5.1
Protein Domains Involved in Protein–Protein Interactions in Signal Transduction

Domain	Recognition Motif
SH_2 domain	Specific phosphotyrosine residue determined by three amino acids
SH_3 domain	Amino acid sequence variations around a basic PXXP site
WW domain	Specific phosphoserine or phosphothreonine residues followed by a proline and some proline-rich protein motifs
PTB domain	Specific phosphotyrosine residue within a NPXpY motif
PH domain	Phosphatidylinositol di- and tri-phosphates
PDZ domain	Carboxyterminal (T/S)XV motif

associated in order to create pathways or networks ensure the specificity of signal transduction. This is achieved either by recruitment of active signaling molecules into multiprotein networks, or by activation of dormant enzymes already positioned close to their substrate.[2] In these processes, protein–protein interactions play central roles, mediated by modular protein-binding domains which are structurally conserved elements with unique molecular specificities.[3] Examples of such domains are src homology 2 and 3 domains (SH_2 and SH_3), pleckstrin homology domains (PH), phosphotyrosine-binding domains (PTB), phosphoserine and phosphothreonine-binding domains (WW),[4] and PDZ domains.[5] The specificity of protein interaction by these domains is summarized in Table 5.1. Many of these interactions occur over short protein sequence motifs, often less than ten amino acids in length, in which the presence of proline residues is critical.[6]

The complexity of signal transduction processes is further enhanced by the capacity of some pathways to cross-talk with one another and, in some instances, to integrate at the level of transcriptional promoters. One such example is the physical interaction between the transcription factors regulated by tumor growth factor (TGF) β and Wnt signaling pathways, which synergistically activate gene expression during embryonic development.[7] The response to signal transduction can further produce new signaling molecules released from the cell that allow a cross-talk with other cells. This is best exemplified by the cyclooxygenase-2 (Cox-2) induction mediated by interleukin-1β in the central nervous system. This inflammatory induction of Cox-2 results in the release of prostanoids which sensitize peripheral nociceptor terminals and produce pain hypersensitivity.[8] The cross-talk effect occurs either within the same cell in absence of any excretion (intracrine effect) or after excretion (autocrine effect), or with other cells that are either neighbors (paracrine effect), or long-distance (endocrine effect).[9] The signal transduction processes are further dependent on the ligand–receptor interactions involved. This is best exemplified by the co-stimulatory signals delivered to T-cells by the B7-1 and B7-2 ligands present on antigen-presenting cells (APC). Signaling by these ligands through the CD28 receptor augments the T-cell response, while signaling through the CTLA-4 receptor attenuates the T-cell response, as a result of distinct structural organization of ligand–receptor complexes.[10,11]

It is quite clear that signal transduction is a cell science in continuous progression. However, the complexity of the mechanisms already elucidated underline the tremendous effects that can be elicited by any deregulation in these cellular processes in pathological conditions. The next chapter in this book (Chapter 6) discusses the implications of signal transduction events and their deregulations in apoptosis, which are beyond the scope of the present chapter. Instead, we will focus our discussion on other signaling pathways and their possible implication in chronic diseases.

5.2 RECEPTORS AND BEYOND

The process of signal transduction within the cell starts with the interaction between a ligand and its specific receptor. The specificity of ligand–receptor interaction is dependent upon both ligand and receptor spatial structures.[12] The receptors can be localized in the cell cytosol, in the nucleus, or on the cell membrane. Once the ligand is docked in the receptor binding site, receptor triggering ensues and results in a series of cellular events as summarized in Figure 5.1. These events range from action

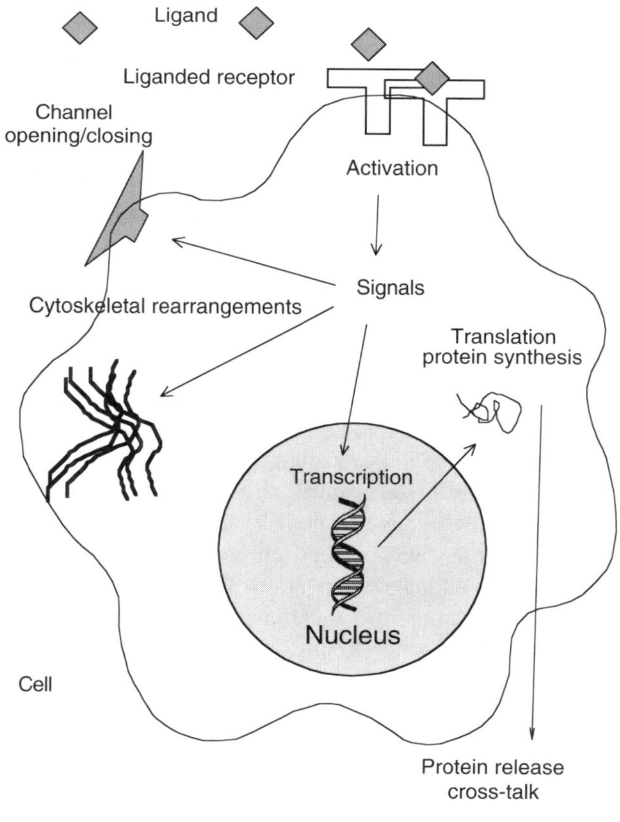

FIGURE 5.1 Cellular events resulting from receptor triggering by a ligand.

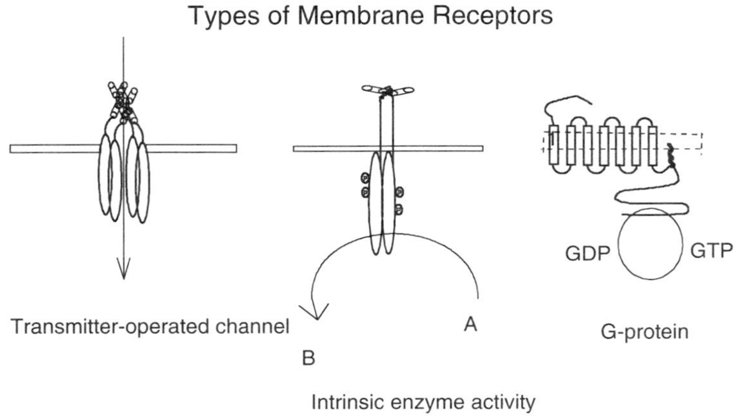

FIGURE 5.2 Schematic representation of the three types of cell membrane receptors.

on cellular channels to translation and release of new cellular messengers. Cytoplasmic-nuclear receptors comprise the members of the steroid and thyroid hormone receptors superfamily.[13] They are intracellular transcription factors existing as inactive apoproteins in the cytosol or the nucleus. Upon binding their respective ligands, they undergo activation which allows them to bind to a responsive DNA element (hormone-responsive element, HRE) and activate transcription of a *cis*-linked gene.

Among the receptors present on the plasma membrane, three types can be distinguished, as summarized in Figure 5.2. These are, respectively, the transmitter-operated channels (formerly known as ligand-gated receptors), the intrinsic activity receptors, and the G-protein coupled receptors, which include the serpentine receptors. These are characterized by seven canonical transmembrane helices.[14] Most membrane receptors require association for activation, and this requirement has been particularly observed with the γ-aminobutyric acid ($GABA_B$) receptor,[15-18] which plays pivotal roles in the central nervous system synaptic transmission. A similar trimeric association has been observed for the tumor necrosis factor (TNF-α) receptor and it was thought that such association took place upon activation by the ligand.[19] However, more recent evidence indicates that the TNF-α receptor, as well as the Fas receptor, functions as a preformed complex, rather than as an individual subunit.[20,21]

The transmitter-operated channels are beyond the scope of this chapter. The receptors with intrinsic enzymatic activity are receptors which contain intrinsic or associated kinase activities and, upon activation, are capable of phosporylating either their cytoplasmic domains, or distinct cytoplasmic proteins mainly on tyrosine, but also on serine or threonine residues, respectively.[22] Examples of such receptors are the insulin receptor and the growth factor receptors. Interferons and the insulin-like growth factor receptors that will be discussed later in this chapter are, respectively, receptors with associated and intrinsic kinase activity. The receptors involved in lymphocyte activation function according to this model. Engagement of the T-cell receptor by MHC-antigen complexes activates membrane localized src kinases which then phosphorylate the linker of activated T-cells (LAT). The phosphorylated tyrosine residues on the cytoplasmic domain of LAT provide a scaffold for the recruitment of the Grb2- (growth factor-receptor bound protein 2) and Grb2-like proteins-Sos (son of sevenless, a guanine nucleotide exchange factor identified in *Drosophila*) complexes and the p85 subunit of phosphatidylinositol 3'-kinase (PI-3K). The translocation of Grb2-Sos in proximity to cytoplasmic membrane-bound Ras results in exchange of GDP for GTP on Ras and the subsequent initiation of the mitogen-activated phosphokinases (MAPK).[23] The recruitment of PI-3K in turn leads to the activation of protein kinases B (c-Akt) and C (PKC), which further enhances the Ras pathway by phosphorylating Raf downstream of Ras.[24]

The G-protein-coupled receptors mediate their intracellular actions by a pathway that involves activation of one or more guanine nucleotide-binding regulatory proteins (G-proteins).[25] These receptors comprise a large and functionally diverse superfamily which includes neurotransmitter, sensory, immune recognition, and hormone receptors.[14] These receptors interact in the cytoplasm with the G-proteins normally latent in a GDP-bound form. Ligand binding activates the G-proteins as the GTP-bound form, which further deactivate by GTP hydrolysis. This results in the release of a phosphate group used by the subsequent kinases in the transduction pathway. This effect is mediated by adenylyl cyclase, which functions as a GTPase-activating protein for the α-subunit of G-proteins and increases the GTP-GDP exchange factor (GEF) activity of heterotrimeric (αβγ) G-protein complexes.[26] These early events of receptor activation on the plasma membrane lead to a series of signaling cascades within the cell which ultimately result in the cellular response to receptor engagement.

5.3 THE SIGNAL TRANSDUCTION CASCADES

Among the signal transduction cascades, the Ras/Rho pathways are probably the best known to date. Ras (rat sarcoma virus; p21) are 21-kDa proteins which play a major role in the signal transduction of growth factor receptors.[27] The Rho (Ras homologous) family encompasses small G-proteins that play dynamic roles in the regulation of the actin cytoskeleton. The Ras pathway occurs in parallel to the Rac (Ras-related C3-botulinum toxin substrate) pathway, which is activated by inflammatory mediators and cellular stress.[28,29] These pathways further activate MAPKs, extracellular-signal-regulated kinases (ERKs), and MAP/ER kinases (MEKs), at several downstream levels, resulting in the activation of nuclear transcription factors such as NFκB, p53, and Elk-1. These pathways are summarized in Figure 5.3.

At the MAPK4 level, the activated receptor recruits Ras and/or PI-3K or Rac, which in turn activate Raf, PKC, Akt, and PAK (p21-activated kinase or $p65^{pak}$). Akt (PKB) constitutes the central point of cross-talk between the phosphatidyl inositol pathway and Ras, as it is able to downregulate Raf by phosphorylating this protein in highly conserved serine residues,[30,31] resulting in inhibition of this signaling cascade. At the second downstream level (MAPK3, i.e., MAPK kinases), Raf and MEKKs (MEK kinases) are activated, which leads to the activation of important transcription factors such as NFκB, as the result of the phosphorylation of its IκB inhibitors. This factor is central to inflammatory processes, as it induces the

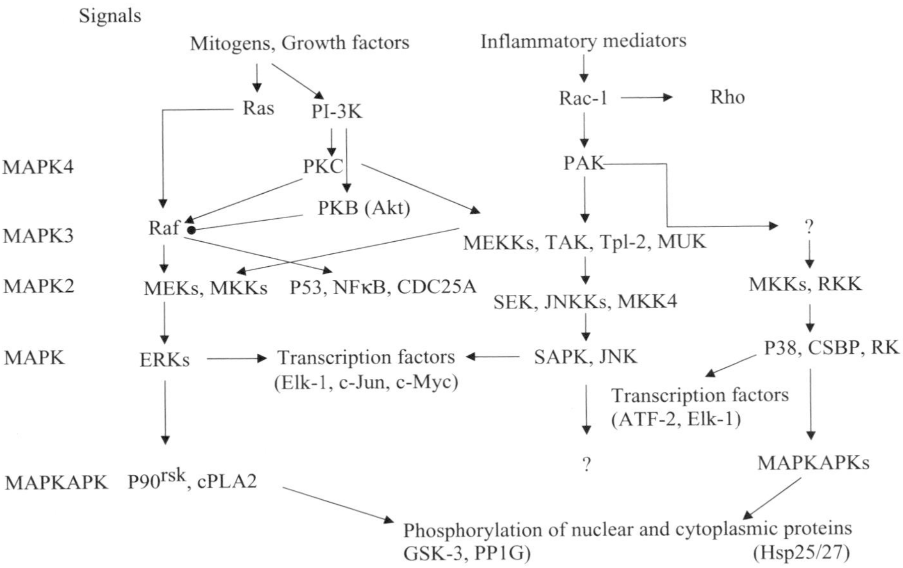

FIGURE 5.3 The mitogen-activated protein kinase cascade.

transcription of key genes such as the interferon β, cox-2, and inducible nitric oxide synthetase (iNOS).[32] The signal transducer and activator of transcription 1 (STAT1), a key element in the interferon signal transduction (see below), inhibits the activation of NFκB[33] upon phosphorylation by the TNF-α receptor signaling pathway, which underlines the apoptotic-related mission of this transducer.

At the MAPK2 level, MEKs and stress-activated protein kinase (SAPK)/ERK (SEKs) are activated, which leads to the MAPK level of the cascade where ERKs, SAPKs, and JNK (Jun N-terminal kinase) are in turn phosphorylated. The resulting activation is particularly sustained under cellular stress and leads to opposed pro-apoptotic (ERK) as well as anti-apoptotic (SAPK/JNK) signals influencing cell survival.[34] This level partially corresponds to a downregulation step since activated SAPK binds to the SH3 domain of Grb2, which in turn forms a heterotrimeric complex by binding the SH2 domain of PI-3K, thereby inhibiting its protein serine kinase activity.[28] The resulting activated kinases further phosphorylate key transcription factors such as c-Jun, Elk-1 and c-Myc. In the last stage of the cascade, activated MAPKAPKs (mitogen-activated protein kinase activated protein kinase), such as the ribosomal S6 kinase $p90^{sk}$ and MAPKAPK 2 and 3, phosphorylate cytoplasmic and nuclear protein, such as heat shock proteins (Hsp 25/27), which regulate actin polymerization, and c-Fos, respectively.[28]

The activation of PI-3K results in the phosphorylation at the 3 position of the inositol ring of phosphoinositides present on the cytoplasmic surface of the cell membrane, generating phosphatidylinositol 3,4,5-triphosphate (PIP_3). This in turn serves as a source of phosphate groups for the activation (phosphorylation) of protein kinase B (c-Akt, PKB) by a PIP_3-dependent kinase (PDK),[35] leading to a cascade of metabolic and cell survival events respectively summarized in Figures 5.4 and

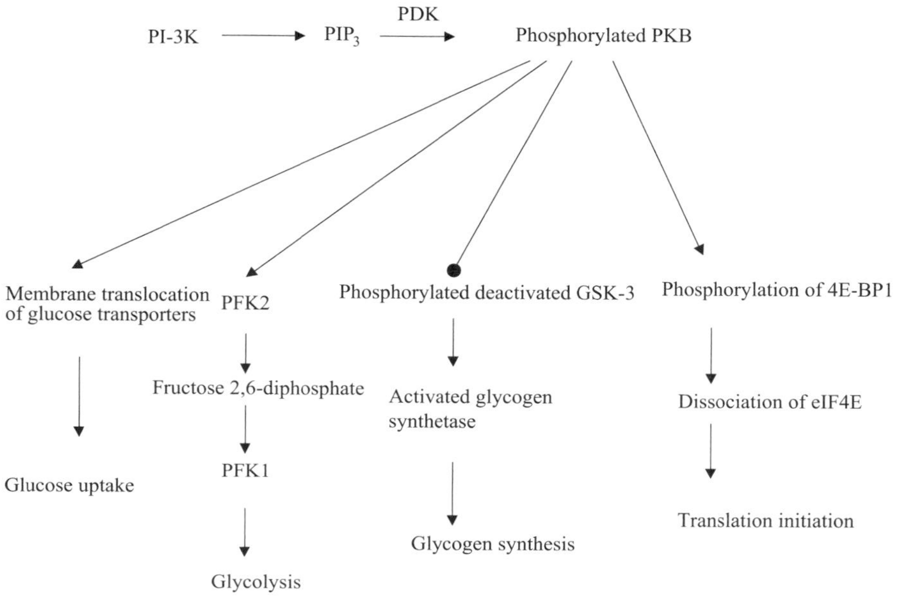

FIGURE 5.4 Cellular metabolic processes regulated by PI-3K and PKB.

5.5. The metabolic events mediated by PI-3K and PKB activation primarily concern cellular energy supply and protein synthesis (Figure 5.4). The activation of PKB results on the one hand in the recruitment of glucose transporters from internal cellular pools to the membrane, which enhances glucose uptake. On the other hand, the phosphorylation of 6-phosphofructose 2-kinase (PFK2) by PKB increases the production of fructose 2,6-biphosphate, which in turn activates phosphofructose 1-kinase (PFK1) and glycolysis. The phosphorylation of glycogen synthetase kinase-3 (GSK-3) inactivates this enzyme, which regulates glycogen synthetase phosphorylation, resulting in increased glycogen synthesis. Finally, the phosphorylation of the eukaryotic initiation factor 4E (eIF4E)-binding protein (4EBP1) results in the dissociation of eIF4E from the complex, which turns on the translation processes.

The PI-3K and PKB pathway is also central in cell protection (Figure 5.5). The pathway regulates several key components of the apoptotic process (Chapter 6). Stimulation of cells by the insulin-like growth factor (IGF-1), or interleukins 2 and 3, blocks the induction of caspase 3, and increases the expression of Bcl2, Bcl_{XL}, and c-Myc, independently of the Ras pathway. The inhibition of caspase 3 limits protein degradation. The conjugation of Bcl2 and c-Myc activities stimulates the progression through the cell cycle. Moreover, PKB phosphorylates Bad, a proapoptotic protein, on a serine pertaining to a sequence motif homologous to GSK-3 and PFK2, resulting in its sequestration by the τ form of 14-3-3 proteins and preventing its binding to, and inhibition of, BclXL,[36] leading to enhanced cell survival.

The sphingomyelin pathway is a third ubiquitous signal transduction pathway interacting with those decribed above. It is mainly induced by cellular stress mediated by Fas, TNF-α, interferon γ, or interleukin-1β receptors.[37] These liganded receptors

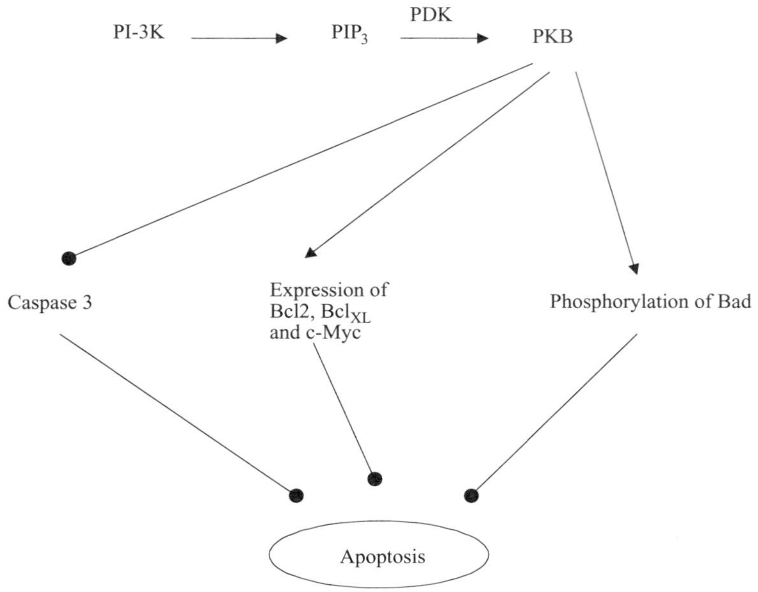

FIGURE 5.5 Cell survival pathways mediated by PI-3K and PKB.

activate various sphingomyelinases (Smases), which are specific forms of activated phospholipase C. Recent studies indicate that initiator caspases play a role in the activation of Smases. These enzymes transform sphingomyelin present in the cell membrane, which is transformed in ceramide, according to the scheme presented in Figure 5.6. Ceramide activates a specific ceramide-activated protein kinase (CAPK/KSR), which activates the MAPK pathway through phosphorylation of Raf. CAPK is also likely to increase mitochondrial membrane permeability by a mechanism mediated by Bad.[37] Also consider a ceramide-activated serine-threonine protein phosphatase (CAPP), which plays a role in the downregulation of *c-myc* expression[38] and dephosphorylates c-Jun, thereby counteracting the proapoptotic activities of JNK.[37] Ceramide also activates the ζ isoform of protein kinase C (PKCζ), which activates NF-κB.

The interrelationship of these various signal transduction pathways and their capacity at inducing interrelated cell survival or death signals underlines the complexity of the cell machinery. Besides the major signal transduction pathway induced by their liganded receptors, interferons also exert pleiotropic cellular effects by inducing the signaling pathways summarized above,[39] which might have severe implications in the context of the interferon dysregulations observed in chronic fatigue syndrome.[40]

5.4 INTERFERON RECEPTORS AND SIGNALS

Interferons (IFNs) are homologous cytokines that play a central protective role during infection by pathogens. They induce a cellular antiviral state and modulate

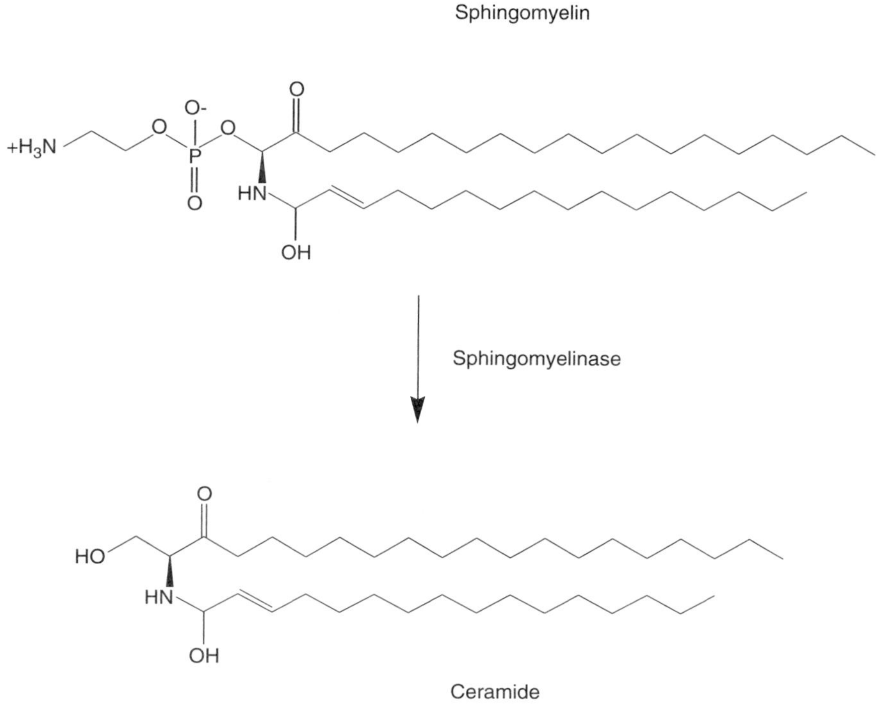

FIGURE 5.6 Ceramide structure and formation from sphingomyelin by sphingomyelinase.

the immune response.[41,42] Type I IFNs (IFN-α, IFN-β, and IFN-ω) are secreted by virus-infected cells. Type II IFN (IFN-γ) is secreted by T-cells under certain conditions of activation and by natural killer (NK) cells.[42] Type I IFNs bind to a cell surface receptor consisting of two transmembrane proteins, IFNAR1 and IFNAR2, which associate upon binding.[43] A peptide motif (FSSLKLNVY) of the extracellular region of IFNAR1 has recently been identified as overlapping a site of the receptor with which both IFN-α and IFN-β interact.[44] So far, there are at least 14 different IFN-α subtypes identified[45] that elicit different biological responses by interaction with the same receptor.[43,45] The major signal transduction pathway of type I IFNs is schematically represented in Figure 5.7.

The cytoplasmic domain of IFNAR1 and IFNAR2 are, respectively, associated with a tyrosine kinase (Tyk2) and a janus kinase (JAK1) activated upon dimerization. These kinases in turn phosphorylate the signal transducers and activators of transcription (STAT) 1 and 2, which heterodimerize by mutual recognition of their phosphorylated SH2 domains.[46] The heterodimer, called interferon-stimulated gene factor-3α (ISGF-3α), recruits a 48-kDa DNA-binding protein (p48, ISGF-3γ) from the cytoplasm, and the trimeric complex (now termed ISGF-3) translocates to the nucleus where it interacts with the interferon-stimulated response element (ISRE) of DNA.[47] Recent evidence[48] indicates that STAT 3 and 5 are also activated by the type I interferons, but their target genes are presently unknown. Activated STAT 3

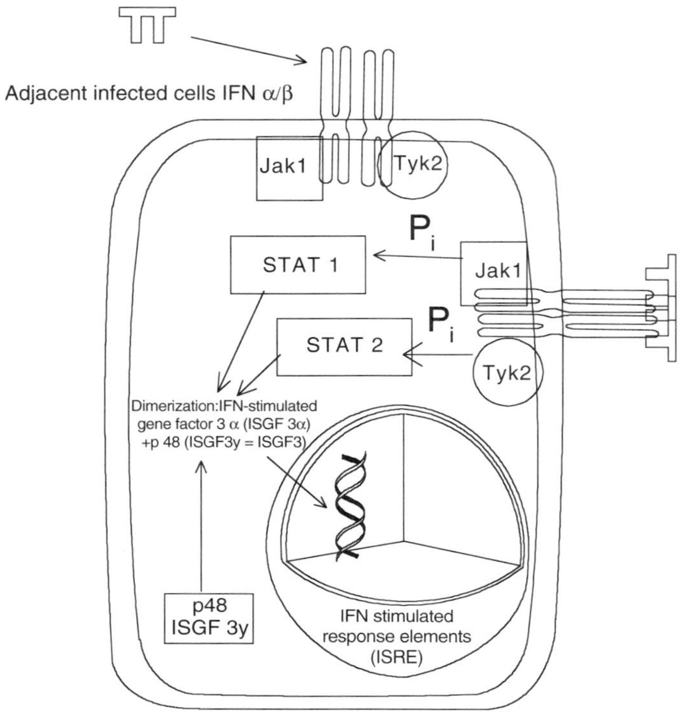

FIGURE 5.7 Major signal transduction pathway by the type I interferons.

has been tentatively identified as the adapter molecule that couples PI-3K to IFNAR1, thereby inducing its activation;[49] however, this role of STAT 3 remains controversial.[50,51]

The link between type I interferons and PI-3K activation, however, is well established and IFN is likely to play the role of an autocrine co-factor.[52] Activation of the MAP kinase p38 by type I IFNs has also been reported.[53] This mechanism of signal transduction is required for IFN-dependent transcriptional activation independently of the STAT signaling pathway.[54] This mechanism is likely to occur directly at MEK/ERK levels of the cascade (Figure 5.3), as it is independent from Ras/Raf.[55] The type II IFN receptor (Figure 5.8) consists of two subunits (α and β), with each chain constitutively associated with a specific janus kinase (JAK1 with the α-chain and JAK2 with the β-chain). The chain association consecutive to ligand binding leads to the transphosphorylation and activation of the JAKs, which in turn phosphorylate STAT1.[42] The activated STAT1 homodimerizes (the complex is termed GAF, which stands for gamma-interferon activation factor) and translocates to the nucleus where it is able to bind specific DNA sequences (called GAS, gamma-activated site) and initiate transcription.[47]

The importance of STAT1 in mediating the action of both type I and II IFNs is no longer questioned, as a lack of its expression is consistently associated with IFN resistance.[56,57] The apparent dysregulation in types I and II IFN pathways in chronic

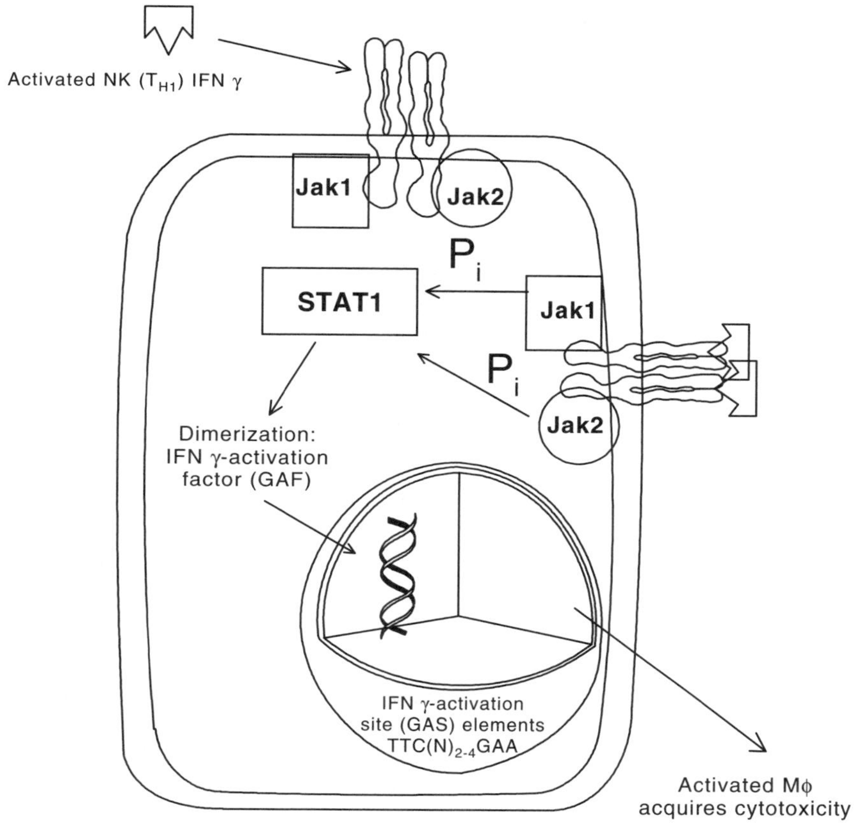

FIGURE 5.8 Major signal transduction pathway by the type II IFN.

fatigue syndrome[58] led us to investigate the expression of STAT1 in PBMCs. We classified the samples according to the ratio of 37- over 80-kDa RNase L (refer to Chapters 1, 2, and 3), which is representative of the proteolytic activity of the PBMC samples. As shown in Figure 5.9, STAT 1 is fully degraded in positive samples, suggesting that it may also be a substrate of the proteases responsible for RNase L cleavage. A degradation of STAT1 in those cells might well constitute the missing link explaining unresponsiveness to IFNs.

While the downstream events of interferon receptor activation are quite well understood, the mechanisms leading to IFN expression are less clear. The induction of type I INF gene expression is known to require the phosphorylation-induced activation of the transcription factors interferon regulatory factors 3 and 7 (IRF).[59] The role of viral ds-RNA has been demonstrated in IFN expression and activation of PKR and has been suggested as a possible means for IRF activation. More recent evidence shows, however, that the activation of IRFs does not require PKR[59,60] and can be stimulated by a viral component other than, or in addition to, ds-RNA, involving a new cellular kinase.[59] This was confirmed independently in a more recent study[61] which further showed that activation and nuclear translocation require

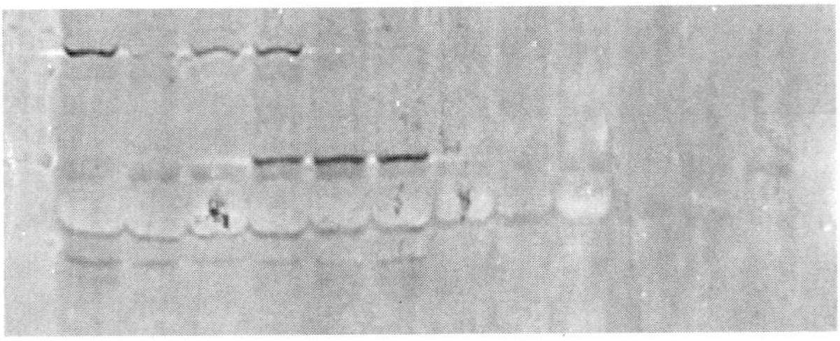

FIGURE 5.9 Cleavage of STAT 1 as a function of RNase L cleavage (expressed as the ratio of 37 over 80-kDa proteins) in PBMC extracts, as detected by immunoblotting.

the phosphorylation of IRF 3 on serine and threonine residues in the C-terminal region. In contrast, cell exposure to stress results in the activation of a MAPKKK-related signaling pathway, which results in the phosphorylation of IRF 3 on the N-terminal part of the sequence. N-terminal phosphorylation, however, is insufficient to promote nuclear translocation, transactivation, or degradation of IRF 3. An increasing body of evidence also indicates that the IRFs take part of a stereospecific enhancer complex which interacts with the positive regulatory domains (PRD) of the type I IFNs. This complex has been termed the enhanceosome.[62] In the case of IFN-β, the PRDs are recognized by an enhanceosome made of the p50/p65 of NFκB, an IRF 3 dimer, and the heterodimeric complex of ATF2/c-Jun. Two different signaling pathways intervene in the activation of the proteins pertaining to the IFN-β enhanceosome. Indeed, IRF 3 and ATF2/c-Jun are activated by the MEKK/JNK pathway, while NFκB is activated by the Iκ kinases.[62] Besides IRF 3 and IRF 7, IRF 1 and 2 are also implicated in the expression of type I IFNs.[63] Seven different IRFs have been identified.

Type II interferon (IFN-γ) is one of the genes regulated by type I IFNs. The regulation of IFN-γ is effected in an autocrine manner and plays a major role in the immune system as a differentiating factor for dendritic cells, which subsequently trigger T-cell-mediated immunity.[64] Consequently, type I interferon link innate and adaptive immunity. The STAT4 is likely to be responsible for the IFN-γ gene expression by type I IFNs and acts upon recruitment by the activated STAT1-STAT2 complex.[65] STAT1 is further suspected to play a regulatory role in this signal transduction process, which is currently less than clear.[66,67] Nevertheless, STAT4 undoubtedly plays a major role in T_H1 differentiation and has been found highly expressed in PBMCs, dendritic cells, and macrophages at sites of T_H1-mediated inflammation, such as in synovial tissue obtained from rheumatoid arthritis patients.[68] STAT4 is also involved in the transduction of signals from the activated interleukin-12 (IL-12) receptor,[69] leading to the production of IFN-γ. Like others,[70] we did not find altered IL-12(p40) serum levels in CFS patients when compared to healthy controls. However, we found significantly elevated levels of IFN-γ in the serum of CFS patients, which might reflect the loss of negative regulation by STAT1 on its expression. These apparently discrepant results might thus reflect, on the one hand, a sustained lack of negative regulation of IFN-γ production elicited by type I IFNs in monocytes, and on the other hand a lack of IL-12 production resulting from an increased sensitivity to glucocorticoids.[71,72] Overall, such immune dysregulation results in a poor activation of NK cell activity,[73] a CFS characteristic pointed out in the recent past.[74]

Besides the cross-talk between type I and type II IFNs evoked above, the genes regulated by these cytokines, either directly or indirectly, are deeply involved in apoptotic regulation and consequently regulate the cell fate. IFN-γ induces the cyclin-dependent kinase (CDK) inhibitor $p21^{WAF1/CIP1}$ via the JAK/STAT pathway.[75] This results in cell growth inhibition. IFN-γ upregulates the IFN-α/β gene expression by inducing the expression of the p48 subunit of ISGF3 (see Figure 5.7). IFN-γ activates a transcriptional element termed GATE (gamma-interferon-activated transcriptional element), which interacts with the gene encoding the CCAAT/enhancer-binding protein-β (C/EBP-β), a pleiotropic transcription factor.[76] The transcriptional activity of C/EBP-β is further enhanced by IFN-γ through activation of ERKs independently of Raf/Ras (Figure 5.3).[77] Type I IFNs in turn induce a primary antiviral response gene coding for IRFs, the ds-RNA-dependent protein kinase (PKR), and the 2',5'-oligoadenylate synthetase (2-5OAS). As a secondary response, the gene coding for RNase L is rapidly expressed.[78,79] All these proteins are involved in the induction of apoptosis (Chapter 6).

PKR plays an important modulating role in the type I IFN pathway by binding STAT1[80] in the absence of stimulation by IFNs and ds-RNA, thereby reducing its transcriptional activity. The observed degradation of STAT1 in the PBMCs of CFS patients (Figure 5.9), could therefore be held responsible for the upregulation of PKR as well as a dysregulated induction of 2-5OAS in these cells (see below).[78] Interferon-α further enhances the transduction cascade by increasing the expression of IRF-7, which modulates the inflammatory response and may be involved in Epstein–Barr virus-associated malignancies.[81] Type I IFNs also activate the expression of several other antiviral genes coding for diverse effector proteins.[82]

Among these, the MxA protein is a 76-kDa GTPase that inhibits the multiplication of several RNA viruses.[83-86] MxA can act in the absence of other interferon-induced proteins and its overexpression induces apoptosis.[83] The GTPase activity of MxA is regulated by its oligomerization,[83,85] which is critical for its viral target recognition.[85] MxA is present in both the cytoplasm and nucleus. In the cytoplasm, the protein prevents the translocation of viral ribonucleoproteins to the nucleus, and in the cell nucleus, the protein directly inhibits the viral polymerase activity.[86] The gene coding for protein p202 (a 52-kDa phosphoprotein) is another gene induced by the type I IFNs that could prove of high interest in CFS. The p202 is a negative regulator of apoptosis induced by p53 and c-myc and the transcription of its gene is directly repressed by p53.[87] The degradation of p53 observed in the PBMCs of CFS patients (Chapter 6) might thus be responsible for an upregulation of this protein and a subsequent apoptotic inhibition. The 17-kDa ISG15 protein, also known as the ubiquitin cross-reactive protein (UCRP), is also induced by type I IFNs.[88-91] Ubiquitin is a ubiquitous protein that targets cellular proteins for degradation by the 26S proteasome.[88] ISG15/UCRP exerts its biological effects by its covalent conjugation to cellular proteins through an enzyme pathway distinct from that of ubiquitin ligation.[88,90] However, ISG15/UCRP is also likely to pertain to the cytokine cascade, as it has been shown to be directly involved in the induction of the proliferation of, and cytolytic activity by, NK cells.[89,91] Finally, IFN-α enhances Fas surface expression on PBMC and T-cells and participates in the stimulation of Fas ligand production.[92]

Many of the secondary responses to the genes expressed under the influence of IFNs are mediated by the activation of the nuclear transcription factor NF-κB.[93] NF-κB activation is required for the expression of iNOS in macrophages under the stimulation of IFN-γ,[94] and such activation is likely to be regulated by STAT6.[95] Similarly, IFN-α/β promotes cell survival through the transcription by NF-κB of genes like Bcl-2, which counteract strongly its own proapoptotic signals.[96] PKR has been shown to be responsible for NF-κB activation by phosphorylation of the I-κB inhibitor,[97,98] its action likely mediated by the I-κB kinase.[99] Recent evidence indicates the existence of a separate PKR-independent NF-κB activation pathway by IFNs.[60,100]

Several compounds linked to the thyroid and steroid receptor superfamily, such as retinoic acid and the antiestrogen tamoxifen, enhance the proapototic signals induced by the type I IFNs mediated by the 2-5A pathway.[101,102] These effects are mediated by a post-transcriptional expression of several genes associated with retinoic acid-IFN-induced mortality (GRIM). The product of one such gene, GRIM12, has been identified as the thioredoxin reductase,[103] an enzyme which, among several oxidized substrates, reduces p53, resulting in its higher capacity to interact with DNA.[104] A similar protein, GRIM19, acts by activating caspase 9.[105] These examples constitute a first mode of interaction between the interferons and the steroid and thyroid receptor superfamily. These mechanisms regulating the pro-anti-apoptotic balance of type I interferons might be severely dysregulated in CFS, since we have observed inactivating cleavages of both p53 and caspase 9 in the PBMCs of these patients (Chapter 6).

5.5 TYPE I INTERFERON-STIMULATED GENES AND THE THYROID RECEPTOR

Noteworthy among the genes induced by IFNs are those coding for the 2-5A synthetase (2-5OAS). These enzymes pertain to a family of proteins induced by type I IFNs through the JAK/STAT pathway.[106] The enzymes are activated by ds-RNA of viral origin, as well as by single-stranded RNA,[107,108] and their activity is likely to be further regulated by an IFN-α/IFN-γ balance.[109] The enzymes bind and polymerize ATP into 2',5'-oligoadenylates (2-5A), which further activate the RNase L (Chapters 1 and 2). Besides two small proteins (p41/46),[110] the family comprises higher molecular weight enzymes, namely p69/71 and p100.[111] The three isoforms exhibit different catalytic characteristics, the p100 protein producing preferentially 2-5A dimers instead of the higher oligomers produced by the smaller isoforms,[112,113] which mediate a more efficient antiviral state.[114] The activation of p41/46 and p69/71 proceeds, respectively, through tetramerization[115] and dimerization[116] of these proteins. Oligomerization is a prerequisite for catalytic activity.[117]

Besides the 2-5OAS, type I interferons also induce three closely related proteins termed 2-5 oligoadenylate synthetase-like proteins (2-5OASL). These proteins are the p30, p56, and p59 OASL.[118,119] Counter to the 2-5OAS, the 2-5OASL do not share the catalytic ATP-polymerizing activity. The 2-5OAS genes have been mapped to chromosomal segment 12q24.1, while 2-5OASL genes are located on chromosomal segment 12q24.2.[120] In order to have a better insight of the homology between the proteins of the OAS/OASL superfamily, we have aligned their sequences using ClustalW in Figure 5.10. Amino acid numbering follows the p100 sequence. The catalytic domain of 2-5OAS is located in the N-terminal part of the proteins. The

```
                    730        740        750        760        770        780
            ....|....| ....|....| ....|....| ....|....| ....|....| ....|....|
p41/46 OAS  ---------- ---------- ---------- MMDLRNTPAK SLDKFIEDYL LPDTCFRMQI
p69/71 OAS  QWLKKEAQTW LTSPNLDN-- ELPAPSWNVQ PAPLFTTPGH LLDKFIKEFL QPNKCFLEQI
p100 OAS    ELLAQEAAAL GMQACFLSRD GTSVQPWDVM PALLYQTPAG DLDKFISEFL QPNRQFLAQV
p30 OASL    ---------- ---------- -------MAL MQELYSTPAS RLDSFVAQWL QPTGVE-GRG
p56 OASL    ---------- ---------- -------MAL MQELYSTPAS RLDSFVAQWL QPTGVE-GRG
p59 OASL    ---------- ---------- -------MAL MQELYSTPAS RLDSFVAQWL QPHREWKEEV
Clustal Co                                      *  **    ** *.   *  *

                    790        800        810        820        830        840
            ....|....| ....|....| ....|....| ....|....| ....|....| ....|....|
p41/46 OAS  DHAIDIICGF LKERCFRGSS ---YP-VCVS KVVKGGSSGK GTTLRGRSDA DLVVFLSPLT
p69/71 OAS  DSAVNIIRTF LKENCFRQS- ------TAKI QIVRGGSTAK GTALKTGSDA DLVVFHNSLK
p100 OAS    NKAVDTICSF LKENCFRNS- ----P-IKVI KVVKGGSSAK GTALRGRSDA DLVVFLSCFS
p30 OASL    ARRCATVEEF LRQEHFQGKR G-LDQDVRVL KVVKVGSFGN GTVLRSTREV ELVAFLTCFH
p56 OASL    ARRCATVEEF LRQEHFQGKR G-LDQDVRVL KVVKVGSFGN GTVLRSTREV ELVAFLSCFH
p59 OASL    LDAVRTVEEF LRQEHFQGKR G-LDQDVRVL KVVKVGSFGN GTVLRSTREV ELVAFLSCFH
Clustal Co        .  * *..   *.               ..*. **     ** *.    .  .** *

                    850        860        870        880        890        900
            ....|....| ....|....| ....|....| ....|....| ....|....| ....|....|
p41/46 OAS  TFQDQLNRRG EFIQEIRRQL EACQRERALS VKFEVQAPRW GN--PRALSF VLSSL-QLGE
p69/71 OAS  SYTSQKNERH KIVKEIHEQL KAFWREKEEE LEVSFEPPKW KA--PRVLSF SLKSK-VLNE
p100 OAS    QFTEQGNKRA EIISEIRAQL EACQQERQ-- FEVKFEVSKW EN--PRVLSF SLTSQTMLDQ
p30 OASL    SFPG-GSQ-- ASQRCSEADM ENHVQSQD-L LDLGLEDLRM EQRVPDALVF TIQTR-GTAE
p56 OASL    SFPG-GSQ-- ASQRCSEADM ENHVQSQD-L LDLGLEDLRM EQRVPDALVF TIQTR-GTAE
p59 OASL    SFQE-AAK-- HHKDVLRLIW KTMWQSQD-L LDLGLEDLRM EQRVPDALVF TIQTR-GTAE
Clustal Co   .                        . .         .  .      *  * *   . .     .

                    910        920        930        940        950        960
            ....|....| ....|....| ....|....| ....|....| ....|....| ....|....|
p41/46 OAS  GVEFDVLPAF DALGQLTGSY KPNPQIYVKL IEECTDLQ-K EGEFSTCFTE LQRDFLKQRP
p69/71 OAS  SVSFDVLPAF NALGQLSSGS TPSPEVYAGL IDLYKSSDLP GGEFSTCFTV LQRNFIRSRP
p100 OAS    SVDFDVLPAF DALGQLVSGS RPSSQVYVDL IHSYSN---- AGEYSTCFTE LQRDFIISRP
p30 OASL    PITVTIVPAY RALGPSLPNS QPPPEVYVSL IKACGG---- PGNFCPSFSE LQRNFVKHRP
p56 OASL    PITVTIVPAY RALGPSLPNS QPPPEVYVSL IKACGG---- PGNFCPSFSE LQRNFVKHRP
p59 OASL    PITVTIVPAY RALGPSLPNS QPPPEVYVSL IKACGG---- PGNFCPSFSE LQRNFVKHRP
Clustal Co   .   ..**.  ***        *  ..*  * *           *  ..  .*.  *** *.  **

                    970        980        990       1000       1010       1020
            ....|....| ....|....| ....|....| ....|....| ....|....| ....|....|
p41/46 OAS  TKLKSLIRLV KHWYQNCKKK L---GKLPPQ YALELLTVYA WERGSM-KTH FNTAQGFRTV
p69/71 OAS  TKLKDLIRLV KHWYKECERK LKPKGSLPPK YALELLTIYA WEQGSG-VPD FDTAEGFRTV
p100 OAS    TKLKSLIRLV KHWYQQCTKI SKGRGSLPPQ HGLELLTVYA WEQGGK-DSQ FNMAEGFRTV
p30 OASL    TKLKSLLRLV KHWYQQ---A HHP------- ---------- ---GS----- ----------
p56 OASL    TKLKSLLRLV KHWYQQYVKS RSPRANLPPL YALELLTIYA WEMGTEEDEN FMLDEGFTTV
p59 OASL    TKLKSLLRLV KHWYQQYVKA RSPRANLPPL YALELLTIYA WEMGTEEDEN FMLDEGFTTV
Clustal Co  **** *.*** ****.                               *
```

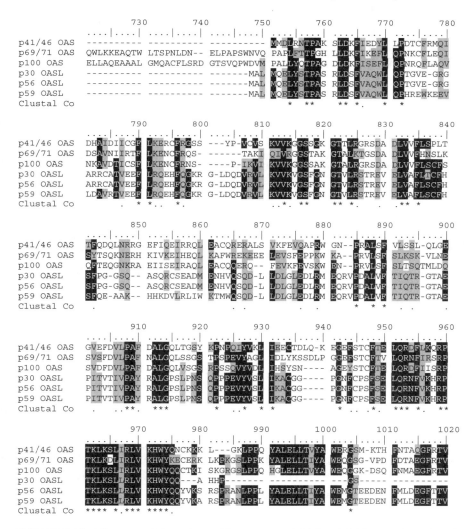

FIGURE 5.10 Alignment of the amino acid sequences of OAS and OASL proteins using the ClustalW program. Identities are in black and similarities (conservative replacements) are shaded. Amino acid numbering follows the p100 OASL sequence. The consensus is indicated on the last line.

```
            ....|....| ....|....| ....|....| ....|....| ....|....| ....|....|
p41/46 OAS  LELVINYQQL CIYWTKYYDF KNPIIEKYLR RQLTKPRPVI LDPADPTGNL GGGDPKGWRQ
p69/71 OAS  LELVTQYQQL CIFWKVNYNF EDETVRKFLL SQLQKTRPVI LDPAEPTGDV GGGDRWCWHL
p100 OAS    LELVTQYRQL CIYWTINYNA KDKTVGDFLK QQLQKPRPII LDPADPTGNL G--HNARWDL
p30 OASL    ---------- ---------- ---------- ---------- ---------- ----------
p56 OASL    MDLLLEYEVI CIYWTKYYTL HNAIIEDCVR KQLKKERPII LDPADPTLNV A--EGIRWDT
p59 OASL    MDLLLEYEVI CIYWTKYYTL HNAIIEDCVR KQLKKERPII LDPADPTLNV A--EGYRWDI
Clustal Co

                  1090       1100       1110       1120       1130       1140
            ....|....| ....|....| ....|....| ....|....| ....|....| ....|....|
p41/46 OAS  LAQEAEAWLN YPCFKNWDGS PVSSWILLAE --SNSTDDET DDP------- ----------
p69/71 OAS  LAKEAKEWLS SPCFKDGTGN PIPPWKVPVK ---------- ---------- ----------
p100 OAS    LAKEAAACTS ALCCMGRNGI PIQPWPVKAA ---------- ---------- ----------
p30 OASL    ---------- ------GRPH P--------- --------QR ---------- ----------
p56 OASL    VALRASQCLK QDCCYDNREN PISSWNVKRA RDIHLTVEQR GYPDFNLIVN PYEPIRKVKE
p59 OASL    VAQRASQCLK QDCCYDNREN PISSWNVKRA RDIHLTVEQR GYPDFNLIVN PYEPIRKVKE
Clustal Co                        *

                  1150       1160       1170       1180       1190       1200
            ....|....| ....|....| ....|....| ....|....| ....|....| ....|....|
p41/46 OAS  RTYQKYGYIG ---------T HEYPHFSHRP S--------- ---------- ----------
p69/71 OAS  -------VI- ---------- ---------- ---------- ---------- ----------
p100 OAS    -------V-- ---------- ---------- ---------- ---------- ----------
p30 OASL    ---------- ---------G -RRVQMGHRC S--------- ---------- ----------
p56 OASL    KIRRTRGYSG LQRLSFQVPG SERQLLSSRC SLAKYGIFSH THIYLLETIP SEIQVFVKNP
p59 OASL    KIRRTRGYSG LQRLSFQVPG SERQLLSSRC SLAKYGIFSH THIYLLETIP SEIQVFVKNP
Clustal Co

                  1210       1220       1230       1240       1250       1260
            ....|....| ....|....| ....|....| ....|....| ....|....| ....|....|
p41/46 OAS  ---------- ---------- ---TLQAAST PQAEEDWTCT IL-------- ----------
p69/71 OAS  ---------- ---------- ---------- ---------- ---------- ----------
p100 OAS    ---------- ---------- ---------- ---------- ---------- ----------
p30 OASL    ---------- ---------- -----EGLPV PET-----GL LL-------- ----------
p56 OASL    DGGTYAYAIN PNSFILGLKQ QIEDQQGLPK KQQQLEFQGQ VLQDWLGLGI YGIQDSDTLI
p59 OASL    DGGSYAYAIN PNSFILGLKQ QIEDQQGLPK KQQQLEFQGQ VLQDWLGLGI YGIQDSDTLI
Clustal Co

                  1270       1280       1290       1300
            ....|....| ....|....| ....|....| ....|....| ....|...
p41/46 OAS  ---------- ---------- ---------- ---------- --------
p69/71 OAS  ---------- ---------- ---------- ---------- --------
p100 OAS    ---------- ---------- ---------- ---------- --------
p30 OASL    ---------- ---------- ---------- ---------- --------
p56 OASL    LSKKKGEALF PAS------- ---------- ---------- --------
p59 OASL    LSKKKGEALF PAS------- ---------- ---------- --------
Clustal Co
```

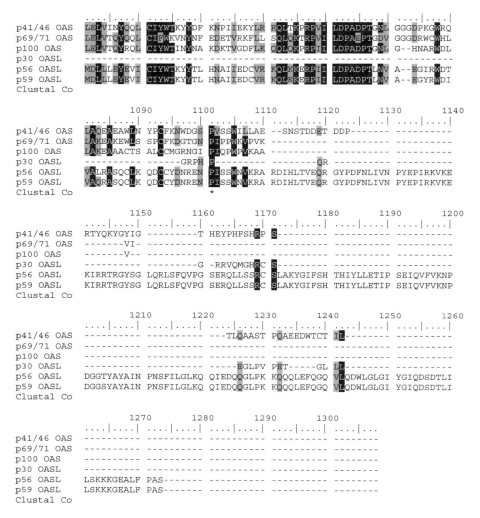

FIGURE 5.10 *Continued.*

consensus among the six proteins in this region is particularly low; the three aspartic acid residues (829, 831, and 905) that have been shown to be critical for the catalytic activity[115] are not conserved in the OASLs. Similarly, the P-loop (residues 813 to 822) is poorly conserved in the 2-5OASL proteins, and — noteworthy — the crucial lysine (residue 820) is replaced by an asparagine. These differences explain why the OASL proteins are devoid of catalytic activity.[118,119] In contrast, all six proteins have conserved the ATP-binding site (residues 951 to 962).[122] A phylogenetic analysis of the six sequences (Figure 5.11) allows us to assign a single common ancestor to these proteins, the tree having eventually diverged into several subfamilies, including either the OAS or the OASL proteins.

```
                    +--p59 OASL
                 +--5
           +-----4  +--p56 OASL
           !     !
        +--3     +-----p30 OASL
        !  !
        !  !        +--p100 OAS
      --1  +--------2
        !           +--p69/71 OAS
        !
        +--------------p41/46 OAS
```

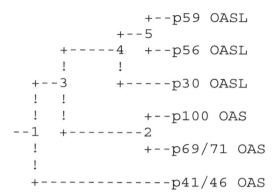

FIGURE 5.11 Phylogenetic tree of the 2-5A synthetase family.

A blast search with the p59OASL sequence has permitted its identification as the thyroid receptor (TR)-interacting protein 14.[119] TRIPS are proteins interacting with the thyroid and retinoid receptor in the presence or absence of the triiodothyronine (T3) ligand.[123] These proteins do not interact with the glucocorticoid receptor. All these proteins interact with the TR through a receptor-binding motif made of a consensus LXXLL sequence.[124,125] The three OASL proteins contain this motif (LKSLL, residues 963 to 967, Figure 5.10) and because of the high level of identity between these proteins, we can consequently argue that p30 and p56 are lower molecular variants of p59/TRIP14. In contrast, the terminal leucine is conservatively replaced by an isoleucine in the OAS proteins. TRIP14/p59 does not present any sequence similarity with other thyroid receptor-interacting proteins such as the TRAPs (TR-associated proteins)[126] or TRUPs (TR-uncoupling proteins).[127] Interestingly, the OAS as well as the OASL proteins are ubiquitously expressed in both the cytoplasm and the cell nucleus,[119,128] and the OASL proteins might thus be allowed to interact with TR in the nucleus.

The thyroid hormone T3 plays an important role in metabolic balance. Its action is mediated by nuclear thyroid hormone receptors (TR), which are members of the steroid and thyroid receptor superfamily and act as transcription factors regulating target gene expression directly through DNA response elements (HRE) (review in Reference 13). Although TR can bind to HRE as a monomer or a homodimer, the major form interacting with HRE is the heterodimer with retinoid X receptor (RXR). Unliganded TR represses transcription and ligand binding causes derepression. A group of coactivator and corepressor proteins mediate repression and activation, respectively, through histone acetylase (HDAC) or acetylase (HAT) domains (Figure 5.12; review in Reference 129). The coactivator proteins interact with TR through LXXLL motifs heavily conserved,[124,125] a motif shared by the OASL proteins. Surprisingly enough, the sequences of p56 and p59 OASL/TRIP proteins also contain two highly conserved overlapping ubiquitin motifs (Figure 5.10, residues 1207 to 1243, respectively $IX_2LKXQIX_6PX_2KQXLX_6LQ$ and $YXIX_5IX_2LKX_2IX_5LX_6LXFXGX_2L$). This suggests that TRIP14 could be capable of binding the TR and targeting it for destruction by the proteasome.[130]

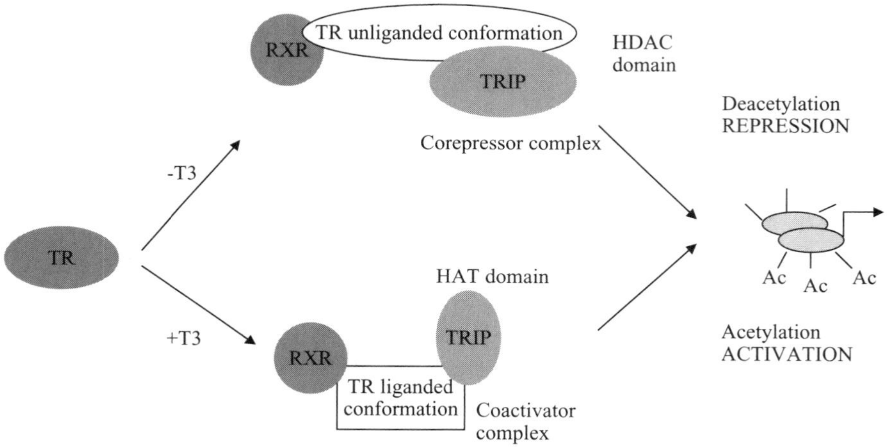

FIGURE 5.12 The mode of activation or repression of the TR on HRE, through either the histone deacetylase (HDAC) domain of corepressors, or the histone acetylase (HAT) domain of coactivators, respectively, in absence or presence of T3.

A cross-talk between different nuclear signaling pathways is not unusual. For instance, p53 modulates the transcriptional activity of TR[131] and the vitamin D receptor represses the basal transcription by the TR.[132] However, a cross-talk between signaling pathways proceeding through membrane receptors like IFNs and nuclear receptors like T3 is less common. The interaction between TRIP15 and the interferon consensus sequence-binding protein (ICSBP) is another example.[133] The TR-interacting protein TRIP15 has recently been identified as a component of a regulatory subunit of the 26S proteasome (COP9/CSN2).[134] The protein is present in both the cytoplasm and the nucleus and possesses an associated kinase activity which specifically phosphorylates signaling molecules such as IκBα and c-Jun. The protein also interacts with ICSBP in that it phosphorylates on a specific serine residue essential for the efficient interaction of ICSBP with IRF-1, leading to transcriptional repression of the ISRE.[133] ICSB (also known as IRF-8) is expressed exclusively in immune cells and is induced by IFN-γ. Ablation of IRF-8/ICSBP expression results in mice deficient in T_H1 mediated immune response, which is attributed to a lack of IL-12 expression.[133]

The induction of the 2-5OASL proteins by IFNs is a complex process which requires not only a proper balance of different IFN subtypes,[135-138] but also the activation of PKC through PI-3K.[139] Chronic fatigue syndrome is characterized by an unexplained long-lasting severe fatigue along with immune dysfunction and hypersensitivity to glucocorticoids, which is likely to occur at the transcriptional level.[72] A strong dysregulation of the interferon 2-5A pathway has also been pointed out as a characteristic of the illness (Chapters 2 and 3). Consequently, one can reasonably consider that the dysregulation of the IFN signaling pathway in these patients can be responsible for a peripheral resistance to thyroid hormones, explaining the extreme fatigue with a normal or subnormal thyroid hormone profile. The dysregulation occurring at this level, characterized by upregulated 2-5OAS enzymes,[140] can thus reasonably be thought to be accompanied by a similar up-regulation of the 2-5OASL/TRIP14 proteins, eventually targeting the TR for destruction by the proteasome. A similar mechanism might also be involved, explaining the resistance to glucocorticoids.

5.6 RNase L AND SIGNAL TRANSDUCTION

Strong indications for the possible involvement of RNase L in signal transduction pathways are progressively emerging. The enzyme has been shown to induce apoptosis by a caspase-dependent mechanism, involving an 18S rRNA cleavage.[141] The enzyme has also been implicated, along with PKR, in the suppression of gene expression from viral and non-viral vectors.[142] RNase L has also been implicated in the regulation of the ds-RNA activation of MAPK and JNK through the cleavage of the 28S rRNA and inhibition of translation.[100] Finally, RNase L has recently been linked to the direct regulation of the IFN pathway.[143] The enzyme cleaves the mRNAs transcribed by ISG15 and by a gene induced as a primary response to interferons that encodes a 43-kDa ubiquitin-specific protease (ISG43). The ISG15 protein is a ubiquitin-like protein which acts as an immunoregulator.[88,89] RNase L negatively regulates these genes, which results, on the one hand, in a decreased ubiquitination of cellular proteins by ISG15, and on the other hand in a decreased deubiquitination by ISG43 of selective ubiquitinated cellular substrates.[143] In the PBMC of CFS patients, RNase L is cleaved and one of the fragments generated (37-kDa) retains catalytic activity, and is likely to be regulated differently than the 80-kDa native enzyme (Chapter 3). Consequently, a time-out of such regulatory action by RNase L, as can be suspected in CFS, can consequently have dramatic biological effects.

Besides the catalytically active fragment, the cleavage of RNase L in PBMC of CFS patients releases other fragments. One of these contains the N-terminal part of the protein. A BLAST search performed with this fragment of RNase L (residues 1 to 362) indicates a high degree of similarity with three human proteins. The sequence alignment is given in Figure 5.13. The overall similarity between these proteins is over 40%. Interestingly, one of these proteins is TRIP9, another member of the family.[123] The two other proteins are unnamed protein products directly submitted in 2000 to Genbank by Japanese scientists involved in the NEDO human cDNA sequencing projects. The protein sequences are translations of clones respectively isolated from teratocarcinoma cells NT2 previously induced by retinoic acid (accession GI 7022441) and primary renal epithelial cells (accession GI 10438501).

Surprisingly, the LXXLL motif[124,125] of TRIP 9 (LDFLL, residues 131 to 135, Figure 5.13) is not matched by any of the other proteins. However, the RNase L fragment contains such a motif later in the sequence (LKILL, residues 270 to 274, Figure 5.13), which is matched by one of the unnamed proteins (LEILL, GI 7022441). The sequence of the second unnamed protein does not contain such a motif. However, unless TRIP 9, the two unnamed proteins alternatively share the repetitive ankyrin motifs with unknown functions of RNase L (Chapter 2), namely GANVN (residues 70 to 75, Figure 5.13), GADVN (residues 174 to 178), GADVNA (residues 279 to 284), and GADVN (residues 316 to 321). Most strikingly, the BLAST search further identifies TRIP 9 as the β-chain of I-κB (accession GI

```
                          10         20         30         40         50         60
                  ....|....| ....|....| ....|....| ....|....| ....|....| ....|....|
RNase Lank   1    ME-SRDHNN- -----PQ--- EGPTSS-SGR R--------- AAVEDNHLLI KAVQNED--V
TRIP9        1    MAGVACLGK- ---------- AADADEWCDT G--------- LGSLGPDAAA PGGPGLG--A
GI7022441    1    MKTFEGFCAL HLAASQGHWK IVQILLEAGA DPN------- ATTLEETTPL FSAVENG-QI
GI10438501   1    MSNTPTHSIA ASISQPQTPT PSPIISPSAM LPIYPAIDID AQTESNHDTA LTLACAGGHE

                          70         80         90        100        110        120
                  ....|....| ....|....| ....|....| ....|....| ....|....| ....|....|
RNase Lank   39   DLVQQLLEGG ANVN-FQEEE G--------- ---------- -G-------- -----WTPLH
TRIP9        39   ELGPGLSWAP LVF-GYVTED G--------- ---------- ---------- -----DTALH
GI7022441    53   DVLRLLLQHG ANVNGSHSMC G--------- ---------- ---------- -----WNSLH
GI10438501   61   ELVQTLLERG ASI-EHRDKK GFTPLILAAT AGHVGVVEIL LGNGADIEAQ SERTKDTPLS

                         130        140        150        160        170        180
                  ....|....| ....|....| ....|....| ....|....| ....|....| ....|....|
RNase Lank   65   NAVQMSREDI VELLLRHGAD PVLRKKN--- GATPFILAAI AGSVKLLKLF LSKGADVNEC
TRIP9        64   LAVIHQHEPF LDFLLGFSAG TEYMDLQNDL GQTALHLAAI LGETSTVEKL YAAGAGLCVA
GI7022441    79   QASFQENAEI IKLLLRKGAN KECQDDF--- GITPLFVAAQ YGKLESLSIL ISSGANVNCQ
GI10438501   120  LACSGGRQEV VELLLARGAN KEHRNVS--- DYTPLSLAAS GGYVNIIKIL LNAGAEINSR

                         190        200        210        220        230        240
                  ....|....| ....|....| ....|....| ....|....| ....|....| ....|....|
RNase Lank   122  D--FYGFTAF MEAAVYGKVK ALKFLYKRGA NVNLRR-KTK ED-------- QER-------
TRIP9        124  E--RRGHTAL HLACRVGAHA CARALLQP-- ---------- ---------- RPR-------
GI7022441    136  A--LDKATPL FIAAQEGHTK CVELLLSSGA DPDLYCNEDS WQLPIHAAAQ MGHTKILDLL
GI10438501   177  TGSKLGISPL MLAAMNGHTA AVKLLLDMGS DINAQI-ETN RNTALTLACF QGRTEVVSLL

                         250        260        270        280        290        300
                  ....|....| ....|....| ....|....| ....|....| ....|....| ....|....|
RNase Lank   163  -LRKG----- ---GATALMD AAEKGHVEVL KILLDEM-GA DVNACDNMGR NALIHALLSS
TRIP9        152  --RPR----- ---------- -------EAP DTYLAQ--GP DRTPDTNH-- TP--VALYPD
GI7022441    194  IPLTNRACDT GLNKVSPVYS AVFGGHEDCL EILLRNGYSP DAQACLVFGF SSP-VCMAFQ
GI10438501   236  LDRKANVEHR AKTGLTPLME AASGGYAEVG RVLLDK--GA DVNAPPVP-- SSRDTALTIA

                         310        320        330        340        350        360
                  ....|....| ....|....| ....|....| ....|....| ....|....| ....|....|
RNase Lank   214  DDSDVEAITH LLLDHGADVN VR-------- ---------G ERGK-----T PLILAVEKKH
TRIP9        183  SD-LEKEEEE SEEDWKLQLE AE-------- ---------N YEGH-----T PLHVAVIHKD
GI7022441    253  KDCEFFGIVN ILLKYGAQIN ELHLAYCLKY EKFSIFRYFL RKGCSLGPWN HIYEFVNHAI
GI10438501   292  ADKGHYKFCE LLIGRGAHID VR-------- ---------N KKGN-----T PLWLAANGGH

                         370        380        390        400        410        420
                  ....|....| ....|....| ....|....| ....|....| ....|....| ....|....|
RNase Lank   252  LGLVQRLLEQ EHIEINDTD- ---------- SDGKTALLLA VELKLKKIAE LLCKRG----
TRIP9        220  VEMVR-LLRD AGADLDKPEP ---------- TCGRSPLHLA VEAQAADVLE LLLRAG---A
GI7022441    313  KAQAKYKEWL PHLLVAGFDP LILLCNSWID SVSIDTLIFT LEFTNWKTLA PAVERM---L
GI10438501   330  LDVVQ-LLVQ AGADVDAAD- ---------- NRKITPLMAA FRKGHVKVVR YLVKEVNQFP

                         430        440        450        460        470        480
                  ....|....| ....|....| ....|....| ....|....| ....|....| ....|....|
RNase Lank   297  ASTDCGDLVM TARR----NY DHSLVKVLLS HGAKEDFHP- ---------- PAEDWKPQSS
TRIP9        266  NPAARMYGGR TPLGSAMLRP NPILARLLRA HGAPEPEGE- ---------- -DEKSGPCSS
GI7022441    370  SARASNAWIL QQHIATVPSL THLCRLEIRS SLKSERLRSD S--------Y ISQLPLPRSL
GI10438501   378  SDSECMRYIA TITDKEMLKK CHLCMESIVQ AKDRQAAEAN KNASILLEEL DLEKLREESR

                         490        500
                  ....|....| ....|....| ....|
RNase Lank   342  HWGAALKDLH R--IYRPMIG KLK--
TRIP9        314  SSDSDGGDEG VSQEERQGSP AGGSG
GI7022441    422  HNYLLYEDVL R-MYEVPELA AIQDG
GI10438501   438  RLALAAKREK RKEKRRKKKK KKK--
```

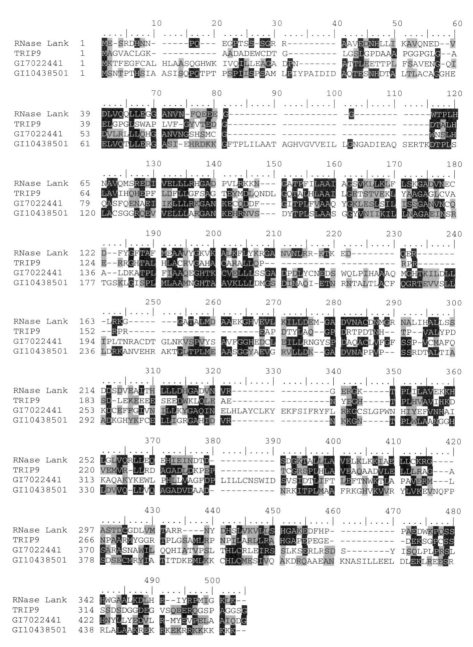

FIGURE 5.13 Sequence alignment (ClustalW) of the ankyrin fragment of RNase L (residues 1 to 362) with human TRIP 9 and the unnamed proteins (accession GI 7022441 and GI 10438501).

```
                       10         20         30         40         50         60
                ....|....| ....|....| ....|....| ....|....| ....|....| ....|....|
RNase L      1  ------FFID EKYK-IADTS EGGIYLGFYE KQE------- ----VAVKTF CEGSPR-AQR
Cdk-6        1  MEKDGLCRAD QQYECVAEIG EGAYGKVFKA RDLKNGGRFV ALKRVRVQTG EEGMPLSTIR
Clustal Co   1           *  ..*   .*.  **        *   .              * *.*   ** *   . *

                       70         80         90        100        110        120
                ....|....| ....|....| ....|....| ....|....| ....|....| ....|....|
RNase L     42  EVSCLQ--SS RENSHLVTFY GSE-SHRGHL FVCVTLCEQT LEACLDVHRG EDVE-NEEDE
Cdk-6       61  EVAVLRHLET FEHPNVVRLF DVCTVSRTDR ETKLTLVFEH VDQDLTTYLD KVPEPGVPTE
Clustal Co  14  **. *.    .  *. ..*  .       *       .**  .  ..  *       .  *  .  *

                      130        140        150        160        170        180
                ....|....| ....|....| ....|....| ....|....| ....|....| ....|....|
RNase L     98  FARNVLSSIF KAVQELHLSC GYTHQDLQPQ NILIDSKKAA HLADFDKS-- ----------
Cdk-6      121  TIKDMMFQLL RGLDFLH-SH RVVHRDLKPQ NILVTSSGQI KLADFGLARI YSFQMALTSV
Clustal Co  24    ....  .  . .  ** *    *.**.** ***. *     .****   .

                      190        200        210        220        230        240
                ....|....| ....|....| ....|....| ....|....| ....|....| ....|....|
RNase L    145  --IKWAGDPQ ---------- ---------- -EVKR----- ----DLEDLG RLVLYVVKKG
Cdk-6      180  VVTLWYRAPE VLLQSSYATP VDLWSVGCIF AEMFRRKPLF RGSSDVDQLG KILDVIGLPG
Clustal Co  39      *   *.                        *. *           *.. **  ...  .  *

                      250        260        270        280        290        300
                ....|....| ....|....| ....|....| ....|....| ....|....| ....|....|
RNase L    174  S--------- --ISFEDLKA QSNEEVVQLS PDEETKDLIH RLFHPGEHVR DCLSDLLGHP
Cdk-6      240  EEDWPRDVAL PRQAFHSKSA QPIEKFVTD- IDELGKDLLL KCLTFNPAKR ISAYSALSHP
Clustal Co  47              .*    * *  *  *     **  ***. .   .  *  .    * **

                      310        320
                ....|....| ....|....| ....|....
RNase L    223  FFWTWESRYR TLR------- ---------
Cdk-6      299  YFQDLERCKE NLDSHLPPSQ NTSELNTA-
Clustal Co  61  .*   *     .*
```

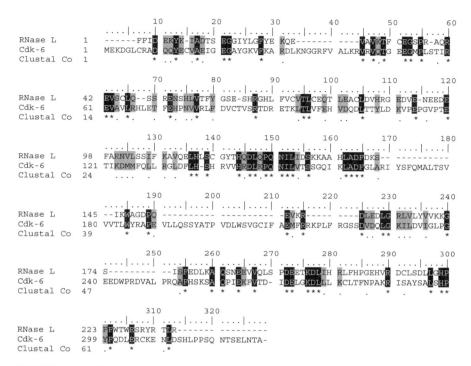

FIGURE 5.14 Alignment (ClustalW) of the sequences of RNase L (intermediary fragment, residues 358 to 596) and of the human serine/threonine kinase Cdk6.

4505385), assigned to chromosome 19q13.1,[144] with which the first 88 residues of RNase L share 67% similarity. The intermediate fragment of RNase L, which contains the kinase homology region, shares up to 50% similarity with the cell division protein kinase 6 (Cdk6, serine/threonine-protein kinase PLSTIRE, Figure 5.14). This kinase blocks apoptosis by driving G1 progression during the cell cycle.[145] Finally, the catalytic domain of RNase L (residues 558 to 741) shares 40% similarity with the catalytic domain of Ire1 (Chapter 2, Figure 2.6).[146]

Despite the strikingly high degree of similarity between the RNase L fragments released by proteolytic cleavage in PBMCs of CFS patients and proteins involved in signal transduction, at the present stage it is too early to draw definitive conclusions regarding their possible implications in the immune dysfunctions associated with the illness.[147] However, the present path is certainly worth further investigation in light of current knowledge of signal transduction.

5.7 THE INSULIN-LIKE GROWTH FACTOR RECEPTOR

Insulin growth factors (IGF-I and II; IGF-I was formerly known as somatomedin) are single chain polypeptides of 70 and 67 residues, respectively, which share 40 to 50% identity with insulin.[148] The IGFs are ubiquitously expressed and produced in large quantities by the liver under the main regulatory influence of growth hormone

(GH). IGF-I production is regulated in peripheral tissues by other factors, such as estrogen in the uterus, follicle-stimulating hormone in the ovary, and parathyroid hormone and estrogens in bone.[149] Indirect determinants such as adrenal androgens (particularly dehydroepiandrosterone, DHEA) and inflammatory cytokines also play a role in the regulation of circulating IGF-I levels.[150] The circulating IGFs are circulating in serum in association with high-affinity binding proteins (insulin-like growth factor-binding proteins, IGFBPs), which sequester them away from their receptors, facilitate their transport and regulate their functions at the cellular level.[150,151] Among these, IGFBP-3 is the dominant binding protein in serum that displays the highest affinity for IGF-I.[151]

IGF-I has been recognized as an important mitogen required by some cell types to progress from the G1 phase to the S phase of the cell cycle,[152] an activity regulated by the IGFBP-3. Its activity has therefore been directly linked to the development and progression of cancer.[152,153] The survival signals given by the IGF-I-receptor interaction are still partly unknown, but are progressively unraveled. The IGF-I receptor (IGF-IR) is a transmembrane tyrosine kinase widely expressed in many cell types.[154] Upon activation, the receptor autophosphorylates and recruits a large docking protein, insulin receptor substrate-2, which in turn recruits PI-3K. This latter (Figure 5.5) further activates PDK and PKB (Akt). PKB subsequently phosphorylates Bad, resulting in its sequestration from the death suppressor Bcl-xL, which remains in a complex with the caspase activator Apaf-1, preventing the formation of the apoptosome.[148] Besides their inhibition of IGF action, the IGFBPs have also been suggested to play a minor stimulating role of IGF action by a putative interaction with specific cell-surface receptors, or a decrease of affinity for IGFs.[155] However, their main regulatory role remains the inhibition of the cell survival signals by IGF-I, and they cause apoptosis in a way which can be IGF-I-dependent or even independent, respectively. Noteworthy in this respect, the proapoptotic action of the tumor suppressor p53 has been shown recently to be mediated by the induction of IGFBP-3.[156]

The IGF-I and IGFBP system has been implicated as mediator in many catabolic illnesses involving GH resistance.[157] Specific regulatory roles have also been devoted to this cell signaling pathway in the modulation of cognitive function,[158,159] cardiac function,[160] steroid hormone actions in the endometrium through paracrine and autocrine mechanisms,[161] and musculoskeletal instability.[162] Finally, IGF-I, in conjunction with prolactin (PRL) and GH, has been implied in immunomodulation.[163] Interestingly, some links are likely to exist between the IFN/2-5A and the IGF-I/GH signaling pathways. On the one hand, IFN-γ is likely to play an important switching role in macrophage differentiation leading to the induction of IGF-I by these cells.[164] On the other hand, an interaction between the low molecular weight form of the 2-5A synthetase and the PRL receptor has been demonstrated, which exerts an inhibitory effect on PRL induction of the interferon-regulatory factor 1 promoter and a reduction of PRL-inducible STAT 1.[165]

Impaired cognitive and cardiac function, musculoskeletal instability, and deficient neuroendocrine-immune communication are common complaints and observations in CFS.[166-168] Several authors have pointed out a dysregulation in the hypothalamic–pituitary–adrenal axis (HPA) in patients suffering from CFS,[166,169] which might thus involve GH and IGF-I resistance. Attempts to detect significant abnormalities in GH and IGF-I function in CFS patients, however, have been far from conclusive. While some authors have found small or significant differences in basal IGF-I levels or in GH response to provoked hypoglycemia,[170,171] many others have reported no clear-cut or even insignificant differences when compared to healthy controls.[172-174] In a study intended to evaluate the efficacy of GH therapy in CFS, it was noted that despite no significant improvement in the quality of life, among 20 patients receiving the therapy, four were able to resume work after a long period of sick leave.[175] These discrepancies further underline the heterogeneity of CFS patient groups resulting from the failure of the current case definition to identify a truly discrete group of individuals suffering from the same pathophysiology.[166]

5.8 CONCLUSIONS AND PROSPECTS

In this chapter, we have reviewed several cell signaling pathways and have pinpointed their possible interactions with the IFN/2-5A pathway. We have provided evidence of extensive cross-talk between these signaling pathways and have indicated several abnormalities in the IFN and HPA signaling systems that might be involved in the pathogenesis of CFS. In particular, a dysregulation of the 2-5OAS induction by IFN, involving a preferential induction of 2-5OASL/TRIP proteins can explain a peripheral resistance to thyroid hormones leading to chronic fatigue. Similarly, impaired immunomodulation and cognitive and cardiac dysfunctions can partially originate from a combined dysregulation in the IFN/2-5A and HPA signaling pathways. Finally, we have also indicated the central roles that RNase L plays in these cell signaling cross-talking systems and have pointed out the significant implications that might have the proteolytic cleavage of this enzyme in the immune cells of CFS patients.

While our current, continuously evolving understanding of the cell signaling cascades strongly supports biological dysregulations of cellular homeostasis in CFS, further experimental evidence will be required in order to link the different pathways possibly involved.

REFERENCES

1. Pawson, T. and Scott, J. D., Signaling through scaffold, anchoring, and adaptor proteins, *Science*, 1997; 278: 2075–80.
2. Schillace, R. V. and Scott, J. D., Organization of kinases, phosphatases, and receptor signaling complexes, *J. Clin. Invest.,* 1999; 103: 761–5.
3. Virkamäki, A., Ueki, K., and Kahn, C. R., Protein–protein interaction in insulin signaling and the molecular mechanisms of insulin resistance, *J. Clin. Invest.,* 1999; 103: 931–43.
4. Lu, P.-J. et al., Function of WW domains as phosphoserine or phosphothreonine-binding modules, *Science*, 1999; 283: 1325–8.
5. Fanning, A. S. and Anderson, J. M., PDZ domains: fundamental building blocks in the organization of protein complexes at the plasma membrane, *J. Clin. Invest.,* 1999; 103: 767–72.

6. Kay, B. K., Williamson, M. P., and Sudol, M., The importance of being proline: the interaction of proline-rich motifs in signaling proteins with their cognate domains, *FASEB J.*, 2000; 14: 231–41.
7. Nishita, M. et al., Interaction between Wnt and TGF-beta signaling pathways during formation of Spemann's organizer, *Nature*, 2000; 403: 781–5.
8. Samad, T. A. et al., Interleukin-1β-mediated induction of Cox-2 in the CNS contributes to inflammatory pain hypersensitivity, *Nature*, 2001; 410: 471–5.
9. Englebienne, P., *Immune and Receptors Assays in Theory and Practice,* CRC Press, Boca Raton, FL, 2000, 32.
10. Schwartz, J.-C. D. et al., Structural basis for co-stimulation by the human CTLA-4/B7-2 complex, *Nature*, 2001; 410: 604–8.
11. Stamper, C. C. et al., Crystal structure of the B7-1/CTLA-4 complex that inhibits human immune responses, *Nature*, 2001; 410: 608–11.
12. Englebienne, P., Molecular basis of ligand-receptor interactions, in *Immune and Receptors Assays in Theory and Practice,* CRC Press, Boca Raton, FL, 2000, 141–79.
13. Tsai, M.-J. and O'Malley, B. W., Molecular mechanisms of action of steroid/thyroid receptor superfamily members, *Annu. Rev. Biochem.,* 1994; 63: 451–86.
14. Englebienne, P., Receptors at work, in *Immune and Receptors Assays in Theory and Practice,* CRC Press, Boca Raton, FL, 2000, 1–21.
15. Jones, K. A. et al., $GABA_B$ receptors function as a heterodimeric assembly of the subunits $GABA_BR1$ and $GABA_BR2$, *Nature*, 1998; 396: 674–9.
16. White, J. H. et al., Heterodimerization is required for the formation of a functional $GABA_B$ receptor, *Nature*, 1998; 396: 679–82.
17. Kaupmann, K. et al., $GABA_B$ receptor subtypes assemble into functional heterodimeric complexes, *Nature*, 1998; 396: 683–7.
18. Kuner, R. et al., Role of heteromer formation in $GABA_B$ receptor function, *Science*, 1999; 283: 74–7.
19. Wallach, D. et al., Tumor necrosis factor receptor and Fas signaling mechanisms, *Annu. Rev. Immunol.,* 1999; 17: 331–67.
20. Chan, F. K.-M. et al., A domain in TNF receptors that mediates ligand-independent receptor assembly and signaling, *Science*, 2000; 288: 2351–4.
21. Siegel, R. M. et al., Fas preassociation required for apoptosis signaling and dominant inhibition by pathogenic mutations, *Science*, 2000; 288: 2354–7.
22. Hubbard, S. R., Mohammadi, M., and Schlessinger, J., Autoregulatory mechanisms in protein-tyrosine kinases, *J. Biol. Chem.,* 1998; 273: 11987–90.
23. Clements, J. L. and Koretzky, G. A., Recent developments in lymphocyte activation: linking kinases to downstream signaling events, *J. Clin. Invest.,* 1999; 103: 925–929.
24. Apli, A. E. et al., Signal transduction and signal modulation by cell adhesion receptors: the role of integrins, cadherins, immunoglobulin-cell adhesion molecules, and selectins, *Pharmacol. Rev.,* 1998; 50: 197–263.
25. Strader, C. D. et al., Structure and function of G protein-coupled receptors, *Annu. Rev. Biochem.,* 1994; 63: 101–32.
26. Scholich, K. et al., Facilitation of signal onset and termination by adenylyl cyclase, *Science*, 1999; 283: 1328–31.
27. Malarkey, K. et al., The regulation of tyrosine kinase signaling pathways by growth factor and G-protein-coupled receptors, *Biochem. J.,* 1995; 309: 361–75.
28. Denhardt, D. T., Signal-transducing protein phosphorylation cascades mediated by Ras/Rho proteins in the mammalian cell, *Biochem. J.,* 1996; 318: 729–47.
29. Panaretto, B. A., Aspects of growth factor signal transduction in the cell cytoplasm, *J. Cell Sci.,* 1994; 107: 747–52.

30. Rommel, C. et al., Differentiation stage-specific inhibition of the Raf-MEK-ERK pathway by Akt, *Science*, 1999; 286: 1738–41.
31. Zimmermaann, S. and Moelling, K., Phosphorylation and regulation of Raf by Akt (protein kinase B), *Science*, 1999; 286: 1741–4.
32. Christman, J. W., Lancaster, L. H., and Blackwell, T. S., Nuclear factor κ B: pivotal role in the systemic inflammatory response syndrome and new target for therapy, *Intensive Care Med.*, 1998; 24: 1131–8.
33. Wang, Y. et al., Stat1 as a component of tumor necrosis factor alpha receptor 1-TRADD signaling complex to inhibit NF-κB activation, *Mol. Cell. Biol.*, 2000; 20: 4505–12.
34. Wang, X. et al., The cellular response to oxidative stress: influences of mitogen-activated protein kinase signaling pathways on cell survival, *Biochem. J.*, 1998; 333: 291–300.
35. Coffer, P. J., Jin, J., and Woodgett, J. R., Protein kinase B (c-Akt): a multifunctional mediator of phosphatidylinositol 3-kinase activation, *Biochem. J.*, 1998; 335: 1–13.
36. Del Peso, L. et al., Interleukin-3-induced phosphorylation of BAD through the protein kinase Akt, *Science*, 1997; 278: 687–9.
37. Mathias, S., Pena, L. A., and Kolesnick, R. N., Signal transduction of stress via ceramide, *Biochem. J.*, 1998; 335: 465–80.
38. Hannun, Y. A., Functions of ceramide in coordinating cellular responses to stress, *Science*, 1996; 274: 1855–9.
39. Platanias, L. C. and Fish, E. N., Signaling pathways activated by interferons, *Exp. Hematol.*, 1999; 27: 1583–92.
40. Vojdani, A. and Lapp, C. W., Interferon-induced proteins that are elevated in blood samples of patients with chemically- or virally-induced chronic fatigue syndrome, *Immunopharm. Immunotoxicol.*, 1999; 21: 175–202.
41. Durbin, J. E. et al., Type I IFN modulates innate and specific antiviral immunity, *J. Immunol.*, 2000; 164: 4220–8.
42. Boehm, U. et al., Cellular responses to interferon-γ, *Annu. Rev. Immunol.*, 1997; 15: 749–95.
43. Piehler, J., Roisman, L. C., and Schreiber, G., New structural and functional aspects of the type I interferon-receptor interaction revealed by comprehensive mutational analysis of the binding interface, *J. Biol. Chem.*, 2000; 275: 40425–33.
44. Eid, P. et al., Localization of a receptor nonapeptide with a possible role in the binding of the type I interferons, *Eur. Cytokine Netw.*, 2000; 11: 560–73.
45. Pattyn, E. et al., Dimerization of the interferon type I receptor IFNaR2-2 is sufficient for induction of interferon effector genes but not for full antiviral activity, *J. Biol. Chem.*, 1999; 274: 34838–45.
46. Stark, G. R. et al., How cells respond to interferons, *Annu. Rev. Biochem.*, 1998; 67: 227–64.
47. Darnell, J. E., Jr., Kerr, I. M., and Stark, G. R., Jak-STAT pathways and transcriptional activation in response to interferons and other extracellular signaling proteins, *Science*, 1994; 264: 1415–21.
48. Su, L. and David, M., Distinct mechanisms of STAT phosphorylation via the interferon α/β receptor, *J. Biol. Chem.*, 2000; 275: 12661–6.
49. Pfeffer, L. M. et al., STAT3 as an adapter to couple phosphatidylinositol 3-kinase to the IFNAR1 chain of the type I interferon receptor, *Science*, 1997; 276: 1418–20.
50. Rani, M. R. R. et al., Catalytically active Tyk2 is essential for interferon-β-mediated phosphorylation of STAT3 and interferon-α receptor-1 (IFNAR-1) but not for the activation of phosphoinositol 3-kinase, *J. Biol. Chem.*, 1999; 274: 32507–11.

51. Uddin, S. et al., Interferon-dependent activation of the serine kinase PI 3'-kinase requires engagement of the IRS pathway but not the Stat pathway, *Biochem. Biophys. Res. Comm.,* 2000; 270: 158–62.
52. Weinstein, S. L. et al., Phosphatidylinositol 3-kinase and mTOR mediate lipopolysaccharide-stimulated nitric oxide production in macrophages via interferon-beta, *J. Leukoc. Biol.*, 2000; 67: 405–14.
53. Uddin, S. et al., Activation of the p38 mitogen-activated protein kinase by type I interferons, *J. Biol. Chem.,* 1999; 274: 30127–31.
54. Uddin, S. et al., The Rac1/p38 mitogen-activated protein kinase pathway is required for interferon α-dependent trnascriptional activation but not serine phosphorylation of Stat proteins, *J. Biol. Chem.,* 2000; 275: 27634–40.
55. Romerio, F., Riva, A., and Zella, D., Interferon-alpha2b reduces phosphorylation and activity of MEK and ERK through a Ras/Raf-independent mechanism, *Br. J. Cancer,* 2000; 83: 532–8.
56. Wong, L. H. et al., Interferon-resistant human melanoma cells are deficient in ISGF3 components, STAT1, STAT2, and p48-ISGF3γ, *J. Biol. Chem.,* 1997; 272: 28779–85.
57. Sun, W. H. et al., Interferon-α resistance in a cutaneous T-cell lymphoma cell line is associated with lack of STAT1 expression, *Blood,* 1998; 91: 570–6.
58. Komaroff, A. L., The biology of chronic fatigue syndrome, *Am. J. Med.,* 2000; 108: 169–71.
59. Smith, E. et al., IRF3 and IRF7 phosphorylation in virus-infected cells does not require double-stranded-RNA-dependent protein kinase R or IκB kinase but is blocked by vaccinia virus E3L protein, *J. Biol. Chem.,* 2001; 276: 8951–7.
60. Iordanov, M. S. et al., Activation of NF-kappaB by double stranded RNA (dsRNA) in the absence of protein kinase R and RNase L demonstrates the existence of two separate dsRNA-triggered antiviral programs, *Mol. Cell. Biol.,* 2001; 21: 61–72.
61. Servant, M. J. et al., Identification of distinct signaling pathways leading to the phosphorylation of interferon regulatory factor 3, *J. Biol. Chem.,* 2001; 276: 355–63.
62. Kim, T. et al., Signaling pathways to the assembly of an interferon-β enhanceosome, *J. Biol. Chem.,* 2000; 275: 16910–7.
63. Lohoff, M. et al., Deficiency in the transcription factor interferon regulatory factor (IRF)-2 leads to severely compromised development of natural killer and T helper type 1 cells, *J. Exp. Med.,* 2000; 192: 325–35.
64. Kadowaki, N. et al., Natural interferon α/β-producing cells link innate and adaptive immunity, *J. Exp. Med.,* 2000; 192: 219–25.
65. Farrar, J. D. et al., Selective loss of type I interferon-induced STAT4 activation caused by a minisatellite insertion in mouse Stat2, *Nature Immunol.,* 2000; 1: 65–9.
66. O'Shea, J. J. and Visconti, R., Type 1 IFNs and regulation of T_H1 responses: enigmas both resolved and emerge, *Nature Immunol.,* 2000; 1: 17–9.
67. Nguyen, K. B. et al., Interferon α/β-mediated inhibition and promotion of interferon γ: STAT1 resolves a paradox, *Nature Immunol.,* 2000; 1: 70–6.
68. Frucht, D. M. et al., Stat4 is expressed in activated peripheral blood monocytes, dendritic cells, and macrophages at sites of Th1-mediated inflammation, *J. Immunol.,* 2000; 164: 4659–64.
69. Fukao, T. et al., Inducible expression of Stat4 in dendritic cells and macrophages and its critical role in innate and adaptive immune responses, *J. Immunol.,* 2001; 166: 4446–55.
70. Visser, J. T., De Kloet, E. R., and Nagelkerken, L., Altered glucocorticoid regulation of the immune response in the chronic fatigue syndrome, *Annu. N.Y. Acad. Sci.,* 2000; 917: 868–75.

71. Visser, J. et al., Differential regulation of interleukin-10 (IL-10) and IL-12 by glucocorticoid *in vitro, Blood*, 1998; 91: 4255–64.
72. Visser, J. et al., Increased sensitivity to glucocorticoids in peripheral blood mononuclear cells of chronic fatigue syndrome patients, without evidence for altered density or affinity of glucocorticoid receptors, *J. Invest. Med.*, 2001; 49: 195–204.
73. Faderl, S. and Estrov, Z., Hematopoietic growth factors and cytokines, *J. Clin. Ligand Assay,* 2000; 23: 169–80.
74. Klimas, N. G. et al., Immunologic abnormalities in chronic fatigue syndrome, *J. Clin. Microbiol.,* 1990; 28: 1403–10.
75. Chin, Y. E. et al., Cell growth arrest and induction of cyclin-dependent kinase inhibitor $p21^{WAF1/CIP1}$ mediated by STAT1, *Science*, 1996; 272: 719–22.
76. Roy, S. K. et al., CCAAT/enhancer-binding protein-beta regulates interferon-induced transcription through a novel element, *J. Biol. Chem.,* 2000; 275: 12626–32.
77. Hu, J. et al., ERK1 and ERK2 activate CCAAAT/enhancer-binding protein-β-dependent gene transcription in response to interferon-γ, *J. Biol. Chem.,* 2001; 276: 287–97.
78. Tnami, M. and Bayard, B. A., Evidence for IRF-1-dependent gene expression deficiency in interferon-unresponsive HepG2 cells, *Biochim. Biophys. Acta,* 1999; 1451: 59–72.
79. Tam, N. W. N. et al., Up-regulation of STAT1 protein in cells lacking or expressing mutants of the double-stranded RNA-dependent protein kinase PKR, *Eur. J. Biochem.,* 1999; 262: 149–54.
80. Wong, A. T.-H. et al., Physical association between STAT1 and the interferon-inducible protein kinase PKR and implications for interferon and double-stranded RNA signaling pathways, *EMBO J.,* 1997; 16: 1291–1304.
81. Lu, R. et al., Regulation of the promoter activity of interferon regulatory factor-7 gene, *J. Biol. Chem.,* 2000; 275: 31805–12.
82. Nicholl, M. J., Robinson, L. H., and Preston, C. M., Activation of cellular interferon-responsive genes after infection of human cells with herpes simplex virus type 1, *J. Gen. Virol.,* 2000; 81: 2215–18.
83. Schumacher, B. and Staeheli, P., Domains mediating intramolecular folding and oligomerization of MxA GTPase, *J. Biol. Chem.,* 1998; 273: 28365–70.
84. Hefti, H. P. et al., Human MxA protein protects mice lacking a functional alpha/beta interferon system against La Crosse virus and other lethal viral infections, *J. Virol.,* 1999; 73: 6984–91.
85. Flohr, F. et al., The central interactive region of human MxA GTPase is involved in GTPase activation and interaction with viral target structures, *FEBS Lett.,* 1999; 463: 24–8.
86. Weber, F., Haller, O., and Kochs, G., MxA GTPase blocks reporter gene expression of reconstituted Thogoto virus ribonucleoprotein complexes, *J. Virol.,* 2000; 74: 560–3.
87. D'Souza, S. et al., The gene encoding p202, an interferon-inducible negative regulator of the p53 tumor supressor, is a target of p53-mediated transcriptional repression, *J. Biol. Chem.,* 2001; 276: 298–305.
88. Narasimhan, J., Potter, J. L., and Haas, A. L., Conjugation of the 15-kDa interferon-induced ubiquitin homolog is distinct from that of ubiquitin, *J. Biol. Chem.,* 1996; 271: 324–30.
89. D'Cunha, J. et al., Immunoregulatory properties of ISG15, an interferon-induced cytokine, *Proc. Natl. Acad. Sci. U.S.A.,* 1996; 93: 211–5.
90. Potter, J. L. et al., Precursor processing of pro-ISG15/UCRP, an interferon-β-induced ubiquitin-like protein, *J. Biol. Chem.,* 1999; 274: 25061–8.

91. Smith, J. K. et al., Oral use of interferon-alpha stimulates ISG15 transcription and production by human buccal epithelial cells, *J. Interferon Cytokine Res.*, 1999; 19: 923–8.
92. Kaser, A., Nagata, S., and Tilg, H., Interferon-α augments activation-induced T cell death by upregulation of Fas (CD95/Apo-1) and Fas ligand expression, *Cytokine*, 1999; 11: 736–43.
93. Grossmann, M. et al., New insights into the roles of Rel/NF-κB transcription factors in immune function, hemopoiesis, and human disease, *Int. J. Biochem. Cell Biol.*, 1999; 31: 1209–19.
94. Heitmeier, M. R., Scarim, A. L., and Corbett, J. A., Double-stranded RNA-induced inducible nitric-oxide synthase expression and interleukin-1 release by murine macrophages requires NF-κB activation, *J. Biol. Chem.*, 1998; 273: 15301–7.
95. Ohmori, Y. and Hamilton, T. A., Interleukin-4/STAT6 represses STAT1 and NF-κB-dependent transcription through distinct mechanisms, *J. Biol. Chem.*, 2000; 275: 38095–103.
96. Yang, C. H. et al., IFNα/β promotes cell survival by activating NF-κB, *Proc. Natl. Acad. Sci. U.S.A.*, 2000; 97: 13631–6.
97. Maran, A. et al., Blockage of NF-κB signaling by selective ablation of an mRNA target by 2-5A antisense chimeras, *Science*, 1994; 265: 789–92.
98. Zamanian-Daryoush, M. et al., NF-κB activation by double-stranded-RNA-activated protein kinase (PKR) is mediated through NF-κB-inducing kinase and IκB kinase, *Mol. Cell. Biol.*, 2000; 20: 1278–90.
99. DiDonato, J. A. et al., A cytokine-responsive IκB kinase that activates the transcription factor NF-κB, *Nature*, 1997; 388: 548–54.
100. Iordanov, M. S. et al., Activation of p38 mitogen-activated protein kinase and c-Jun NH_2-terminal kinase by double-stranded RNA and encephalomyocarditis virus: involvement of RNase L, protein kinase R, and alternative pathways, *Mol. Cell. Biol.*, 2000; 20: 617–27.
101. Pelicano, L. et al., Retinoic acid enhances the expression of interferon-induced proteins: evidence for multiple mechanisms of action, *Oncogene*, 1997; 15: 2349–59.
102. Lindner, D. J. et al., The interferon-β and tamoxifen combination induces apoptosis using thioredoxin reductase, *Biochim. Biophys. Acta,* 2000; 1496: 196–206.
103. Hofmann, E. R. et al., Thioredoxin reductase mediates cell death effects of the combination of beta interferon and retinoic acid, *Mol. Cell. Biol.*, 1998; 18: 6493–504.
104. Mustacich, D. and Powis, G., Thioredoxin reductase, *Biochem. J.*, 2000; 346: 1–8.
105. Angell, J. E et al., Identification of GRIM-19, a novel cell death-regulatory gene induced by the interferon-β and retinoic acid combination, using a genetic approach, *J. Biol. Chem.*, 2000; 275: 33416–26.
106. Yokosawa, N., Kubota, T., and Fuji, N., Poor induction of interferon-induced 2',5'-oligoadenylate synthetase (2-5 AS) in cells persistently infected with mumps virus is caused by decrease of STAT-1α, *Arch. Virol.*, 1998; 143: 1985–92.
107. Desai, S. Y. et al., Activation of interferon-inducible 2',5' oligoadenylate synthetase by adenoviral VAI RNA, *J. Biol. Chem.*, 1995; 270: 3454–61.
108. Hartman, R. et al., Activation of 2'-5' oligoadenylate synthetase by single-stranded and double-stranded RNA aptamers, *J. Biol. Chem.*, 1998; 273: 3236–46.
109. Fish, E. N. et al., The interaction of interferon-alpha and -gamma: regulation of (2-5)A synthetase activity, *Virology*, 1988; 165: 87–94.
110. Benech, P. et al., Structure of two forms of the interferon-induced (2'-5') oligo A synthetase of human cells based on cDNAs and gene sequences, *EMBO J.*, 1985; 4: 2249–56.

111. Hovanessian, A. G. et al., Characterization of 69- and 100-kDa forms of 2-5A-synthetase from interferon-treated human cells, *J. Biol. Chem.,* 1988; 263: 1959–69.
112. Marié, I. et al., 69-kDa and 100-kDa isoforms of interferon-induced (2'-5') oligoadenylate synthetase exhibit differential catalytic parameters, *Eur. J. Biochem.,* 1997; 248: 558–66.
113. Rebouillat, D. et al., The 100-kDa 2',5'-oligoadenylate synthetase catalyzing preferentially the synthesis of dimeric pppA2'p5'A molecules is composed of three homologous domains, *J. Biol. Chem.,* 1999; 274: 1557–65.
114. Marié, I., Rebouillat, D., and Hovanessian, A. G., The expression of both domains of the 69/71 kDa 2',5' oligoadenylate synthetase generates a catalytically active enzyme and mediates an anti-viral response, *Eur. J. Biochem.,* 1999; 262: 155–65.
115. Sarkar, S. N. et al., The nature of the catalytic domain of 2'-5'-oligoadenylate synthetases, *J. Biol. Chem.,* 1999; 274: 25535–42.
116. Sarkar, S. N. et al., Enzymatic characteristics of recombinant medium isozyme of 2'-5' oligoadenylate synthetase, *J. Biol. Chem.,* 1999; 274: 1848–55.
117. Ghosh, A. et al., Enzymatic activity of 2'-5'-oligoadenylate synthetase is impaired by specific mutations that affect oligomerization of the protein, *J. Biol. Chem.,* 1997; 272: 33220–6.
118. Rebouillat, D., Marié, I., and Hovanessian, A. G., Molecular cloning and characterization of two related and interferon-induced 56-kDa and 30-kDa proteins highly similar to 2'-5' oligoadenylate synthetase, *Eur. J. Biochem.,* 1998; 257: 319–30.
119. Hartmann, R. et al., p59OASL, a 2'-5' oligoadenylate synthetase-like protein: a novel human gene related to the 2'-5' oligoadenylate synthetase family, *Nucleic Acids Res.,* 1998; 26: 4121–7.
120. Kumar, S. et al., Expansion and molecular evolution of the interferon-induced 2'-5' oligoadenylate synthetase gene family, *Mol. Biol. Evol.,* 2000; 17: 738–50.
121. Ghosh, A. et al., Effects of mutating specific residues present near the amino terminus of 2'-5'-oligoadenylate synthetase, *J. Biol. Chem.,* 1997; 272: 15452–8.
122. Kon, N. and Suhadolnik, R. J., Identification of the ATP binding domain of recombinant human 40-kDa 2',5'-oligoadenylate synthetase by photoaffinity labeling with 8-azido-[α-^{32}P]ATP, *J. Biol. Chem.,* 1996; 271: 19983–90.
123. Lee, J. W. et al., Two classes of proteins dependent on either the presence or absence of thyroid hormone for interaction with the thyroid receptor. *Mol. Endocrinol.* 1995; 9: 243–54.
124. Ko, L., Cardona, G. R., and Chin, W. W., Thyroid hormone receptor-binding protein, an LXXLL motif-containing protein, functions as a general coactivator, *Proc. Natl. Acad. Sci. U.S.A.,* 2000; 97: 6212–7.
125. Takeshita, A. et al., Thyroid hormone response elements differentially modulate the interactions of thyroid hormone receptors with two receptor binding domains in the steroid receptor coactivator-1, *J. Biol. Chem.,* 1998; 273: 21554–62.
126. Yuan, C.-X. et al., The TRAP220 component of a thyroid hormone receptor-associated protein (TRAP) coactivator complex interacts directly with nuclear receptors in a ligand-dependent fashion, *Proc. Natl. Acad. Sci. U.S.A.,* 1998; 95: 7939–44.
127. Burris, T. P. et al., A nuclear hormone receptor-associated protein that inhibits transactivation by the thyroid hormone and retinoic acid receptor, *Proc. Natl. Acad. Sci. U.S.A.,* 1995; 92: 9525–9.
128. Besse, S. et al., Ultrastructural localization of interferon-inducible double-stranded RNA activated enzymes in human cells, *Exp. Cell Res.,* 1998; 239: 379–92.
129. Zhang, J. and Lazar, M. A., The mechanism of action of thyroid hormones, *Annu. Rev. Physiol.,* 2000; 62: 439–66.

130. Ciechanover, A., The ubiquitin-proteasome pathway: on protein death and cell life, *EMBO J.*, 1998; 17: 7151–60.
131. Yap, N., Yu, C.-L., and Cheng, S.-Y., Modulation of the transcriptional activity of thyroid hormone receptors by the tumor suppressor p53, *Proc. Natl. Acad. Sci. U.S.A.*, 1996; 93: 4273–7.
132. Yen, P. M. et al., Vitamin D receptors repress basal transcription and exert dominant negative activity on triiodothyronine-mediated transcriptional activity, *J. Biol. Chem.*, 1996; 271: 10910–6.
133. Cohen, H. et al., Interaction between interferon consensus sequence-binding protein and COP9/signalosome subunit CSN2 (Trip15), *J. Biol. Chem.*, 2000; 275: 39081–9.
134. Seeger, M. et al., A novel protein complex involved in signal transduction possessing similarities to 26S proteasome subunits, *FASEB J.*, 1998; 12: 469–78.
135. Sanceau, J. et al., IFN-beta induces serine phosphorylation of Stat-1 in Ewing's sarcoma cells and mediates apoptosis via induction of IRF-1 and activation of caspase-7, *Oncogene*, 2000; 19: 3372–83.
136. Yu, F. and Floyd-Smith, G,. Protein synthesis-dependent and -independent induction of p69 2'-5'-oligoadenylate synthetase by interferon-α, *Cytokine*, 1999; 11: 744–50.
137. Floyd-Smith, G., Wang, Q., and Sen, G. C., Transcriptional induction of the p69 isoform of 2',5'-oligoadenylate synthetase by interferon-β and interferon-γ involves three regulatory elements and interferon-stimulated gene factor 3, *Exp. Cell. Res.*, 1999; 246: 138–47.
138. Coccia, E. M. et al., Activation and repression of the 2-5A synthetase and p21 gene promoters by IRF-1 and IRF-2, *Oncogene*, 1999; 18: 2129–37.
139. Yu, F. and Floyd-Smith, G., Protein kinase C is required for induction of 2',5'-oligoadenylate synthetases, *Exp. Cell Res.*, 1997; 234: 240–8.
140. Suhadolnik, R. J. et al., Upregulation of the 2-5A synthetase/RNase L antiviral pathway associated with chronic fatigue syndrome, *Clin. Infect. Dis.*, 1994: 18: S96–104.
141. Rusch, L., Zhou, A., and Silverman, R. H., Caspase-dependent apoptosis by 2',5'-oligoadenylate activation of RNase L is enhanced by interferon-β, *J. Interferon Cytokine Res.*, 2000; 20: 1091–100.
142. Terenzi, F. et al., The antiviral enzymes PKR and RNase L suppress gene expression from viral and non-viral based vectors, *Nucleic Acids Res.*, 1999; 27: 4369–75.
143. Li, X.-L. et al., RNase L destabilization of interferon-induced mRNAs, *J. Biol. Chem.*, 2000; 275: 8880–8.
144. Okamoto, T. et al., Assignment of the IkappaB-beta gene NFKBIB to human chromosome band 19.q13.1 by *in situ* hybridization, *Cytogenet. Cell Genet.*, 1998; 82: 105–6.
145. Russo, A. L. et al., Structural basis for inhibition of the cyclin-dependent kinase Cdk6 by the tumor suppressor p16^{INK4a}, *Nature*, 1998; 395: 237–43.
146. Urano, F., Bertolotti, A., and Ron, D., IRE1 and efferent signaling from the endoplasmic reticulum, *J. Cell Sci.*, 2000; 113: 3697–702.
147. De Freitas, E. et al., Retroviral sequences related to human T-lymphotropic virus type II in patients with chronic fatigue immune dysfunction syndrome, *Proc. Natl. Acad. Sci. U.S.A.*, 1991; 88: 2922–6.
148. O'Connor, R., Fennelly, C., and Krause, D., Regulation of survival signals from the insulin-like growth factor-I receptor, *Biochem. Soc. Trans.*, 2000, 28: 47–51.
149. Le Roith, D., What is the role of circulating IGF-I? *Trends Endocrinol. Metab.*, 2001; 12: 48–52.

150. Rosen, C. J., Serum insulin-like growth factors and insulin-like growth factor-binding proteins: clinical implications, *Clin. Chem.,* 1999; 45: 1384–90.
151. Kostecka, Z. and Blahovec, J., Insulin-like growth factor binding proteins and their functions, *Endocrine Regul.,* 1999; 33: 90–4.
152. Giovannucci, E. Insulin-like growth factor-I and binding protein-3 and risk of cancer, *Horm. Res.,* 1999; 51, suppl. 3: 34–41.
153. Yu, H. and Rohan, T., Role of the insulin-like growth factor family in cancer development and progression, *J. Natl. Cancer Inst.,* 2000; 92: 1472–89.
154. Adams, T. E. et al., Structure and function of the type 1 insulin-like growth factor receptor, *Cell. Mol. Life Sci.,* 2000; 57: 1050–93.
155. Baxter, R. C., Insulin-like growth factor (IGF)-binding proteins: interaction with IGFs and intrinsic bioactivities, *Am. J. Physiol. Endocrinol. Metab.,* 2000; 278: E967–76.
156. Grimberg, A., P53 and IGFBP-3: apoptosis and cancer protection, *Mol. Genet. Metab.,* 2000; 70: 85–98.
157. Von Laue, S. and Ross, R. J., Inflammatory cytokines and acquired growth hormone resistance, *Growth Horm. IGF Res.,* 2000; 10: S9-14.
158. Lobie, P. E. et al., Growth hormone, insulin-like growth factor I and the CNS: localization, function and mechanism of action, *Growth Horm. IGF Res.,* 2000; 10: S51–6.
159. Van Dam, P. S. et al., Growth hormone, insulin-like growth factor I and cognitive function in adults, *Growth Horm. IGF Res.,* 2000; 10: S69–73.
160. Ren, J., Samson, W. K., and Sowers, J. R., Insulin-like growth factor I as a cardiac hormone: physiological and pathophysiological implications in heart disease, *J. Mol. Cell. Cardiol.,* 1999; 31: 2049–61.
161. Rutanen, E. M., Insulin-like growth factors and insulin-like growth factor binding proteins in the endometrium. Effect of intrauterine levonorgestrel delivery, *Human Reprod.,* 2000; 15, suppl. 3: 173–81.
162. Rosen, C. J., IGF-I and osteoporosis, *Clin. Lab. Med.,* 2000; 20: 591–602.
163. Dorshkind, K. and Horseman, N. D., The roles of prolactin, growth hormone, insulin-like growth factor-I, and thyroid hormones in lymphocyte development and function: insights from genetic models of hormone and hormone receptor deficiency, *Endocr. Rev.,* 2000; 21: 292–312.
164. Winston, B. W. et al., Cytokine-induced macrophage differentiation: a tale of two genes, *Clin. Invest. Med.,* 1999; 22: 236–55.
165. McAveney, K. M. et al., Association of 2',5'-oligoadenylate synthetase with the prolactin (PRL) receptor: alteration in PRL-inducible stat1 (signal transducer and activator of transcription 1) signaling to the IRF-1 (interferon-regulatory factor 1) promoter, *Mol. Endocrinol.,* 2000; 14: 295–306.
166. Komaroff, A. L. and Buchwald, D. S., Chronic fatigue syndrome: an update, *Annu. Rev. Med.,* 1998; 49: 1–13.
167. Kavelaars, A. et al., Disturbed neuroendocrine-immune interactions in chronic fatigue syndrome, *J. Clin. Endocrinol. Metab.,* 2000; 85: 692–6.
168. Streeten, D. H., Thomas, D., and Bell, D. S., The roles of orthostatic hypotension, orthostatic tachycardia, and subnormal erythrocyte volume in the pathogenesis of chronic fatigue syndrome, *Am. J. Med. Sci.,* 2000; 320: 1–8.
169. Demitrack, M. A. and Crofford, L. J., Evidence for and pathophysiologic implications of hypothalamic-pituitary-adrenal axis dysregulation in fibromyalgia and chronic fatigue syndrome, *Annu. N.Y. Acad. Sci.,* 1998; 840: 684–97.

170. Allain, T. J. et al., Changes in growth hormone, insulin, insulin-like growth factors (IGFs), and IGF-binding protein-1 in chronic fatigue syndrome, *Biol. Psych.,* 1997; 41: 567–73.
171. Moorkens, G. et al., Characterization of pituitary function with emphasis on GH secretion in the chronic fatigue syndrome, *Clin. Endocrinol.,* 2000; 53: 99–106.
172. Buchwald, D., Umali, J., and Stene, M., Insulin-like growth factor-I (somatomedin-C) levels in chronic fatigue syndrome and fibromyalgia, *J. Rheumatol.,* 1996; 23: 739–42.
173. Berwaerts, J., Moorkens, G., and Abs, R., Secretion of growth hormone in patients with chronic fatigue syndrome, *Growth Horm. IGF Res.,* 1998; 8: 127–9.
174. Cleare, A. J. et al., Integrity of the growth hormone/insulin-like growth factor system is maintained in patients with chronic fatigue syndrome, *J. Clin. Endocrinol. Metab.,* 2000; 85: 1433–9.
175. Moorkens, G., Wynants, H., and Abs, R., Effect of growth hormone treatment in patients with chronic fatigue syndrome: a preliminary study, *Growth Horm. IGF Res.,* 1998: 8: 131–3.

6 Immune Cell Apoptosis and Chronic Fatigue Syndrome

Marc Frémont, Anne D'Haese, Simon Roelens, Karen De Smet, C. Vincent Herst, and Patrick Englebienne

CONTENTS

6.1 INTRODUCTION

Apoptosis (from the Greek meaning "fall of the leaves") is a phenomenon which occurs at any site in the body where cells are dividing. Apoptosis is a physiological energy requiring mechanism of cell death which is accomplished by a specialized cellular machinery. Apoptosis is under genetic control and may be initiated by an internal clock; therefore it has also been called "programmed cell death" or "cell suicide." Programmed cell death occurs during normal cellular differentiation and development of multicellular organisms and is involved in maintenance of tissue homeostasis, protection against pathogens, and ageing. The amount of programmed cell death is enormous. Millions of cells, most of them perfectly healthy, die in this way every minute in an adult human being. Inappropriate induction or regulation of apoptosis has severe implications for the organism and can lead to various chronic

0-8493-1046-6/02/$0.00+$1.50

and sometimes deadly pathologies including (auto-)immune diseases, neurological disorders, and cancer.[1-4]

Not all cell death is accidental. Based on the characteristic look of cells when they die, Kerr et al.[5] originally described two forms of cell death, necrosis and apoptosis. More recently, another programmed cell death form, failing to fulfill all the criteria of apoptosis, has been described and termed paraptosis.[6]

Necrosis involves groups of cells that die accidentally as a result of acute injury (ischemia, hyperthermia, irradiation, and metabolic toxins).[7] Necrosis can be thought of as murder. Morphologic changes consist of swelling of cells and organelles, early loss of membrane integrity, while the nucleus remains intact. The cells burst and spill their cytosolic contents into the extracellular space over their neighbors and cause an inflammatory reaction.[4,8]

Apoptosis affects a single cell and is considered suicide; the cell activates a death program and kills itself.[9] Morphologic changes consist of plasma membrane blebbing, cytoplasmic and nuclear condensation, cell shrinkage, chromatin aggregation, DNA fragmentation, and disassembly into membrane-enclosed vesicles called apoptotic bodies. The fragmentation and degradation of genomic DNA is a critical apoptotic event, which results in an irreversible loss of viability. Macrophages and viable neighboring cells recognize some ligands on the surface of the apoptotic bodies. The apoptotic bodies are rapidly engulfed and phagocytized by these cells, preventing inflammatory damage to surrounding cells that would result from leakage of intracellular contents.[3,8] Apoptosis occurs in a predictable, reproducible sequence of events and can be completed within 30 to 60 min. Therefore the deaths easily go unnoticed.[3,9]

Distinct stages of apoptosis and their corresponding controlling genes have been identified for the first time in the nematode worm *Caenorhabditis elegans*. When a specific set of genes, identified as "*ced*" (cell death defective), is inactivated by mutation, 131 cells (of the 1090 present in mature *C. elegans*) genetically programmed to die will survive instead. Four major genes have been identified that regulate the programmed cell death, namely ced-9, egl-1, ced-4, and ced-3.[10] In mammals, two major classes of proteins, which are homologues to the nematode ced proteins, are involved in apoptosis: the Bcl-2 family and the caspase family.

Apoptosis is controlled by external or internal death-inducing signals.[4,11] Apoptosis is induced by multiple physical and biochemical agents or by a failure to meet the requirements of cell cycle check points, respectively.[12] Physical agents include x-ray, γ- and UV-radiations; chemical agents include cellular toxins, hormones, natural signaling molecules such as TGF-β, and two death factors: tumor necrosis factor-α (TNF-α) and Fas ligand (FasL).[13] FasL and TNF-α binding to their respective receptors Fas (APO-1 or CD95) and TNFR1 induce apoptosis. Both activated receptors, Fas and TNFR1, recruit intracellular proteins containing signal-transducing death factor domains: Fas-associated protein with death domain (FADD) and TNFR1-associated death domain protein (TRADD). The death domain proteins may in turn activate caspases.[4,11,13]

Besides this Fas/TNFR-1 death receptor pathway, apoptosis can also occur through the mitochondrial pathway. The Bcl-2 family of proteins controls the release of mitochondrial apoptogenic factors, cytochrome c (cyt c) and apoptosis-inducing factor (AIF), which activate downstream executional phases, including the activation of caspases.[14] A large number of Bcl-2 family members has been identified in mammals, which can be grouped in prosurvival members (such as Bcl-2, Bcl-X, Bcl-w, Mcl-1, Bfl-1, and Boo), proapoptotic members (such as Bax, Bad, Bak, Mtd), and BH3-only proapoptotic members, i.e., members containing only the Bcl-2 homolog domain 3 (such as Bik, Bid, Bim). Prosurvival members and proapoptotic members interact with one another to suppress the activity of their cognate members and also function independently to directly regulate the apoptotic mitochondrial changes.[1,15,16] Recently, evidence was provided that additional apoptotic pathways are triggered by the endoplasmatic reticulum (ER). Stress to the ER results in caspase-12 activation.[17,18]

The central component of the specialized machinery accomplishing apoptosis is a proteolytic system involving a family of cytosolic proteases. All have cysteine in their active site and cleave their target proteins after aspartic acid residues. They are termed by coining as caspases (*C*ys *Asp* prote*ases*). They are the homologues of the *C. elegans* protein ced-3.[3,19] The first identified member of the caspase family was the human interleukin 1β-converting enzyme (ICE, caspase-1), although caspase-1 has no identified obvious role in cell death.[3,20]

All caspases are constitutively present within cells as large inactive precursors, the procaspases (30- to 55-kDa). They are tightly regulated at the post-translational level by cleavage at Asp-X sites to produce the active (proteolytic) heterodimers.[3,4,21] Procaspases contain three basic domains: a prodomain (NH_2-terminal), a large subunit (17- to 22-kDa), and a small subunit (10- to12-kDa).[2] Caspases have overlapping substrate specificities that suggest at least partially overlapping functions.[22] Caspases cut off contacts with other cells, reorganize the cytoskeleton, shut down DNA replication and repair, interrupt splicing, activate DNases which destroy DNA, disrupt the nuclear structure, induce the cell to display signals that mark it for phagocytosis, and disintegrate the cell into apoptotic bodies.[3] Several distinct mechanisms activate the caspases.[21] Pro-apoptotic Bcl-2 family members promote changes in mitochondrial membrane potential which result in the release of cyt c. Released cyt c subsequently binds to apoptosis protease-activating factor-1 (Apaf-1), the recently described ced-4 homologue, leading to its activation. Apaf-1 activates caspase-9, which in turn cleaves and activates effector caspases (like caspase-3), which can then go on to cleave and activate other effector caspases (caspase cascades) or other cellular substrates (multiple cytoplasmic and nuclear proteins).[2,3,9,19,21] In contrast, death signals mediated by Fas/TNFR1 receptors can usually activate caspases directly, bypassing the need for the mitochondrial release of cyt c, thereby escaping the regulation by Bcl-2 family proteins.[16]

Besides the caspases, calpains are other Ca^{2+}-dependent cysteine proteases that bind to membranes upon activation and cleave substrate proteins in a limited manner. Two calpain species are known to exist ubiquitously: a form highly sensitive to Ca^{2+} (μ-calpain), and a form only slightly sensitive to this cation (m-calpain).[23] Calpains are implicated in apoptosis and necrosis, signal-transduction pathways, as well as cytoskeletal reorganization in response to elevations in intracellular calcium concentration $[Ca^{2+}]_i$.[24-26]

Apoptotic induction also involves promoting genes such as the tumor suppressor gene p53 and the proto-oncogene c-myc. p53 triggers apoptosis if DNA repair mechanisms fail. It activates death genes, such as Bax, or downregulates survival genes such as Bcl-2.[4,27] When c-myc is expressed, withdrawal of cellular growth factors results in apoptosis.[4]

Another regulator of apoptosis is the double-stranded (ds)-RNA-activated serine/threonine protein kinase, PKR. This is an interferon (IFN)-inducible enzyme of widespread occurrence in eukaryotic organisms activated by several cellular stress conditions, including viral infection (ds-RNA), cytokine treatment, and growth factor deprivation.[28-31] Once activated, the enzyme phosphorylates the alpha subunit of eukaryotic initiation factor (eIF2α), thereby inhibiting translation initiation.[31] PKR has also been shown to play a variety of important roles in the regulation of gene transcription and signal transduction pathways through the activation of several transcription factors, such as nuclear factor NF-κB, p53 or the signal transducers and activators of transcription (STATs).[30,32] PKR is also implicated in regulating uninfected cell proliferation and transformation and may function as a tumor suppressor and inducer of apoptosis.[29-31] Among the genes upregulated in response to PKR are Fas, Bax, and p53.[32]

Finally, the hallmark of apoptosis is the proteolytic inactivation of the DNA repair enzyme poly (ADP-ribose) polymerase (PARP) catalyzed by several caspases.[22,33] Apoptosis is also associated with the cleavage of other substrates, including DNA-protein kinase C, protein kinase C-δ, and structural proteins of the nucleus or cytoskeleton.[22,34] Disassembly of the nuclear lamina, through cleavage of nuclear lamins by proteases (caspases and a Ca^{2+}-dependent serine proteases), is required for packaging the condensed chromatin into apoptotic bodies.[35,36] The proteolytic action of calpains or caspases has also been found to mediate cleavage of many proteins of the cytoskeleton, including actin, growth arrest specific protein Gas2, and the actin-binding proteins gelsolin and fodrin.[37,38]

6.2 RNase L AND APOPTOSIS IN CHRONIC FATIGUE SYNDROME

As we have seen earlier in this book (Chapters 1 and 2), the 2-5A system includes the 2',5' oligoadenylate (2-5A) synthetases (2-5OAS), which produce 2-5A in response to ds-RNA and interferons (IFNs). The 2-5A are unusual and very specific activators of RNase L, which upon activation cleaves single-stranded (ss)-RNA with a moderate specificity for UU and UA sequences. It is now well established that the 2-5A system is involved in the antiviral action of IFN.[39] Inhibition of RNase L by an analog of 2-5A can reduce the anti-encephalomyocarditis virus (EMCV) effect of IFN,[40] and more recently a similar result was obtained with a dominant negative mutant of RNase L that is able to suppress the effect of IFN on EMCV.[41] Conversely, expression of a recombinant RNase L can inhibit the replication of viruses such as vaccinia virus.[42]

Ribosomal RNA (rRNA) is a substrate for RNase L, and activation of the enzyme in virus-infected cells is associated (at least for viruses like EMCV, vaccinia, or reoviruses) with a specific rRNA cleavage; this cleavage is sometimes used as an indicator for RNase L activity.[43-46] Inhibition of RNase L prevents rRNA cleavage,[40] whereas expression of recombinant RNase L results in its induction.[42] Interestingly, a similar rRNA cleavage can be observed in several types of cells undergoing apoptosis; this was the first indication that RNase L could be involved in apoptotic pathways.[47] A disappearance of 28S rRNA also occurs during γ-ray-induced apoptosis in human lymphocytes,[48] as well as in colon carcinoma cells undergoing IFN- or TNF-α-induced apoptosis.[49] In this latter case, the increased degradation of 28S rRNA is associated with an increased activity of 2-5OAS, suggesting that the rRNA degradation is caused by a 2-5A-mediated activation of RNase L.

Definitive evidence that RNase L is involved in the regulation of apoptosis was provided from expression of the recombinant enzyme in mammalian cells, where it is able to induce apoptosis.[50] This induction can be enhanced by coexpression of 2-5OAS or by increasing the intracellular amount of viral ds-RNA. Interestingly, RNase L-induced apoptosis also occurs in PKR-defective cells and can be blocked by Bcl-2, suggesting that RNase L activates the mitochondrial pathway of apoptosis. Similar results were obtained in cultured cells: overexpression of RNase L in NIH3T3 cells causes apoptosis, as does activation of endogenous RNase L by 2-5A.[51] Overexpression of RNase L in a stable transfected cell line greatly enhances both the cell growth arrest by IFNs and the proapoptotic activity of staurosporine. It also suppresses the replication of vaccinia, ECM, and influenza viruses.[52] Finally, transfection of a trimer form of 2-5A in ovarian cancer cell line results in rRNA cleavage and induction of apoptosis.[53] This effect is enhanced by prior treatment with IFN-β; apoptosis is associated with cyt c release (a further indication that RNase L-induced apoptosis proceeds through the mitochondrial pathway) and can be blocked by a caspase-3 inhibitor.

Conversely, inhibition of RNase L has antiapoptotic effects: in NIH3T3 cells, a dominant negative RNase L can suppress poliovirus- or ds-RNA-induced apoptosis.[51] Other evidence comes from RNase L-defective mice.[54] In these animals, the antiviral effect of interferon-alpha is impaired, but apoptotic functions are also deficient: they present very large thymuses, indicating that apoptosis did not work properly during embryonic development. Thymocytes and fibroblasts derived from these animals are insensitive to treatment with a surprisingly large number of proapoptotic agents, including anti-CD3, anti-Fas, staurosporine, and TNF-α. These results were confirmed in NIH3T3 cells where transfection of a dominant negative RNase L conferred resistance to staurosporine-induced apoptosis.[55]

The involvement of RNase L in the regulation of apoptosis is established, but what is the significance of this apoptotic function, and what triggers RNase L-induced apoptosis? Apoptosis is certainly one of the mechanisms used for achieving the antiviral effects of RNase L. This is demonstrated by observations that a dominant negative RNase L can suppress poliovirus- or ds-RNA-induced apoptosis in NIH3T3 cells[51] or, conversely, that apoptosis induced by expression of a recombinant RNase L can be enhanced by increasing the intracellular amount of viral ds-RNA.[50] Third evidence comes, most unusually, from a plant system. Plants are normally devoid of both 2-5OAS and RNase L. A transgenic tobacco expressing both 2-5OAS and RNase L shows an increased resistance to viral infection. Analysis of infected plants suggests an involvement of apoptosis in the response to viruses: leaves show delimited lesions at the place of infection (infected cells die but the virus does not spread further), whereas inoculated plants expressing either RNase L or 2-5OAS only, or none of them, develop typical systemic infections.[56]

Other evidence suggests that the role of RNase L as a regulator of apoptosis goes far beyond mediating the antiviral effect of IFNs. For example, RNase L-deficient cells become insensitive to a very large number of apoptotic inducers, suggesting that RNase L is involved in several different apoptotic pathways, and not only in IFN-induced pathways. Abnormalities observed in RNase L-deficient mice (enlarged thymuses) strongly suggest a role in the regulation of apoptosis occurring during embryonic development, which is of course a function unrelated to antiviral mechanisms. But how could RNase L be activated in the absence of virus? It has been suggested that RNA formed during the apoptotic degeneration of the nucleus could trigger the activation; these RNAs would present complex secondary structures that could act like viral ds-RNA and activate the 2-5OAS. In this case, RNase L activation would be a secondary event in apoptosis, possibly necessary for its completion but not responsible for its induction. RNase L could contribute to apoptosis by degradation of mRNAs coding for prosurvival factors (like Bcl-2 or Bcl-xL, since we have seen that release of cyt c is required for RNase L-mediated apoptosis), or by global inhibition of protein synthesis via rRNA degradation.

As discussed in Chapters 2 and 3, an unusual form of RNase L is found in the PBMCs of CFS patients: the native 83-kDa enzyme is cleaved and a low molecular weight form of 37-kDa can be detected. This cleavage may certainly have consequences on the ability of RNase L to regulate apoptosis. As we will discuss later in this chapter, apoptotic activity (measured by the activity of various caspases) in PBMCs of CFS patients is related to the extent of RNase L cleavage. There is an increased activity of caspase-3 and caspase-8 in samples where RNase L is cleaved (ratio 37/83-kDa $\times$ 10 above 2); however, at very high ratios (over 20) the activities of these caspases return to normal. In samples characterized by very high ratios, activities of caspases-2 and -9 are even inhibited. As described in Chapters 2 and 3, the 37-kDa fragment is still able to bind 2-5A and has a catalytic activity toward poly(U); its presence is actually associated with an upregulated activity of RNase L. One possible cause, or at least a contributing factor, of the increased apoptotic activity in samples with RNase L ratios below 20 could therefore be this upregulation of RNase L activity, which could induce apoptosis or make the cells more sensitive to apoptosis induced by factors like cytokines or ds-RNA.

But how can we explain the blockade of apoptosis observed in samples characterized by very high RNase L ratios? In order to answer this question, a further investigation of the exact role of the 37-kDa enzyme will be required. It may actually have a different catalytic specificity, or another cellular localization, than the native enzyme, and degrade some RNAs that would normally remain unaltered; for example, an accumulation of this fragment could block apoptosis by degrading mRNAs coding for proapoptotic factors. Accumulating RNase L fragments may also interact (independently of their catalytic activity) with various apoptotic regulators and interfere with their functions. Finally, it is also possible that the blockade of apoptosis is not due directly to RNase L but to the dysregulation of other apoptotic regulators: further in this chapter we will discuss the possible involvement of another apoptosis-linked protease, calpain, as well as the role of PKR and other associated factors such as p53, eIF2α, or NF-κB, which may all play a role in the dysregulation of apoptotic pathways in the PBMCs of CFS patients.

6.3 CASPASES IN CHRONIC FATIGUE SYNDROME

The suspected activation of calpain as one possible mechanism for RNase L cleavage into the 37-kDa fragment (Chapters 2 and 3), suggests a possible involvement of apoptotic dysregulation in CFS, which we attempted to characterize by addressing the caspases in PBMCs. Caspases are a group of cysteine proteases, which specifically cleave proteins after an aspartate residue and play a central role in apoptosis.[57] In normal, nonapoptotic cells, these caspases are present in the cytoplasm as inactive zymogens. Upon apoptotic stimulation, the zymogen form of the caspases is cleaved into a large and a small subunit. These subunits reorganize to an active tetramer, consisting of 2 heterodimers, each made of a large and a small subunit.[58,59] The cleavage of the zymogen may be either autolytic (e.g., caspases-8 and -9)[60,61] or accomplished by other caspases in a caspase-cascade (e.g., caspase-3 activated by capases-8 and -9).[21,62]

The caspase family of proteases may be divided into three subgroups. A first group includes the inducer caspases (caspases-2, -8, -9, -10, -12), which induce apoptosis by amplifying the apoptotic signal exerted on the cell.[18,58] This amplification occurs by cleavage, and thus activation of the effector caspases (caspases-3, -6, -7), a second group of caspases.[58] These effector caspases in turn execute apoptosis by degrading hundreds of regulatory proteins and by activating endonucleases and other proteases.[59] The third group of caspases, the cytokine processors (caspases-1, -4, -5, -11, -13, -14), interact with the cytokines that they cleave from precursors.[58] In this discussion, we will address caspases-2, -8, -9, and -12 as inducer caspases and caspases-3 and -6 as effector caspases.

The different inducer caspases initially have distinct independent pathways of activation, which converge in caspase-3 activation, allowing the death program to proceed.[21,62,63] In case of deficiency in one pathway, some functions can be taken over by another pathway.[21,64,65] Briefly, there are three important pathways leading to cell death. In the first pathway (receptor pathway), binding of TNF-α or FasL to their respective receptors leads to the formation of the Fas-associated death domain (FADD) or the TNF-α-receptor-associated death domain (TRADD) on the cytoplasmic part of the receptor.[66,67] This death domain recruits procaspase-8 that it cleaves and activates.[68] In the second pathway (the mitochondrial pathway), the activation of c-Jun,[69] Bax,[70] or differences in intracellular calcium concentrations[71] lead to the release of cyt c from the mitochondria. Binding of cyt c to Apaf-1 leads to the formation of an apoptosome, which binds and activates procaspase-9.[61] The third pathway, only recently discovered, proceeds by ER stress which induces the release of calcium in the cytoplasm.[17] This imbalance of calcium concentration activates calpains, which in turn cleave and activate procaspase-12. An overview of these three pathways is displayed in Figure 6.1.

The activated inducer caspases in turn activate caspase-3,[21,62,63] which plays a key role in the proteolytic cleavages and morphologic changes that occur in the execution phase of apoptosis.[72,73]

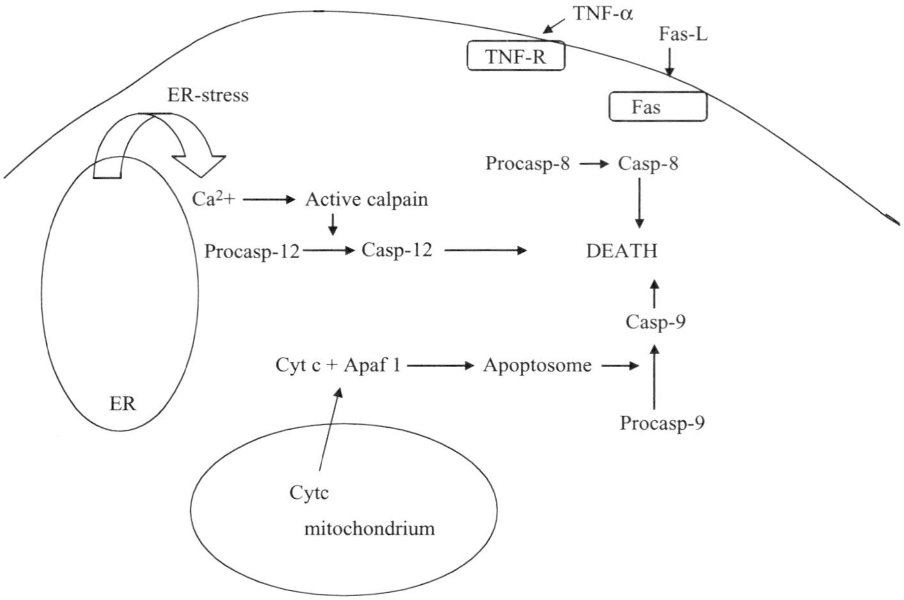

FIGURE 6.1 A general overview of the caspase activating pathways.

6.3.1 Inducer Caspases

The pathways leading to death via caspase-8 are activated via the death receptors: Fas (APO-1/CD95) and TNF-α receptor (TNFR). These transmembrane receptors have an intracellular domain, the death domain (DD), which is essential for their signaling. Binding of TNF-α and FasL to their respective receptors provides the signal for cell suicide.[66] The receptors form multimers[64,74-76] which induce the formation of a complex termed the death-inducing signaling complex (DISC);[75] for further activation, this multimeric complex needs to be enlarged.[67] Upon formation of the multimeric complex, some enzymes dock on the DD in a specific way. These proteins are known as FADD/Mort 1 or TRADD and are respectively specific for TNFR or Fas.[67] Procaspase-8 is then recruited to the DISC. The interaction occurs between the NH_2-terminal part of FADD and the prodomain of procaspase-8, through an 80-amino acid sequence termed the death effector domain (DED).[68] Recruitment of procaspase-8 at the DISC induces its cleavage into p43/p41 and p18/p10 subunits.[77,78] This activation is likely to occur through self-processing of the clustered caspase-8 as the result of an intrinsic proteolytic activity.[60]

Once activated, caspase-8 further leads to the execution of apoptosis, either in a direct or an indirect pathway.[62,64,79] In the direct pathway, caspase-8 cleaves procaspase-3, thus activating this major executioner protease.[62,79] In the indirect pathway, caspase-8 cleaves Bid to tBid (truncated Bid),[64] which then binds to Bax. This results in Bax activation and the subsequent release of cyt c from the mitochondria.[70] At this level, the caspase-8 pathway may also be interconnected with the caspase-9 pathway. An overview of the caspase-8 pathway is given in Figure 6.2. Caspase-

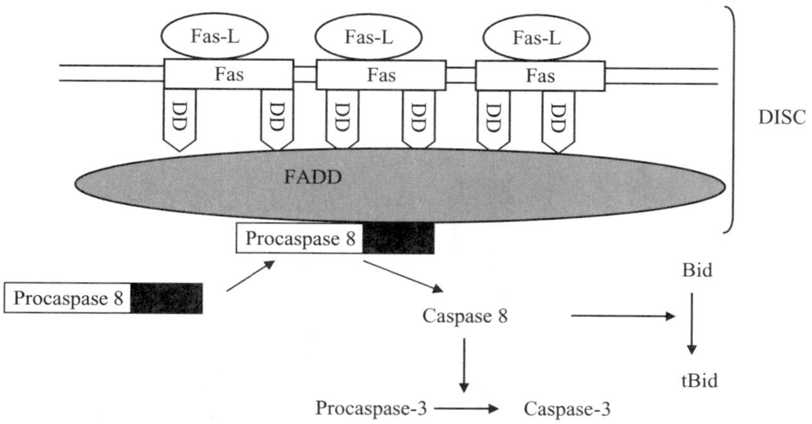

FIGURE 6.2 Caspase-8 activation. Upon stimulation of Fas by Fas-L, the receptors multimerize, bringing the death domains (DD) into contact. These DDs are an ideal docking place for FADD, which has enzymatic activity. This complex of the DD multimer and FADD is called DISC. At the DISC, procaspase-8 is recruited and experiences autolytic cleavage. The activated caspase-8 then leads to apoptosis by cleaving caspase-3 or Bid. In this latter case, tBid induces cyt c release from the mitochondria and caspase-9 activation.

2 is likely to be activated in a similar way, but the importance of its recruitment on the DISC for activation is less clear.[59,80,81]

We measured caspase-8 activity with a commercially available kit (Biosource) in PBMC extracts of CFS patients (Figure 6.3) classified according to their RNase L ratio, which reflects the extent of cleavage of the protein. We observed a significant induction of caspase-8 in samples with ratios between 2 and 20 (i.e., when up to 60% of the 83-kDa RNase L is cleaved). The increase in activity is statistically significant ($p<0.01$) when compared to normal controls. In samples characterized by higher RNase L ratios, the caspase activity returns to normal. This activation of caspase-8 may be due to TNF-α or FasL (cfr. supra). No data are available in the literature on FasL and contradictory data regarding the serum levels of TNF-α in CFS patients have been reported. Some authors mention an elevation of TNF-α in CFS.[82,83] Like others,[84,85] we did not find any differences in serum TNF-α levels between CFS and controls using a commercial ELISA kit (Biosource). This would suggest that the activation of caspase-8 in PBMC of CFS patients could be induced by FasL. The activity of caspase-2 (Figure 6.4) increases in samples with low or intermediate RNase L ratios, but the change is not significant. However, the decrease in caspase-2 activity observed in PBMC samples characterized by high RNase L ratios (>10) correlates with the ratios. In samples characterized by very high RNase L ratios (>20), the activity is significantly lower ($p<0.01$) than that of the negative controls, suggesting a downregulation of caspase-2 activation. Caspase-2 is known to be required for the execution of apoptosis *in vivo*, as demonstrated in the monocytic tumor cell line THP-1. However, its full role in programmed cell death remains unclear.[86]

The release of cyt c from the mitochondria (Figure 6.5) is central to the activation of caspase-9 and can be stimulated by different inducers, namely Bax, a proapoptotic

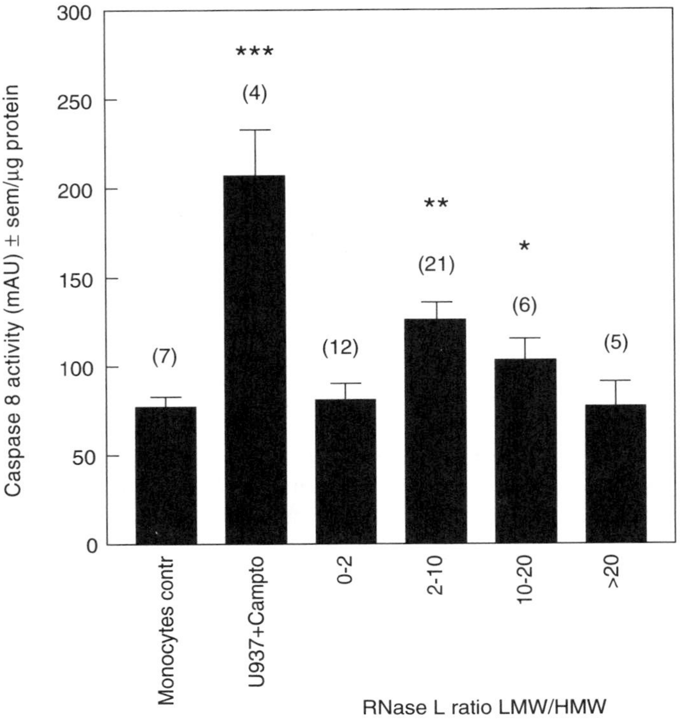

FIGURE 6.3 Evolution of caspase-8 activity in PBMC extracts of CFS patients, classified according to the RNase L ratio. Normal monocytes were used as negative controls. The U937 cell line induced to apoptosis with camptothecin was used as the positive control. Bars represent data ± standard error. The level of significance vs. the negative controls (* = $p<0.05$; ** = $p<0.01$; *** = $p<0.005$; **** = $p<0.001$) is shown above the bars. The number of samples tested is given into brackets for each group. A significant activation of caspase-8 was observed in samples with an intermediate and high RNase L ratio (2 to 20). Caspase-8 activity is back to normal in samples with very high RNase L ratios (>20).

member of the Bcl-2 family;[69-71] c-Jun, the transcription factor;[69] and by changes in the intracellular calcium concentration.[71,87] As mentioned previously, the binding of tBid induces a conformational change which activates Bax.[70] This further results in the formation of oligomeric Bax, which provokes the release of cyt c from the mitochondria.[71]

A second possibility for the induction of cyt c loss by the mitochondria is changes in intracellular calcium concentrations. According to Schild and coworkers,[87] only subtle changes in intracellular calcium concentrations lead to the organized release of cyt c and apoptosis. In contrast and according to Gogvadze and coworkers,[71] larger differences in intracellular calcium concentrations lead also to apoptosis. But with large differences in intracellular calcium concentrations, injury, swelling, and rupture of the mitochondrial membrane occur. According to Gogvadze et al.,[71] this might happen under pathological, less controlled conditions such as necrosis.[87] The third

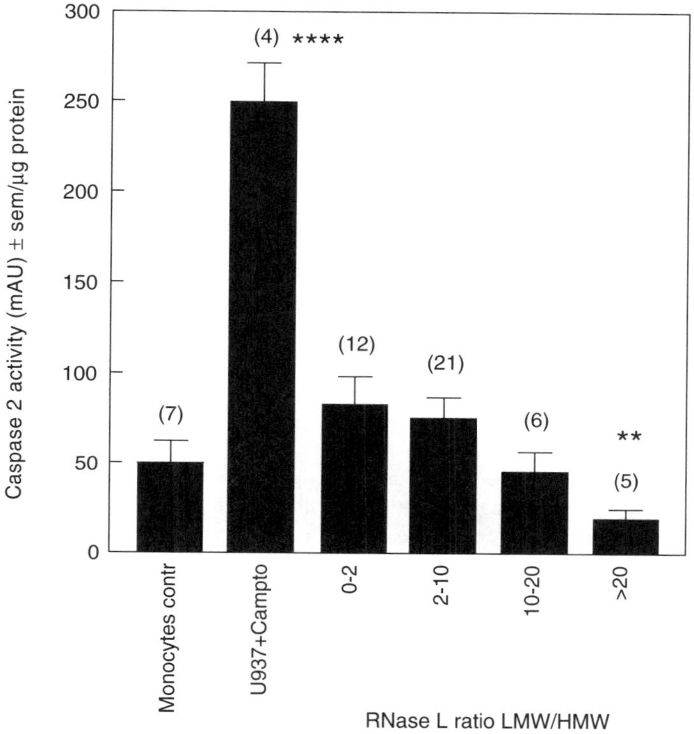

FIGURE 6.4 Evolution of caspase-2 activity in PBMC extracts of CFS patients classified according to the RNase L ratio. Positive and negative controls as well as indications are the same as in Figure 6.3. Counter to caspase-8, no significant activation of caspase-2 was observed. In samples with very high ratios, however, a significant downregulation of caspase-2 activity was noted.

possibility for cyt c release by the mitochondria is an increase in the expression of the transcription factor c-Jun.[69] The release of cyt c alone, however, is insufficient to complete apoptosis. Other signaling proteins, namely Bcl-2 and Bcl $-x_L$, can prevent activation of apoptosis by binding either directly to cyt c[88,89] or to Apaf-1,[90] respectively. For full activation, cyt c must bind to Apaf-1.[21] This is believed to be only possible if Apaf-1 hydrolyzes deoxy-ATP/ATP into deoxy-ADP/ADP, but this has been questioned recently.[91] Binding of cyt c to Apaf-1 promotes the formation of a multimer, the apoptosome which recruits procaspase-9[61,92] through the caspase recruitment domain (CARD) of Apaf-1.[21] Procaspase-9 is activated on the apoptosome by autocatalytic cleavage from which it is then released. Activated caspase-9 cleaves downstream caspases,[61,92] including the two effector caspases -3 and -7, which in turn cleave other caspases in the caspase cascade, including caspase-6,[21] or their respective substrates.[59] An overview of the caspase-9 pathway is provided in Figure 6.5.

We have also measured the enzymatic activity of caspase-9 in PBMC samples, classified according to their RNase L ratio (Figure 6.6). Counter to what was

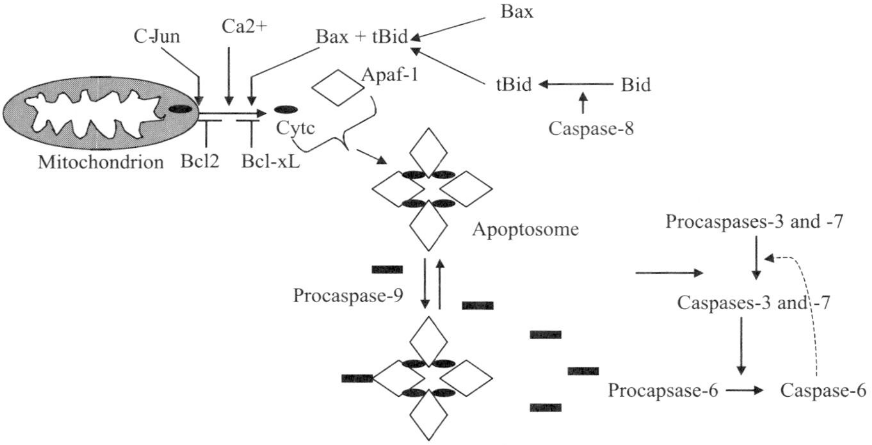

FIGURE 6.5 Schematic representation of the caspase-9 activation mechanism. The release of cyt c from the mitochondria is induced by either c-Jun, augmentation in $[Ca^{2+}]_i$, or the Bax-tBid complex respectively. The released cyt c can either be blocked by binding to Bcl2 or Bcl-x_L or lead to apoptosis by binding to Apaf-1, respectively. The complex formed between cyt c and Apaf-1 is called the apoptosome. Procaspase-9 is recruited on the apoptosome and transformed to active caspase-9 by autolytic cleavage. Caspase-9 is then released from the apoptosome. Activated caspase-9 cleaves procaspases-3 and -7, forming the active enzymes. Activated caspase-3 cleaves in turn procaspase-6. Activated caspase-6 is capable of cleaving procaspase-3, giving an amplification feedback loop.

observed with caspase-8, no significant increase in activity could be observed in samples with moderate RNase L ratios. However, caspase-9 activity, like caspase-2, decreases very significantly in PBMC extracts characterized by higher RNase L cleavage, suggesting a similar downregulation and a blockade of the mitochondrial pathway in these samples. We went on to analyze the molecular forms of the enzyme in PBMC by immunoblotting using an antibody from Santa Cruz for detection. As shown in Figure 6.7, the antibody detected a 20-kDa band appearing progressively with increasing RNase L ratios. This band is likely to match the inactive molecular form produced by m-calpain cleavage.[93]

The pathways leading to death via caspase-12 have only very recently been discovered and still raise many questions.[18] ER stress is likely to be involved in the activation of caspase-12.[17,63,94] Amyloid-β (Aβ)[17] and unfolded protein response (UPR)[64,96] have been suggested to induce the ER stress. Upon ER stress signals, procaspase-12 is suspected to be released from TNF receptor-associated factor 2 (TRAF2), with which it is associated in unstressed cells, forming homodimers.[94] ER stress signals can also induce a release of calcium from the ER[17] that leads to the activation of calpain.[95] m-Calpain has been proposed to cleave the caspase-12 into 35-kDa active fragments.[94,95] How caspase-12 further proceeds to apoptosis is still unclear,[18] but the cleavage of caspase-3 may be involved.[63]

We have analyzed the molecular forms of caspase-12 in PBMC of CFS patients by immunoblotting, using an antibody from oncogene. As shown in Figure 6.8, the

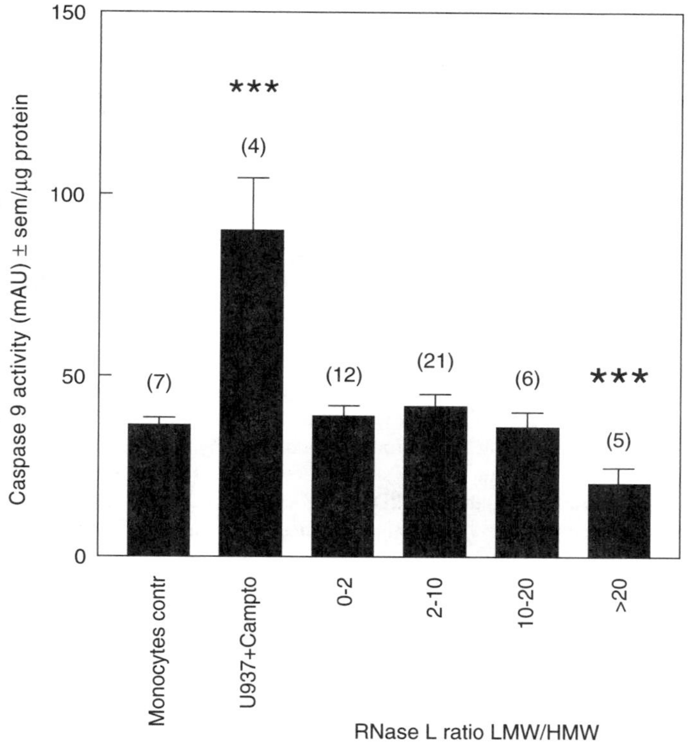

FIGURE 6.6 Evolution of caspase-9 activity in PBMC extracts of CFS patients classified according to the RNase L ratio. Positive and negative controls, as well as indications, are the same as in Figure 6.3. As for caspase-2, no significant activation could be detected, but a significant downregulation of activity was noted in samples with the highest levels of RNase L cleavage.

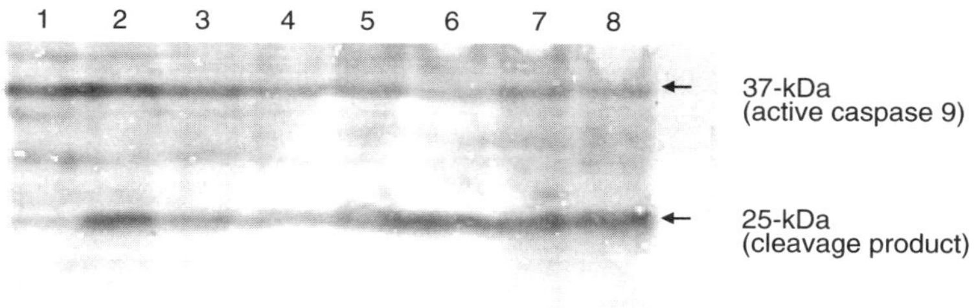

FIGURE 6.7 Molecular forms of caspase-9 in PBMC samples, detected by immunoblotting. The RNase L ratios of these samples were respectively (lane 1 to 8): 0.95; 6.3; 11; 19; 23; 33; 44; 55. The increased caspase-9 cleavage is obvious from the lane 2 sample onward. Increasing RNase L ratios parallel the progressive appearance of a 25 kDa band corresponding to an inactive form of the enzyme. The 37 kDa form disappears accordingly.

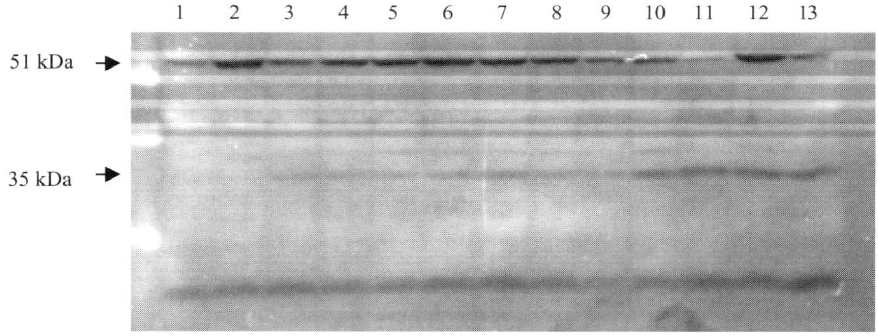

FIGURE 6.8 Molecular forms of caspase-12 in PBMC samples with increasing RNase L ratios, detected by immunoblotting. The RNase L ratios of these samples were respectively (lane 1 to 13): 0.3; 0.8; 1.1; 2.6; 3.2; 3.3; 5.4; 7.5; 10.1; 10.5; 11.2; 12.1; 21. The 51-kDa procaspase-12 and the 35-kDa activated form are indicated by arrows. The increased caspase-12 activation is obvious from the lane 3 sample onward. Increasing RNase L ratios parallel the progressive appearance of this 35 kDa band corresponding to the active form of the enzyme. This activated caspase-12 form is generated by m-calpain.

antibody detected the native 51-kDa form in all samples and the 35-kDa active form in samples with a RNase L ratio above one. The intensity of the 35-kDa fragment increases with increasing RNase L ratios (which reflect the proteolytic activity in the cells), suggesting an increasing m-calpain activity.[95] Counter to what was observed with caspases-2, -8, and -9, caspase-12 is thus likely to activate in parallel with the level of RNase L cleavage in PBMC samples.

6.3.2 Effector Caspases

One of the most important enzymes in the apoptotic process is caspase-3.[72] This enzyme is activated in all apoptotic programs,[21,62,63] and cleaves an enormous amount of substrates.[59] Hence caspase-3 plays a key role in the proteolytic cleavage and morphologic changes occurring in the execution phase of apoptosis.[72,73,96] Caspase-3 is activated by direct cleavage of procaspase-3 by caspase-8[62,79] and caspase-9;[21] caspase-12 has also been proposed to activate caspase-3.[63] Besides the activation of caspase-3 by upstream caspases, another activation mechanism has been recently described, which involves cleavage and activation by m-calpain.[97]

Once activated, caspase-3 leads to apoptosis by cleavage of its substrates. Poly (ADP-ribose) polymerase (PARP) and actin are among its targets.[98-102] The cleavage of PARP results in a decrease in the synthesis of poly (ADP-ribose),[59] which limits the possibilities for DNA repair.[103] Consequently, PARP is not only a death substrate but also plays a positive role in apoptosis, its cleavage preventing the cell from repairing DNA damage,[103] which is essential for the completion of the death program.[104] The cleavage of actin, on the other hand, has dramatic consequences on cellular movement and maintenance of the cytoskeleton (see Section 6.5).

Although the precise role and importance of caspase-6, another effector caspase in apoptosis, has not yet been established,[96] it is probably involved in an amplification

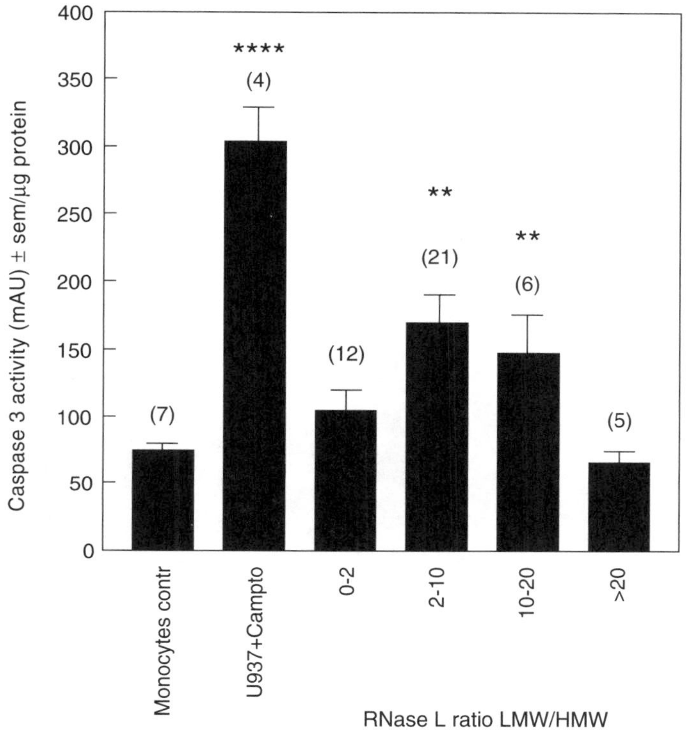

FIGURE 6.9 Evolution of caspase-3 activity in PBMC extracts of CFS patients classified according to the RNase L ratio. Positive and negative controls, as well as indications, are the same as in Figure 6.3. Similarly to caspase-8, a significant activation of caspase-3 was observed for samples with an intermediate and high RNase L ratio (2 to 20). Caspase-3 activity is back to normal in samples with very high RNase L ratios (>20).

loop, in which caspase-3 activates caspase-6 which in turn further activates caspase-3[86] (Figure 6.5).

We have measured the enzymatic activity of both caspases-3 and -6 in various PBMC samples, classified according to their RNase L ratio. We observed a significant induction of caspase-3 enzymatic activity in samples with RNase L ratios ranging from 2 to 20 (Figure 6.9), reflecting the activation of caspase-8. In samples characterized by higher RNase L ratios (>20), caspase-3 activity was not different from normal controls. With caspase-6 (Figure 6.10), we observed a progressive induction of activity with increasing RNase L ratios, which was significant only for samples with very high RNase L ratios.

6.3.3 Caspases and RNase L Cleavage

Our observations of caspase activities in PBMC samples as a function of the extent of RNase L cleavage (RNase L ratio) are likely to unravel abnormalities in apoptotic processes that are worth discussing. When we compared the caspase activities of

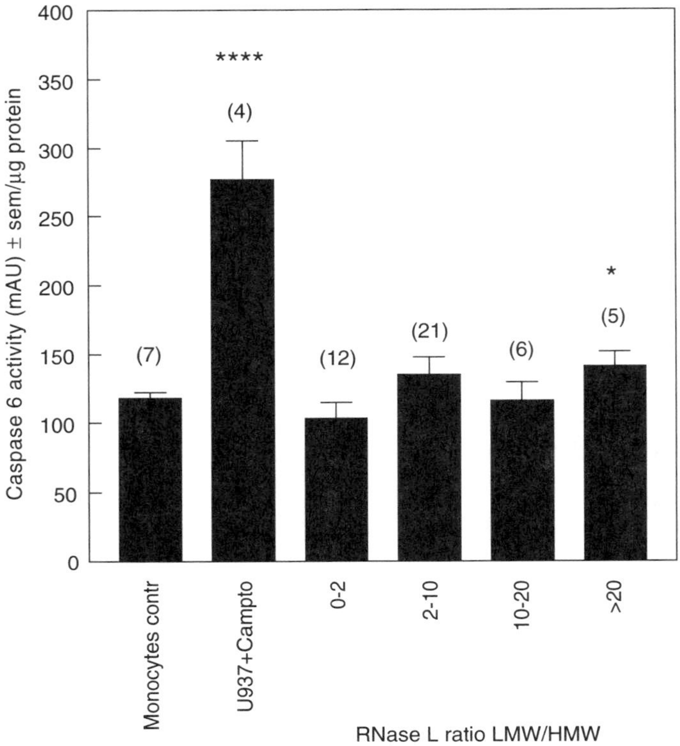

FIGURE 6.10 Evolution of caspase-6 activity in PBMC extracts of CFS patients classified according to the RNase L ratio. Positive and negative controls, as well as indications, are the same as in Figure 6.3. A slight but significant activation can be noted in samples with a very high RNase L ratio.

samples with increased RNase L ratios (>2) with those of samples with either low ratios or of normal control cell lines, no significant differences were observed. We then reasoned that the ratio between LMW and HMW RNase L reflects a proteolytic activity present within the cells; accordingly, we classified the samples as a function of the extent of cleavage: low (0 to 2), intermediate (2 to 10), high (10 to 20), or very high (>20) ratios. At this stage, significant differences were observed as summarized in Figure 6.11.

First, the significant activation of caspase-8 in samples with intermediate and high ratios is paralleled by the activation of caspase-3. These activities return to normal in samples with very high RNase L ratios, while surprisingly, caspase-6 activation becomes slightly but significantly elevated. Counter to that, neither caspase-2 nor -9 is activated in PBMC samples with intermediate or high RNase L ratios. In samples with very high ratios, however, both are very significantly downregulated and it is suspected that this results from a strong activation of calpain or of a similar protease (see below). These results therefore suggest that RNase L cleavage in PBMCs could not only serve as a biological marker of CFS, but could also be indicative of

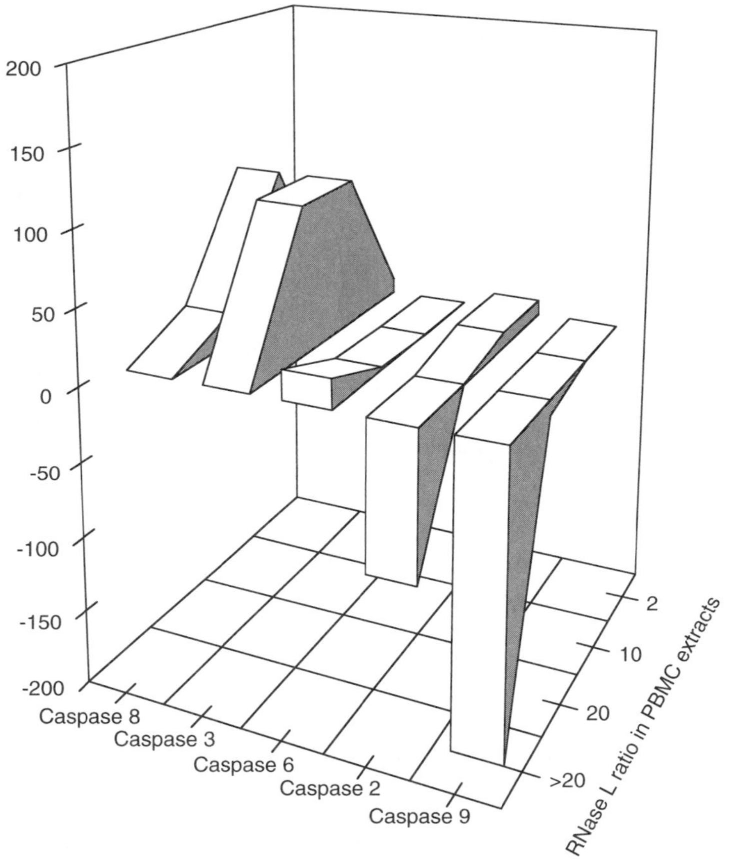

FIGURE 6.11 Summary of the activation of caspases in the PBMC extracts as a function of RNase L ratios. The inverse of activity difference significance vs. the normal controls (y axis) is plotted vs. RNase L ratio for each individual caspase.

the stage of immune cell dysregulation: induction of the death-receptor apoptotic pathway in samples with intermediate and high ratio on the one hand; blockade of this pathway and strong downregulation of the mitochondrial apoptotic pathway in samples with very high RNase L ratios on the other hand. The RNase L ratio measured in PBMC could consequently have major implications in the detection of immune dysfunction and also in the follow-up of any possible therapy.

6.4 CALPAINS IN CHRONIC FATIGUE SYNDROME

Calpains are a family of neutral cysteine proteases that require calcium for their activation.[105,106] Within this family of proteases, one distinguishes two groups: ubiquitous and tissue-specific calpains.[23,107,108] In our discussion, we focus on PBMC, and thus on the ubiquitous calpains. Among these, two major forms have been described:

μ- and m-calpain.[23] Both occur *in vivo* as heterodimers, made of a specific large 80-kDa subunit (CAPN1 and CAPN2 for μ- and m-calpain, respectively) and a common 30-kDa small subunit (CAPN4).[23,105] Both forms are found in many cell types. As far as human blood cells are concerned, μ-calpain is found in all, whereas m-calpain is found only in platelets and lymphocytes and monocytes.[109] The endogenous inhibitor of calpains, calpastatin, is also expressed in these cells and regulates the enzymes.[110]

Requirements in calcium concentration for activation constitute the major difference between the two ubiquitous forms of calpain. It is reported that μ-calpain requires micromolar ranges of calcium, while m-calpain requires millimolar ranges for activation, at least *in vitro*.[23,111-113] The concentration of calcium required by m-calpain is consequently above the physiological calcium concentrations. If m-calpain were to have any function in normal cells, then mechanisms must exist to dramatically reduce its calcium requirements.[112] It was shown that, *in vivo*, calpains undergo an extra activation in the presence of phospholipids present in the cell membrane.[23] For m-calpain particularly, a mechanism was identified that drastically decreases the needs of calcium for activation to approximately 10 μM, which is in physiological ranges. This mechanism is effected by the interaction between m-calpain and acyl-CoA-binding protein, a 20-kDa homodimer.[114]

Although the exact effect of calcium remains speculative, it is generally believed that, in the presence of calcium, calpain subunits are autolytically cleaved into a 76-kDa fragment from the large subunit, and into a 18-kDa fragment from the small subunit, which would give rise to the activated form of the enzyme.[111,115,116] The autolytic cleavage of the large subunit occurs as a two-step process, leading first to an intermediary 78-kDa fragment and finally to the fully cleaved 76-kDa. Both fragments are active, but they differ in their substrate specificity.[117] Although quite generally accepted, the autolytic activation processs is suggested by some authors as one way, by which calpains could be activated, allowing the explanation of their long metabolic half-lives.[118] Moreover, it has been reported that the native 80-kDa subunit could be active in its monomeric form.[119] Calpain activation is suggested to result from the dissociation of the heterodimers in the presence of calcium, with autolysis occurring later.[105,119]

Another intriguing aspect of calpain activation rests with the inhibitor calpastatin. Calpastatin cleavage has been reported in parallel to calpain activation,[120,121] inducing an enhanced enzymatic activity. Therefore, when examining cells for calpain activity, one should consider not only the expression of calpain or the occurrence of autolytically cleaved forms, but also the ratio between calpain and native calpastatin.[121] Counter to other enzymes, such as caspases, calpains have no strict sequence requirements for substrate cleavage;[24] calpains cleave a certain structure instead of a sequence.[122]

Because we suspect m-calpain to be responsible for the cleavage of several proteins in the PBMC of CFS patients, we measured the intracellular calcium concentrations using Arsenazo III and related these to the RNase L cleavage (Chapter 2). We found a significant linear relationship between the intracellular calcium concentrations and the extent of RNase L cleavage, as expressed by the ratio between 37- and 83-kDa forms of the enzyme (Figure 6.12). We detected a progressive increase in calcium concentration up to RNase L ratios of 20. At this level, the

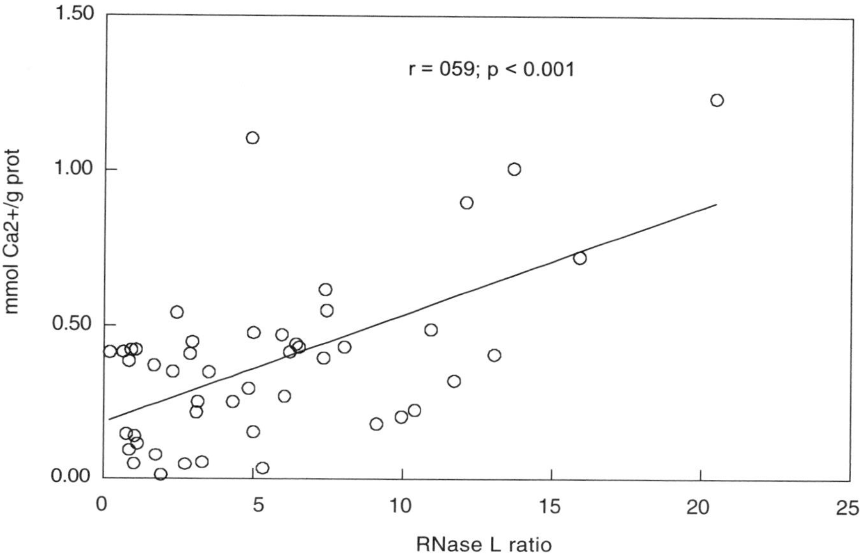

FIGURE 6.12 Quantification of intracellular Ca^{2+} with Arsenazo III in PBMC cytoplasmic extracts (n = 47) as a function of RNase L ratios. A significant linear correlation between the intracellular calcium concentrations and the extent of RNase L cleavage could be detected.

intracellular calcium concentrations are in the millimolar range. Above this ratio, the intracellular calcium concentrations decrease and do not correlate anymore with the extent of RNase L cleavage. As discussed above, an increase in intracellular calcium concentration is necessary for activation of calpains, particularly m-calpain. Consequently, this progressive increase in intracellular calcium concentration supports the possible activation of m-calpain in these cells.

Different reports have been published regarding the relationship between calpains and caspases in apoptosis.[25,95,123] According to one group of authors,[123] calpains act upstream of caspases, which they activate. Other authors[24,25,124] suggest the reverse; in their model, calpains act in the execution phase of apoptosis and are activated by caspases. Recently, strong supporting data for the former model have been published.[93,95,97] All these authors demonstrated the direct cleavage and in most cases activation of caspases by m-calpain, strongly suggesting that calpains, or at least m-calpain, would rather function upstream of the caspases. Interestingly, m-calpain was demonstrated to directly cleave caspase-9 into an inactive fragment.[93] Accordingly, caspase-9-dependent apoptosis is prevented, but not all apoptotic pathways. It is also suggested that apoptosis is alternatively modulated by intracellular calcium signals[93]: activation (cleavage) of caspase-12 by calcium-dependent m-calpain results in apoptosis through activation of the ER stress pathway.[95] m-Calpain is likely to interfere with the activation of caspase-3; cleavage of caspase-3 by m-calpain facilitates its further activation.[97]

As discussed previously (cfr. *supra* and Chapter 2), our results suggest an increased m-calpain activity in the PBMCs of CFS patients. The intracellular calcium concentrations increase linearly as a function of RNase L cleavage (Figure 6.12). We also observed the progressive cleavage of caspase-12 into its activated form and the cleavage of caspase-9 into its inactive form as a function of RNase L cleavage. All these proteins are m-calpain substrates, and the cleavage products detected match those produced by recombinant m-calpain *in vitro*. This strongly suggests an important role for m-calpain in the cleavage of different proteins in the PBMC of CFS patients.

6.5 ACTIN IN APOPTOSIS AND CHRONIC FATIGUE SYNDROME

The cytoskeleton is a dynamic network of intracellular proteinaceous structural elements responsible for cell shape, motility, migration, polarity (orientation), and maintenance of intercellular contacts involved in tissue architecture. The cytoskeleton has the capacity to assemble or disassemble continuously. Actin, the most abundant protein in many animal cells, is the main building block of the cytoskeleton and the motility system in nonmuscle eukaryotic cells. The actin cytoskeleton is a highly active, dynamic, and complex three-dimensional structure that is reshaped and reformed during the cell cycle and in response to extracellular signals.[125] The reorganization of actin filaments and the formation of new actin-containing structures are often associated with motile activities of nonmuscle cells, including cytokinesis, cell locomotion, and growth cone extension.[126-128]

Despite the fact that actin can be present as a globular monomer (G-actin) of 43-kDa, physiological conditions favor the assembly of polymerized (F) actin from monomeric actin. This polymerization starts with the formation of a thermodynamically unstable trimer (nucleation or rate-limiting step) that rapidly grows into a filament by addition of new monomers. The two ends of the filament (the pointed negative end, that grows slower, and the barbed positive end) are structurally and functionally different.[129] Actin polymerization occurs at discrete nucleation sites (also called the transient storage sites of G-actin) located at the plasma membrane, where monomeric actin molecules are present in complexes with sequestering proteins. Actin from these complexes can be rapidly incorporated into motile peripheral regions of the cell, such as lamellipodia and microspikes.[130,131] Actin within punctate structures is predominantly in the nonfilamentous state.[132] Part of the monomeric actin pool diffuses freely in the cytoplasm or concentrates near active assembly sites or sites of secondary messenger release, in discrete beadlike structures. Those structures are found in locomoting cells behind the lamellipodia and may be precursors for the cortical actin network which undergoes continuous turnover in locomoting cells.[132]

Several host-signaling pathways (tyrosine phosphorylation, lipid metabolism, and activation of small G proteins) lead to actin rearrangement.[133,134] The polymerization is controlled by actin-sequestering and capping proteins. The best known are profilin which binds actin 1:1,[135,136] actin depolymerizing factor (ADF), cofilin, and thymosin-β4, a small 5-kDa peptide found in all cell types in extremely high concentrations.[137] Villin and gelsolin bind monomeric actin and cap the barbed ends of actin filaments.[138] These proteins possess a binding site for phosphoinositides, and local changes in the lipids at the cell membrane result in the release of free filamentous ends. GTP-ases, such as Ras and Rho-related proteins, regulate and affect the organization of the actin cytoskeleton in eukaryotic cells.[139] Bacteria are also capable of inducing major changes in the actin cytoskeleton.[140] Agents disrupting actin filaments increase,[141] while actin stabilization protects against, cell permeability.[142] The cytoskeleton affects the nature of cell–cell and cell–substrate interactions by association with cell adhesion molecules. It provides the driving force for cell movements and surface remodeling at the plasma membrane.

More than 100 actin-binding proteins have been identified of which many are expressed in the same cells. We will limit our discussion to deoxyribonuclease I (DNase I), which is related to apoptosis, and to the serum-binding proteins, namely gelsolin and vitamin D-binding protein (DBP). DNase I degrades double-stranded DNA to 5' oligonucleotides by hydrolysis of the 3'-phosphodiester bond. DNase I forms a 1:1 complex with monomeric G-actin. Actin binding inhibits the nuclease activity by sterically blocking the active site around glutamate-13.[143,144] Gelsolin and villin are structurally related actin-binding proteins. Cytoplasmic and plasma gelsolin are very similar, the plasma gelsolin sequence longer by 25 amino acids at the N-terminus.[145] The gelsolin family of capping proteins is characterized by 3 or 6 repeats of a 125–150 amino acid sequence motif. The different segments bind to different sites on actin monomers and filaments.[146] Under the control of Ca^{2+} and membrane polyphosphoinositides, plasma gelsolin severs actin filaments, caps the barbed ends, and nucleates actin polymerization.[138,147] Gelsolin is a substrate for caspases and the cleaved gelsolin may be one physiological effector of morphological change during apoptosis.[148] Overexpression of full-length gelsolin inhibits the loss of mitochondrial membrane potential and cyt c release from the mitochondria, resulting in the lack of activation of caspase-3, -8, and -9.[149] DBP[150] has a molecular weight of 53-kDa. DBP is also referred to as group-specific component or Gc.[151] Besides vitamin D, this major plasma protein complexes actin monomers[152] with high affinity ($K_a = 10^{10}$ M^{-1} at 4°C).[153] Gc is synthesized by the liver and cleared from the circulation by the liver and kidneys. Gc is widely distributed in the cytosol of many tissues and is detected on the membrane of PBMCs.[154] It also effects depolymerization and sequesters actin extremely efficiently in the monomeric state.[155,156]

On the death of cells, whether the result of injury, disease, or natural ageing, G- and F-actin are released into extracellular fluids, including blood plasma. The ionic strength conditions in those fluids favor actin polymerization, an event that would result in increased plasma viscosity.[157] Gelsolin and Gc rapidly break down actin polymers and clear actin from the circulation, a process called the actin scavenging system.

The cellular actin scaffold is continuously reorganized in response to a variety of signals, including apoptotic signals.[37] An elevated intracellular calcium concentration, which occurs during specific apoptotic stages, induces the activation of the severing and capping activities of gelsolin toward actin filaments. In such conditions, more actin filaments with shorter lengths are generated.[147] During apoptosis, several cytoskeletal proteins (such as Gas2, gelsolin, beta-catenin, fodrin, actin, PAK-2) are cleaved by the caspases.[33,37,158-162] Actin is also at least partially cleaved by caspases

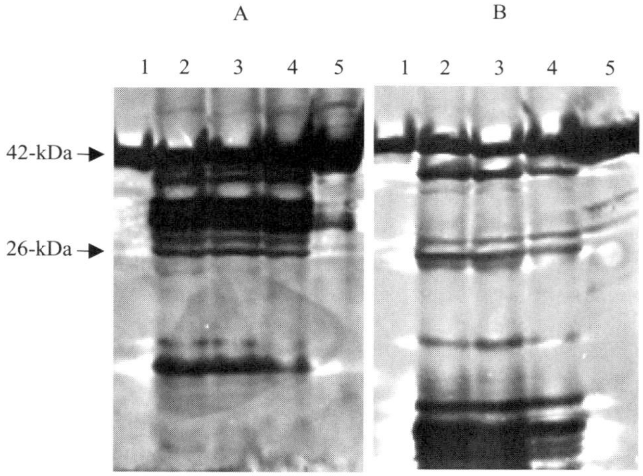

FIGURE 6.13 G-Actin fragmentation pattern in the cytosol of PBMCs of CFS patients and controls. Immunoblotting of purified G-actin (lane 1) and PBMC extracts with decreasing RNase L ratios (lanes 2 to 5) was performed using anti-actin antibodies specific for N- (part A) and C-terminal (part B) domains of G-actin. (Reproduced with permission from Roelens, S., Herst, C. V., and D'Haese, A., *J. Chronic Fatigue Syndrome,* 2001, in press.)

in multiple cell lines (Hela and A431), depending on the stage of apoptotic development.[102] An actin fragment of 15-kDa generated by caspase cleavage can specifically induce morphological changes resembling those of apoptosis. Villa et al.[163] reported that actin proteolysis during neuronal apoptosis requires a protease of the calpain family. Calpain represents one of the few nonlysosomal proteolytic systems of mammalian cells.[164] Actin proteolysis by calpain generates fragments of 40- and 41-kDa and is correlated to DNA fragmentation. In PBMC extracts of CFS patients (Figure 6.13), we detected fragmented G-actin by immunoblotting.[165] Purified G-actin could be cleaved *in vitro* by caspase-3, m-calpain, and PBMC extracts from CFS patients (Figure 6.14), generating fragments of similar sizes.

During the early stages of apoptosis in Jurkat cells, cell nuclei and the cytoskeleton are first modified.[166] The actin cytoskeleton collapses and shrinks, leading to changes in the cell shape. Spano et al.[167] report an initial disassembly of stress fibers, associated with the reorganization of the actin cytoskeleton and changes at the cell surface level when apoptosis is induced *in vitro* in different cell lines (HL-60 and Chang liver cells). Actin accumulates in cellular compartments where annexin V is abundantly present. In later stages of apoptosis, a strong reduction of actin and of annexin V concentrations occurs, probably as a result of proteolytic cleavage. The cleaved actin is released from the cell into the serum. In patients with CFS, we found an increase in fragmented actin in the serum (Reference 165; Figure 6.15). The increase in the presence of a 26-kDa fragment correlates with the fragmentation of both actin and RNase in the PBMC extracts (Figure 6.16). In the human T-cell lymphoma cell line Jurkat membrane blebbing and caspase-3 activation occur in a later stage of apoptosis.[166] Activated caspase-3 produces beta-actin cleavage

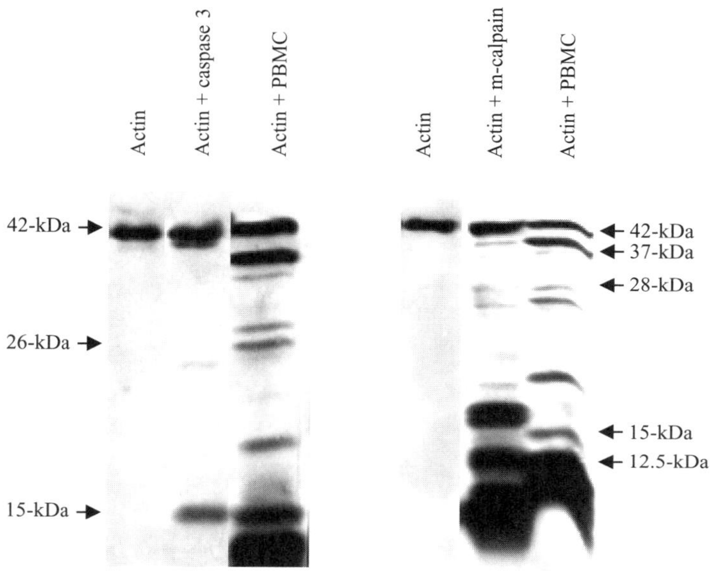

FIGURE 6.14 Possible proteases responsible for G-actin cleavage. Purified G-actin was incubated *in vitro* with either a PBMC extract containing the RNase L 37-kDa fragment or with the apoptotic proteases caspase-3 and m-calpain. Immunoblotting was performed with an antibody specific for C-terminal G-actin. The purified proteases generate G-actin fragments of the same sizes as those generated by the PBMC extract.

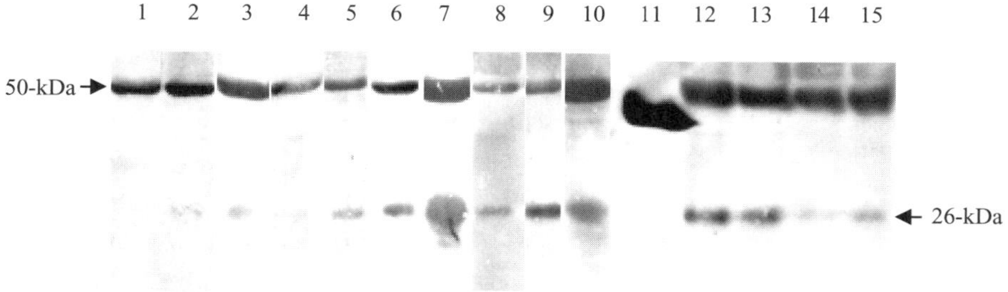

FIGURE 6.15 Detection of G-actin fragments in serum by immunoblotting, using an antibody specific for the N-terminal end. Serum samples corresponding to PBMC samples respectively negative (lanes 1 to 5) or positive (lanes 6 to 10 and 12 to 15) for the presence of the 37-kDa truncated RNase L were electrophoresed and blotted. Lane 11 is the immunoblot of purified G-actin used as a control. The 26-kDa actin fragment is present exclusively in the serum samples corresponding to the PBMC positive for the presence of cleaved RNase L.

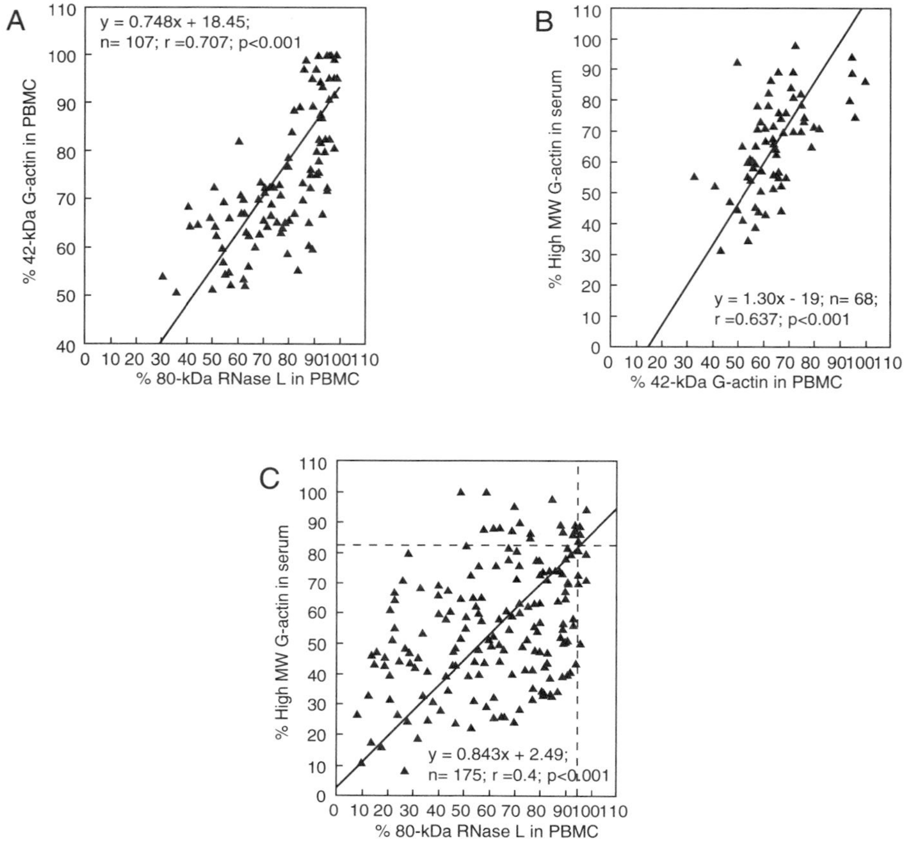

FIGURE 6.16 Correlations between the extent of RNase L cleavage, G-actin cleavage in PBMC extracts, and the presence of the G-actin fragment in serum, respectively. The results are expressed as the percentage of native protein present in the samples. The dotted lines delineate the normal cutoffs. Part A: Correlation between RNase L and G-actin cleavage in PBMC extracts ($n = 107$, $p<0.001$, $r = 0.71$). Part B: Correlation between the presence of G-actin fragments in PBMC and matching serum samples ($n = 68$, $p<0.001$, $r = 0.64$). Part C: Correlation between the presence of RNase L fragments in PBMC and G-actin fragment in matching serum samples ($n = 175$, $p<0.001$, $r = 0.40$).

fragments.[168] Overexpression of the antiapoptotic protein Bcl-2 attenuates neurodegradation, and delays activation of the caspases and fragmentation of beta-actin. Brancolini et al.[37] showed that cytoskeleton reorganization during apoptosis could also be caused by a caspase-like proteolytic cleavage of Gas2 (growth arrest-specific protein).

The roles of actin fragmentation in the apoptotic process are critical. The truncation of actin and fodrin, a spectrin-like protein,[160] by caspase-3[33] helps reduce the stiffness of the cytoskeleton, which permits the formation of apoptotic bodies without bursting the cytoplasmic membrane. Fodrin is suggested to be associated with the maintenance of lipid asymmetry. By binding DNase I, intact actin inhibits

its activity. DNase I is the nuclease responsible for the degradation of DNA during apoptosis in dexamethasone-treated rat thymocytes.[104] Actin cleavage activates DNase I, thereby allowing the cleavage of genomic DNA.[144] The specific proteolytic fragments of actin appear at an apoptotic stage prior to massive protein degradation.[163] Finally, the organelle membrane potential decreases as a consequence to actin cleavage, which leads to the release of cyt c from mitochondria. The subsequent cleavage of procaspase-9 to its active form and its association with Apaf-1 in the cytosol induce caspase-3 activation, leading to proteolytic degradation and completion of apoptosis.[169]

Actin plays a critical role in the development of the immune response and during phagocytosis. The central event in T-cell activation is the interaction of the T-cell receptor (TCR) with the antigenic peptide presented by the major histocompatibility complex (MHC) of the antigen-presenting cell (APC). During T-cell activation, the engagement of costimulatory molecules is crucial for the development of an effective immune response.[170] Wülfing et al.[171] describe an active, cytoskeletal mechanism that drives receptor accumulation at the T-cell-APC interface. The T-cell cortical actin cytoskeleton reorients toward the T-cell-APC interface soon after T-cell activation as the result of interactions between the receptor pairs LFA-1-ICAM-1 and CD28-B7.[171] Phagocytosis of an apoptotic cell by a macrophage implies the recognition of the apoptotic cell, as mediated by receptors (e.g., CD14, scavenger receptor A) on the macrophage membrane and ligands (including sugars and phosphatidylserine) on the apoptotic cell membrane. These signals induce actin polymerization and remodeling around the phagocytic cup in order to allow internalization of apoptotic bodies.

During this event, actin cross-linking proteins (such as paxillin, talin, alpha-actinin) are enriched at the leading edge of the cell and other actin-binding proteins concentrate at the other, trailing edge.[172] Rac and Cdc42 control actin polymerization into lamellipodial and filopodial membrane protrusions[173,174] and are required for the accumulation of F-actin in the phagocytic cups. Cdc42 participates in the formation of filopodia and in the activation of Rac. Rac, in turn, stimulates membrane ruffling and induces Rho, leading to the formation of focal adhesions, stress fibers, and receptor clustering, a prerequisite for efficient particle binding and internalization. Soon after the internalization, F-actin is depolymerized from the phagosome and the vacuole membrane becomes accessible to early endosomes, a process mediated by annexins.[175] The cleavage of G-actin as observed in the PBMC of CFS patients is in line with the apoptotic dysfunctions described above and undoubtedly contributes to the immune dysregulations as a part of the illness.

6.6 PKR AND APOPTOTIC REGULATION IN CHRONIC FATIGUE SYNDROME

RNase L is an important apoptotic effector activated indirectly, via the 2-5OAS, by ds-RNA. These properties are shared by another enzyme, the ds-RNA-activated protein kinase (PKR). Present in a latent state in most cell types, this 68-kDa serine/threonine kinase is induced by IFN-α[176] and activated by ds-RNA[177] or ss-RNA presenting stem-loop structures.[178] These types of RNAs are normally characteristic of viral RNAs, meaning that in most cases PKR will be activated upon viral infection. PKR activation is associated with autophosphorylation and dimerization of the enzyme.[179,180] Activated PKR is able to phosphorylate the α-subunit of the eukaryotic initiation factor 2; phosphorylated eIF2α blocks the exchange of GDP with GTP in the inactive eIF2GDP complex by sequestering the guanine-nucleotide exchange factor eIF2B. This eventually leads to inhibition of translation, blocking the multiplication of viruses.[181,182]

PKR can be activated by stimuli other than ds-RNA, like calcium ionophores, lipopolysaccharides, IFNs, or TNF-α, suggesting that it has other functions than just an antiviral activity.[183] Actually, PKR appears to have important roles in regulating cellular growth control[184] and tumor suppression.[185-189] But the most relevant function of PKR in the context of CFS is most probably its role as an apoptotic regulator. The first evidence for this apoptotic function came from the observation that expression of PKR in HeLa cells resulted in apoptosis, whereas expression of an inactive mutant had no effect.[190] Overexpression of PKR was then shown to induce apoptosis in U937 cells,[191] NIH3T3 cells,[192] or embryonic kidney cells,[193] and to make cells more sensitive to apoptosis induced by ds-RNA[194] or influenza virus.[195] On the one hand, expression of a catalytically inactive PKR mutant confers resistance to apoptosis induced by influenza virus infection in HeLa cells[196] or mouse fibroblasts,[195] as well as to Fas-induced apoptosis in both embryonic kidney[193] and HeLa cells.[196] On the other hand, inactive PKR mutants also make 3T3L1 cells resistant to ds-RNA-induced apoptosis[194] and NIH3T3 cells resistant to apoptosis induced by TNF-α or serum deprivation.[192] Similarly, expression of a PKR antisense RNA in U937 cells renders them resistant to TNF-α-induced apoptosis.[191] Finally, PKR-null mouse fibroblasts are resistant to TNF-α- or LPS-induced apoptosis.[197]

Our interest in PKR was of course triggered by the fact that this protein is involved in the same pathways and regulated by similar mechanisms as RNase L; considering the dysregulation of RNase L activity in CFS, it was logical to investigate PKR levels and activity in CFS patients. It has been shown already that PKR is upregulated in peripheral blood lymphocytes of CFS patients at both mRNA and protein levels.[198] It was suggested that this upregulation was linked to the elevated proportion of apoptotic cells among this population: apoptosis in these cells could be partially inhibited by 2-aminopurine, an inhibitor of PKR.

We have been looking at the level of PKR protein in cytoplasmic extracts of PBMCs from CFS patients. Our results are not in full agreement with the previous observation since, in our samples, the level of 68-kDa protein decreases when the ratio of RNase L (a biological marker of CFS reflecting the extent of cleavage of the protein) is increasing (Figure 6.17). However, when looking at the level of eIF2α phosphorylation, we observed that detection of phosphoserine (corresponding to the phosphorylation and inactivation of eIF2α) was stable or even increasing with RNase L cleavage and, therefore, was still elevated in patients having a relatively low level of PKR protein (Figure 6.18). These two observations may appear contradictory, but several studies have already shown that absolute amounts of PKR in cells do not necessarily reflect the levels of activity. For example, following poliovirus infection, PKR is significantly degraded; at the same time phosphorylation of eIF2α is

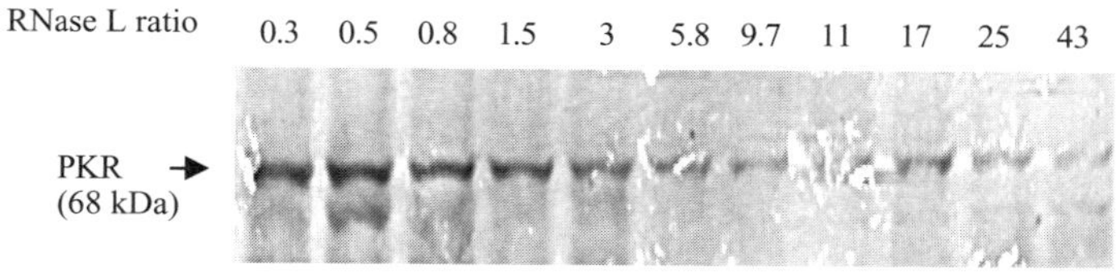

FIGURE 6.17 Immunodetection of native PKR in PBMC cytoplasmic extracts, using a polyclonal anti-PKR antibody (Santa Cruz). The RNase L ratios correspond to the extent of RNase L cleavage in the respective extracts (37 kDa LMW form/80 kDa native form). A decrease of 68-kDa PKR is observed, which parallels the increase of RNase L ratios.

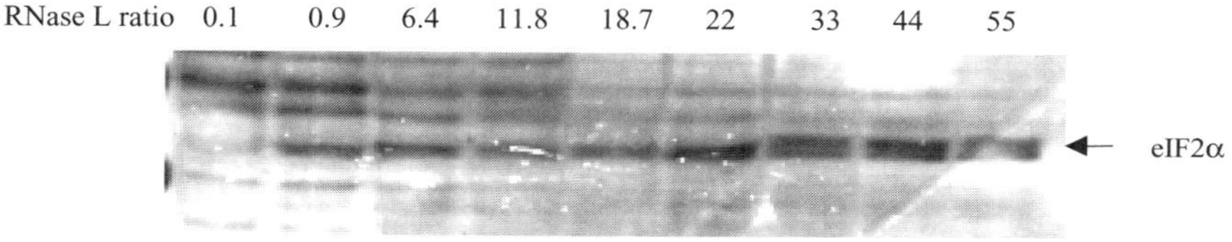

FIGURE 6.18 Phosphorylation of eIF2α related to RNase L ratios detected by immunoblotting using antiphosphoserine antibody (Sigma) on cytoplasmic extracts. Increasing levels of phosphorylated eIF2α are detected. This indicates an increased activity of PKR since global levels of eIF2α do not increase (Figure 6.19).

increased.[199] Conversely, in breast carcinoma cells, a high level of PKR is detected but only minimal activity can be measured; PKR in these cells is refractory to stimulation by ds-RNA.[28]

What are the consequences of an elevated PKR activity? We can certainly speculate that PKR is, at least in part, responsible for the induction of apoptosis observed at intermediate or high RNase L ratios (ratios between 2 to 20). But since phosphorylation of eIF2α is still detected at higher ratios, then what would cause the inhibition of apoptosis at ratios over 20? To try to find an answer, we must consider the different pathways involved in PKR-mediated apoptosis and look for dysregulations of factors normally involved in these pathways.

One possible mechanism for PKR-induced apoptosis is that inhibition of eIF2α pathway leads to cell death. In NIH3T3 cells, TNF-α-induced apoptosis correlates with increased phosphorylation of eIF2α and cells transfected with a variant of eIF2α, resistant to phosphorylation but still able to enhance translation, are resistant to ds-RNA- or TNF-α-mediated apoptosis. Furthermore, expression of an eIF2α variant, which mimics phosphorylated eIF2α, induces apoptosis.[192] Similarly, coexpression of PKR with a dominant negative form of eIF2α reverses both the PKR-mediated translational block and PKR-mediated apoptosis.[200] Yet one study has shown that expression of the human proto-oncogene Bcl-2 could block the PKR-induced apoptosis without blocking the PKR-induced inhibition of translation.[201] This last result would indicate that PKR-induced apoptosis is not, or not only, due to the blockade of protein synthesis.

Blocking of eIF2 function could participate in the initiation of apoptosis at RNase L ratios over 2. What then causes apoptosis to be blocked at very high ratios? Since

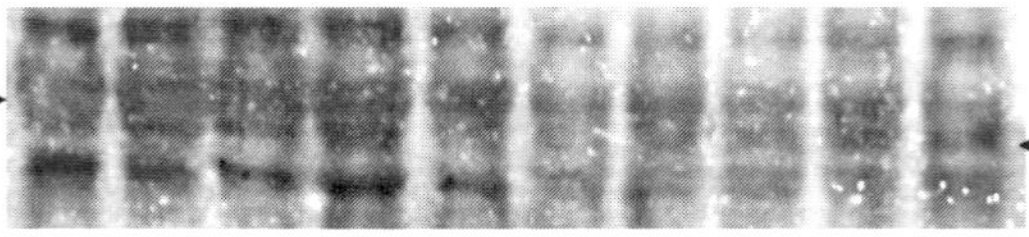

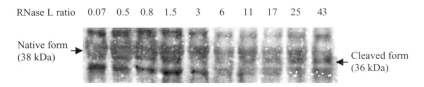

FIGURE 6.19 Progressive cleavage of eIF2a occurring at elevated RNase L ratios. Immunodetection of eIF2α in cytoplasmic extracts from PBMCs with an anti-eIF2 antibody (Santa Cruz).

eIF2α phosphorylation is probably not the only pathway involved, it is possible that apoptotic inhibition comes from the disruption of another pathway. However, looking carefully at eIF2α expression in PBMCs, one notices that, if it is still detectable at elevated RNase L ratios, the level of native 38-kDa eIF2α decreases when a smaller form of approximately 36-kDa appears (Figure 6.19). This observation is reminiscent of the caspase-mediated cleavage of eIF2α occurring in HeLa cells[202] as well as in Saos-2 cells in response to ds-RNA or TNF-α.[203] Consequences of this cleavage on eIF2α function and on the apoptotic process are not yet clear. Interestingly, after cleavage, exchange of GDP bound to eIF2 is very rapid and no longer dependent upon eIF2B,[202] meaning that cleavage could relieve the inhibition of eIF2α activity. Indeed, PKR-mediated translational suppression is repressed when the cleaved product of eIF2α is expressed in Saos-2 cells, even though PKR can still phosphorylate this cleaved product.[203] Along this line of observation, we suggest that apoptosis in PBMCs of CFS patients is first induced when eIF2α is phosphorylated by PKR, but that this induction is blocked when eIF2α starts to be cleaved, generating a fragment whose activity is insensitive to phosphorylation.

Most results suggest an involvement of PKR in TNF-α- or Fas-mediated apoptosis. Both TNF-α and Fas cause apoptosis via recruitment of FADD (Fas-associated death domain-containing protein), which activates caspase-8, which in turn activates other caspases such as caspase-3. While ds-RNA induces apoptosis in fibroblasts overexpressing PKR, it is totally ineffective on FADD-defective fibroblasts.[194] Involvement of FADD-caspase-8 pathway is confirmed by the fact that caspase-8 activity is upregulated during PKR-mediated apoptosis and that inhibitors of caspase-8 can block PKR-mediated apoptosis.[32] It seems likely that PKR could increase the expression of the Fas receptor, making the cells more sensitive to Fas-induced apoptosis;[197,204] however, neither this Fas receptor nor the TNF-α receptor appear to be required for PKR-induced apoptosis.[32] Consequently, PKR could act by a novel mechanism involving FADD independently of the death receptors. The pattern of caspase-8 activation in the PBMCs as a function of RNase L ratio (Figure 6.3; increased caspase-8 activity for RNase L ratios below 20 followed by a sharp decrease at higher ratios) is in good agreement with the involvement of this FADD-caspase-8 mechanism. However, what regulates caspase-8 activity and particularly what causes its sudden inhibition is still unknown.

A major mechanism of PKR action proceeds via the transcription factor NF-κB. NF-κB regulates the expression of genes involved in immune and inflammatory responses. It is normally sequestered in the cytoplasm by its interaction with the inhibitor of κB (I-κB); serine phosphorylation of I-κB causes its degradation by the proteasome, unmasking a nuclear localization sequence on NF-κB that allows its translocation to the nucleus where it activates gene transcription.[205] Activation of NK-κB can be induced by PKR: in addition to eIF2α, PKR is also able to directly phosphorylate I-κB-α.[206] This can result in apoptotic induction, since coexpression of PKR with a repressor form of I-κB-α leads to inhibition of apoptosis, while translation remains blocked. Furthermore, treating cells with proteasome inhibitors that block I-κB-α degradation prevents PKR-induced apoptosis.[200] PKR can also activate NF-κB by a more indirect mechanism involving the activation of I-κB kinase (IKK), a kinase that phosphorylates I-κB and causes its degradation.[207] Interestingly, this activation of IKK does not require the kinase activity of PKR.[208]

Induction of apoptosis via NF-κB may appear surprising since NF-κB is usually considered a cell-survival factor. For example, it is essential in preventing TNF-α-induced cell death[209] and mediates the promotion of cell survival by IFN-α/β.[210] Furthermore, it has been shown that a subunit of NF-κB, RelA, is able to inhibit ds-RNA-induced apoptosis.[211] However, other results have shown a possible pro-apoptotic activity of NK-κB: overexpression of another NF-κB subunit, c-rel, leads to apoptosis in chick bone marrow cells;[212] inhibition of NF-κB activity abrogates virus-induced apoptosis in AT-3 cells[213] and prevents induction of apoptosis by DNA-damaging agents in T lymphocytes.[214]

Is NF-κB active in the PBMCs of CFS patients? We have been looking at the presence of I-κB in the cytoplasm, an indicator of NF-κB activity, since disappearance of I-κB means that NF-κB is released and activated. We could detect I-κB in cytoplasmic extracts of PBMCs from patients having an RNase L ratio lower than 10, but over this value I-κB is no longer detectable (Figure 6.20). This may indicate an activation of NF-κB, possibly consistent with an induction of survival factors and a blockade of apoptosis. However, due to the controversial effects of NF-κB regarding induction or inhibition of apoptosis, it is difficult to predict the consequences of NF-κB activation in these samples.

Another potentially interesting mechanism for PKR-mediated apoptosis is linked to p53. The tumor suppressor p53 is involved in the cellular response to DNA damage or stress factors such as UV radiations, chemotherapeutic drugs, or aberrant growth signals. All these factors result in the stabilization of p53 protein, which then can stop cellular proliferation by inhibiting progression through the cell cycle, or even induce apoptosis. The effect of p53 on apoptosis can be transcription-dependent:

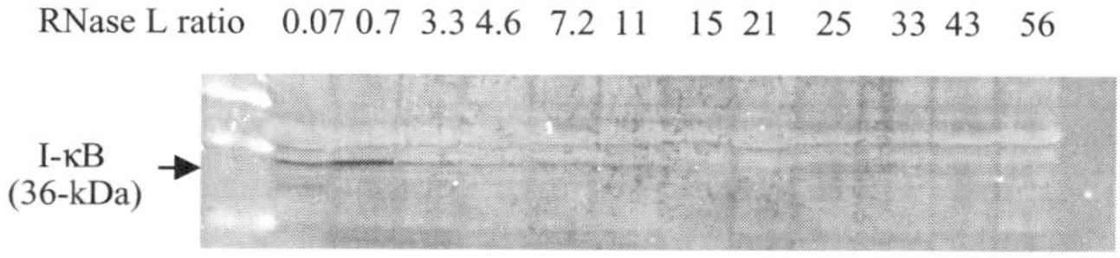

FIGURE 6.20 Immunodetection of I-κB in PBMC cytoplasmic extracts as a function of the RNase L ratio, using anti-I-κB antibody (Santa Cruz). Clearly distinct at very low RNase L ratios, the I-κB protein decreases but remains detectable until ratios around 10, then disappears progressively.

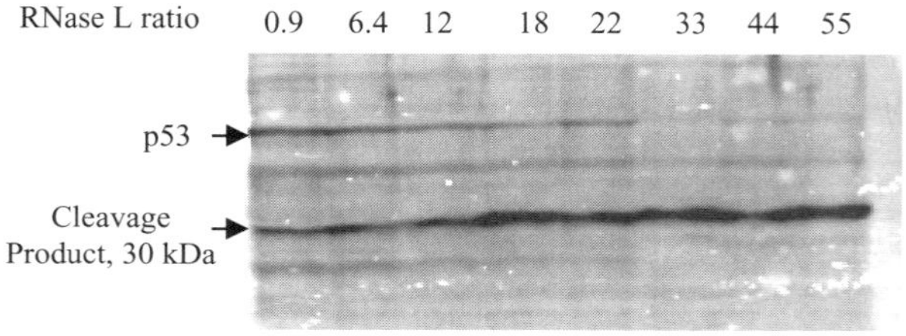

FIGURE 6.21 Immunodetection of p53 in cytoplasmic extracts as a function of the RNase L ratio, using an anti-p53 polyclonal antibody (Santa Cruz). A cleavage product of 30-kDa is already visible at low RNase L ratios and becomes prominent at ratios around 20, where the native 53-kDa form disappears.

p53 is a sequence-specific transcription factor which activates the expression of pro-apoptotic genes such as Bax, NOXA, and p53AIP1, or decreases the expression of antiapoptotic Bcl-2.[215-217] In addition, p53 can also induce apoptosis via a transcription-independent mechanism involving caspase-8 and caspase-9.[218] It has been shown that PKR is able to induce the expression of p53 in promonocytic U937 cells, and that this induction is required for TNF-α-induced apoptosis; conversely, over-expressing p53 in PKR-deficient U937 cells makes them susceptible to TNF-α-induced apoptosis.[219] Consistent with this observation is the fact that PKR activation also causes an upregulation of Bax,[194] which could happen via p53 induction.

We have been looking at p53 expression in PBMCs of CFS patients and were able to detect p53 consistent with an induction of apoptosis. However, the expression of p53 is related to the RNase L ratio: p53 is present in cells where the ratio 37/80-kDa RNase L does not exceed 20 (though its level tends to decrease with increasing ratios). Over this value p53 disappears, and the antibody reveals only a prominent band at about 30-kDa, which is probably a cleavage product of p53 (Figure 6.21). Again, this observation is consistent with a blockade of apoptosis at elevated ratios; a cleavage of p53 may make the cells unable to respond to factors which normally cause p53-mediated apoptosis.

The cause of p53 disappearance is still unknown. The amount of p53 protein in cells is determined more by its degradation rate than by its synthesis rate. Therefore an increased degradation is likely, particularly since we observe the accumulation of a 30-kDa product in the cells. p53 degradation normally proceeds through the ubiquitin–proteasome pathway, a mechanism regulated by the protein MDM2.[220] This protein binds p53 and stimulates the addition of a ubiquitin group to its C-terminus, which leads to its degradation; overactivity of MDM2 is associated with many cancers such as sarcomas, brain, breast, and lung cancers. However, p53 may also be a substrate for m-calpain. Calpain inhibitors have been shown to increase the half-life of p53 *in vivo*,[221-223] whereas calcium ionophores induce its degradation.[224] Most interestingly, this last author reported that p53 cleavage by calpain generates a 30-kDa fragment, which may correspond to the 30-kDa band that we observe in our patients' cells. The calpain activity responsible for the cleavage of RNase L may therefore also cause the degradation of p53, with consequences on apoptotic dysregulation.

A last possibility for PKR to regulate apoptosis is by its many interactions with the pathways linked to IFNs, and particularly its ability to regulate the interferon regulatory factor-1 (IRF-1). This transcription factor, induced by IFN-γ, is a transcriptional activator of IFN-β, but it also regulates many other genes with implications on cell growth, proliferation, and apoptosis.[225] Even if in certain cell types IRF-1 functions as a suppressor of apoptosis,[226] in most cases it acts as an apoptotic activator by inducing the caspase cascade in response to IFN-γ.[227-232] PKR can upregulate IRF-1,[197,229,233] but in turn IRF-1 can also induce PKR; this induction is required for the tumor suppressor activity of IRF-1.[188] Interestingly, IRF-1 can also induce expression of FasL, sensitizing the cells to apoptosis.[234] Considering the important cross-talks between IRF-1 and PKR pathways, it is clear that PKR activation can affect IRF-1-mediated mechanisms, particularly induction of apoptosis. Studying IRF-1 regulation and expression in PBMCs of CFS patients could also give indications about mechanisms resulting in induction and repression of apoptosis in these cells.

In conclusion, most known pathways of PKR-mediated apoptosis appear to be altered in the immune cells of CFS patients, always in a way consistent with the initial observation of an induction of apoptosis followed by its repression. This tends strongly toward an important role of PKR in the dysregulations affecting the PBMCs of CFS patients. It is likely that PKR exerts its regulation on apoptosis via not one, but several, mechanisms and that all the alterations we have detected in PBMCs contribute to the apoptotic dysfunction that may be a significant aspect of the disease. Of course, factors other than PKR are also involved; we have seen earlier in this chapter that the reported activation of RNase L in CFS probably also contributes to apoptotic induction, and that its cleavage could dysregulate many cellular pathways and finally block apoptosis. Of interest is the fact that calpain, which is probably a main effector of RNase L cleavage, may also cleave p53 and therefore act through two different pathways. However, what causes the activation of calpain remains to be determined. Finally, it must be noted that a dysregulation of PKR functions may have consequences not limited to apoptotic dysfunctions. Perturbation of its activity in cells providing for the innate immunity would make the cells more sensitive to viral infections. It was also reported that activation of PKR leads to immunoglobulin E class switching in B-cells,[235] a typical immune dysregulation experienced by CFS patients.[236]

6.7 CONCLUSIONS AND PROSPECTS

From the data presented in this chapter, we can conclude that the PBMCs of CFS patients are characterized by important dysregulations of apoptotic activity. The level of apoptosis in these cells is actually related to the extent of RNase L cleavage: at RNase L ratios comprised between 2 and 20, we observed an increased activity of caspase-3 and -8, whereas at ratios higher than 20 caspase activities return to normal (caspase-3, -8) or even seem to be inhibited (caspase-2, -9). Investigating the expression and activity of different apoptotic regulators (p53, PKR, NF-κB, …) confirmed this observation: expression of these factors is consistent with an induction of apoptosis at moderate and high RNase L ratios, and then with a blockade of apoptosis at very high ratios.

This relation between extent of RNase L cleavage and alteration of apoptotic pathways is interesting, and of course one question is whether RNase L cleavage is a cause or a consequence of these apoptotic dysregulations. For moderate ratios (below 20), a relation between RNase L cleavage and induction of apoptosis is possible: as we have already discussed, generation of a cleaved fragment with higher catalytic activity could contribute to apoptotic induction, and, conversely, activation of proteases during apoptosis could contribute to the generation of the fragment. For RNase L ratios over 20, however, the apoptotic blockade is more difficult to explain. Since we have less apoptosis in these cell populations, it is not possible to explain the extensive RNase L cleavage by an activation of apoptotic proteases (caspases or calpain). As we have seen in Chapter 2, RNase L cleavage at these ratios is more likely caused by other proteases like elastase or cathepsin G, which are involved in inflammatory processes but not in apoptosis. Conversely, it is difficult to understand how RNase L cleavage by itself could cause this apoptotic blockade, either directly or by influencing the expression or activity of other apoptotic factors. We certainly need to further investigate the function of RNase L in apoptotic regulation, and the consequences of the generation of RNase L fragments which potentially have different catalytic activities, specificities, or cellular localization than the native enzyme.

Another interesting possibility would be the involvement of various cellular pathogens in the apoptotic dysregulations. Due to their altered immune functions, CFS patients often develop opportunistic infections (viruses, chlamydia, mycoplasma), whose ability to interfere with apoptotic pathways (and usually block them) is well known. Viruses can inhibit apoptosis by different means, like production of Bcl-2 homologs, inhibition of p53, caspases or PKR,[237] or inhibition of Jak-Stat and eIF2 pathways.[238] Intracellular bacterial pathogens like *Chlamydia* can inhibit caspase-3 activation as well as cyt c release; *Rickettsia* and *Mycobacterium* can activate NF-κB.[239] Once established, these infections could contribute to the maintenance or aggravation of CFS by interfering with various cellular mechanisms (cellular defense such as apoptosis, but possibly also metabolic pathways, cytokine production, etc.), and finally make the pathological process more difficult to reverse. The cleavage of G-actin, which parallels apoptotic induction and RNase L cleavage in PBMCs, has undoubtedly a major impact on immune cell performances, particularly in antigen presentation and phagocytosis. These reduced performances may further contribute to the immune dysfunctions characteristic of CFS.

Regulation of apoptosis is a highly complex process. Three major apoptotic pathways have been described: the mitochondrial pathway involving cyt c release and caspase-9, the death receptors pathway involving FADD and caspase-8, and the less characterized ER pathway involving activation of caspase-12. But these pathways are actually interdependent: they are connected at multiple levels by cross-talks and common regulators which can act together to cause either cell death or cell survival or, conversely, have opposite effects. Predicting the effects of the alteration of one apoptotic factor such as RNase L is therefore a difficult task. However, having established a link between RNase L cleavage and dysregulations in apoptotic functions is already an interesting result. Obviously, the RNase L ratio is a marker which allows distinguishing different subsets of CFS patients who may need different therapeutic approaches. It will be interesting to see whether this relation between RNase L cleavage and apoptotic dysregulation is also found in other pathologies such as cancer and chronic viral infections.

REFERENCES

1. Adams, J. M. and Cory, S., The Bcl-2 protein family: arbiters of cell survival, *Science,* 281, 1322, 1998.
2. Fussenegger, M. and Bailey, J. E., Molecular regulation of cell-cycle progression and apoptosis in mammalian cells: implications for biotechnology, *Biotechnol. Prog.,* 14, 807, 1998.
3. Thornberry, N. A. and Lazebnik, Y., Caspases: enemies within, *Science,* 281, 1312, 1998.
4. Valente, M. and Calabrese, F., Liver and apoptosis, *Ital. J. Gastroenterol. Hepatol.,* 31, 73, 1999.
5. Kerr, J. F., Wyllie, A. H., and Currie, A. R., Apoptosis: a basic biological phenomenon with wide-ranging implications in tissue kinetics, *Br. J. Cancer,* 26, 239, 1972.
6. Sperandio, S., de Belle, I., and Bredesen, D. E., An alternative, nonapoptotic form of programmed cell death, *Proc. Natl. Acad. Sci. U.S.A.,* 97, 14376, 2000.
7. Obaya, A. J., Mateyak, M. K., and Sedivy, J. M., Mysterious liaisons: the relationship between c-Myc and the cell cycle, *Oncogene,* 18, 2934, 1999.
8. Wyllie, A. H., Apoptosis: an overview, *Br. Med. Bull.,* 53, 451, 1997.
9. Alberts, B. et al., *Molecular Biology of the Cell,* 3rd ed., Garland Publishing Inc., New York; 1994; chap. 21.
10. Ellis, R. E., Yuan, J. Y., and Horvitz, H. R., Mechanisms and functions of cell death, *Annu. Rev. Cell Biol.,* 7, 663, 1991.
11. Nagata, S., Apoptosis by death factor, *Cell,* 88, 355, 1997.
12. Hasmall, S. C. and Roberts, R. A., The perturbation of apoptosis and mitosis by drugs and xenobiotics, *Pharmacol. Ther.,* 82, 63, 1999.
13. Roberts, L. R., Adjei, P. N., and Gores, G. J., Cathepsins as effector proteases in hepatocyte apoptosis, *Cell Biochem Biophys.,* 30, 71, 1999.
14. Brown, R., The bcl-2 family of proteins, *Br. Med. Bull.,* 53, 466, 1997.
15. Tsujimoto, Y. and Shimizu, S., VDAC regulation by the Bcl-2 family of proteins, *Cell Death. Differ.,* 7, 1174, 2000.
16. Yin, X. M., Signal transduction mediated by Bid, a pro-death Bcl-2 family of proteins, connects the death receptor and mitochondria apoptosis pathways, *Cell Res.,* 10, 161, 2000.
17. Nakagawa, T. et al., Caspase-12 mediates endoplasmic-reticulum-specific apoptosis and cytotoxicity by amyloid-beta, *Nature,* 403, 98, 2000.
18. Mehmet, H., Caspases find a new place to hide, *Nature,* 403, 29, 2000.
19. Li, J. et al., Nitric oxide suppresses apoptosis via interrupting caspase activation and mitochondrial dysfunction in cultured hepatocytes, *J. Biol. Chem.,* 274, 17325, 1999.
20. Jacobson, M. D., Weil, M., and Raff, M. C., Programmed cell death in animal development, *Cell,* 88, 347, 1997.
21. Slee, E. A. et al., Ordering the cytochrome c-initiated caspase cascade: hierarchical activation of caspases-2, -3, -6, -7, -8, and -10 in a caspase-9- dependent manner, *J. Cell Biol.,* 144, 281, 1999.

22. Talanian, R. V., Quinlan, C., Trautz, S., et al., Substrate specificities of caspase family proteases, *J. Biol. Chem.,* 272, 9677, 1997.
23. Sorimachi, H., Saido, T. C., and Suzuki, K., New era of calpain research. Discovery of tissue-specific calpains, *FEBS Lett.,* 343, 1, 1994.
24. Wood, D. E. and Newcomb, E. W., Caspase-dependent activation of calpain during drug-induced apoptosis, *J. Biol. Chem.,* 274, 8309, 1999.
25. Knepper-Nicolai, B., Savill, J., and Brown, S. B., Constitutive apoptosis in human neutrophils requires synergy between calpains and the proteasome downstream of caspases, *J. Biol. Chem.,* 273, 30530, 1998.
26. Sato-Kusubata, K., Yajima, Y., and Kawashima, S., Persistent activation of Gs-alpha through limited proteolysis by calpain, *Biochem J.,* 347, 733, 2000.
27. Sherr, C. J., Cancer cell cycles, *Science,* 274, 1672, 1996.
28. Savinova, O., Joshi, B., and Jagus, R., Abnormal levels and minimal activity of the dsRNA-activated protein kinase, PKR, in breast carcinoma cells, *Int. J. Biochem. Cell Biol.,* 31, 175, 1999.
29. Gale, M., Jr. et al., Antiapoptotic and oncogenic potentials of hepatitis C virus are linked to interferon resistance by viral repression of the PKR protein kinase, *J. Virol.,* 73, 6506, 1999.
30. Clemens, M. J. and Elia, A., The double-stranded RNA-dependent protein kinase PKR: structure and function, *J. Interferon Cytokine Res.,* 17, 503, 1997.
31. Clemens, M. J., PKR-a protein kinase regulated by double-stranded RNA, *Int. J. Biochem. Cell Biol.,* 29, 945, 1997.
32. Gil, J. and Esteban, M., The interferon-induced protein kinase (PKR) triggers apoptosis through FADD-mediated activation of caspase 8 in a manner independent of Fas and TNF-alpha receptors, *Oncogene,* 19, 3665, 2000.
33. Cryns, V. L. et al., Specific cleavage of alpha-fodrin during Fas- and tumor necrosis factor-induced apoptosis is mediated by an interleukin-1beta-converting enzyme/Ced-3 protease distinct from the poly(ADP-ribose) polymerase protease, *J. Biol. Chem.,* 271, 31277, 1996.
34. Lazebnik, Y. A. et al., Cleavage of poly(ADP-ribose) polymerase by a proteinase with properties like ICE, *Nature,* 371, 346, 1994.
35. Lazebnik, Y. A. et al., Studies of the lamin proteinase reveal multiple parallel biochemical pathways during apoptotic execution, *Proc. Natl. Acad. Sci. U.S.A.,* 92, 9042, 1995.
36. Atencia, R., Asumendi, A., and Garcia-Sanz, M., Role of cytoskeleton in apoptosis, *Vitam. Horm.,* 58, 267, 2000.
37. Brancolini, C., Benedetti, M., and Schneider, C., Microfilament reorganization during apoptosis: the role of Gas2, a possible substrate for ICE-like proteases, *EMBO J.,* 14, 5179, 1995.
38. Franko, J., Pomfy, M., and Prosbova, T., Apoptosis and cell death (mechanisms, pharmacology, and promise for the future), *Acta Medica. (Hradec Kralove),* 43, 63, 2000.
39. Silverman, R. H.; *Ribonucleases: Structures and Functions,* Academic Press, Inc., New York; 1997; chap. 16.
40. Watling, D. et al., Analogue inhibitor of 2-5A action: effect on the interferon-mediated inhibition of encephalomyocarditis virus replication, *EMBO J.,* 4, 431, 1985.
41. Hassel, B. A. et al., A dominant negative mutant of 2-5A-dependent RNase suppresses antiproliferative and antiviral effects of interferon, *EMBO J.,* 12, 3297, 1993.
42. Diaz-Guerra, M., Rivas, C., and Esteban, M., Inducible expression of the 2-5A synthetase/RNase L system results in inhibition of vaccinia virus replication, *Virology,* 227, 220, 1997.

43. Silverman, R. H. et al., Control of the ppp(a2'p)nA system in HeLa cells. Effects of interferon and virus infection, *Eur. J. Biochem.,* 124, 131, 1982.
44. Silverman, R. H. et al., rRNA cleavage as an index of ppp(A2'p)nA activity in interferon-treated encephalomyocarditis virus-infected cells, *J. Virol.,* 46, 1051, 1983.
45. Nilsen, T. W., Maroney, P. A., and Baglioni, C., Synthesis of (2'-5')oligoadenylate and activation of an endoribonuclease in interferon-treated HeLa cells infected with reovirus, *J. Virol.,* 42, 1039, 1982.
46. Goswami, B. B. and Sharma, O. K., Degradation of rRNA in interferon-treated vaccinia virus-infected cells, *J. Biol. Chem.,* 259, 1371, 1984.
47. Houge, G. et al., Fine mapping of 28S rRNA sites specifically cleaved in cells undergoing apoptosis, *Mol. Cell Biol.,* 15, 2051, 1995.
48. Delic, J., Coppey-Moisan, M., and Magdelenat, H., Gamma-ray-induced transcription and apoptosis-associated loss of 28S rRNA in interphase human lymphocytes, *Int. J. Radiat. Biol.,* 64, 39, 1993.
49. Chapekar, M. S. and Glazer, R. I., The synergistic cytocidal effect produced by immune interferon and tumor necrosis factor in HT-29 cells is associated with inhibition of rRNA processing and (2',5') oligo (A) activation of RNase L, *Biochem. Biophys. Res. Commun.,* 151, 1180, 1988.
50. Diaz-Guerra, M., Rivas, C., and Esteban, M., Activation of the IFN-inducible enzyme RNase L causes apoptosis of animal cells, *Virology,* 236, 354, 1997.
51. Castelli, J. C. et al., A study of the interferon antiviral mechanism: apoptosis activation by the 2-5A system, *J. Exp. Med.,* 186, 967, 1997.
52. Zhou, A. et al., Impact of RNase L overexpression on viral and cellular growth and death, *J. Interferon Cytokine Res.,* 18, 953, 1998.
53. Rusch, L., Zhou, A., and Silverman, R. H., Caspase-dependent apoptosis by 2',5'-oligoadenylate activation of RNase L is enhanced by IFN-beta, *J. Interferon Cytokine Res.,* 20, 1091, 2000.
54. Zhou, A. et al., Interferon action and apoptosis are defective in mice devoid of 2',5'-oligoadenylate-dependent RNase L, *EMBO J.,* 16, 6355, 1997.
55. Castelli, J. C. et al., The role of 2'-5' oligoadenylate-activated ribonuclease L in apoptosis, *Cell Death. Differ.,* 5, 313, 1998.
56. Mitra, A. et al., A mammalian 2-5A system functions as an antiviral pathway in transgenic plants, *Proc. Natl. Acad. Sci. U.S.A.,* 93, 6780, 1996.
57. Alnemri, E. S. et al., Human ICE/CED-3 protease nomenclature, *Cell,* 87, 171, 1996.
58. Wolf, B. B. and Green, D. R., Suicidal tendencies: apoptotic cell death by caspase family proteinases, *J. Biol. Chem.,* 274, 20049, 1999.
59. Earnshaw, W. C., Martins, L. M., and Kaufmann, S. H., Mammalian caspases: structure, activation, substrates, and functions during apoptosis, *Annu. Rev. Biochem.,* 68, 383, 1999.
60. Muzio, M. et al., An induced proximity model for caspase-8 activation, *J. Biol. Chem.,* 273, 2926, 1998.
61. Zou, H. et al., An APAF-1.cytochrome c multimeric complex is a functional apoptosome that activates procaspase-9, *J. Biol. Chem.,* 274, 11549, 1999.
62. Kang, J. J. et al., Cascades of mammalian caspase activation in the yeast *Saccharomyces cerevisiae, J. Biol. Chem.,* 274, 3189, 1999.
63. Bitko, V. and Barik, S., An endoplasmic reticulum-specific stress-activated caspase (caspase-12) is implicated in the apoptosis of A549 epithelial cells by respiratory syncytial virus, *J. Cell Biochem.,* 80, 441, 2001.
64. Daniel, P. T., Dissecting the pathways to death, *Leukemia,* 14, 2035, 2000.

65. Zheng, T. S. et al., Deficiency in caspase-9 or caspase-3 induces compensatory caspase activation, *Nat. Med.,* 6, 1241, 2000.
66. Grell, M., Krammer, P. H., and Scheurich, P., Segregation of APO-1/Fas antigen- and tumor necrosis factor receptor-mediated apoptosis, *Eur. J. Immunol.,* 24, 2563, 1994.
67. Boldin, M. P. et al., A novel protein that interacts with the death domain of Fas/APO1 contains a sequence motif related to the death domain, *J. Biol. Chem.,* 270, 7795, 1995.
68. Siegel, R. M. et al., Death-effector filaments: novel cytoplasmic structures that recruit caspases and trigger apoptosis, *J. Cell Biol.,* 141, 1243, 1998.
69. Eilers, A. et al., c-Jun and Bax: regulators of programmed cell death in developing neurons, *Biochem Soc. Trans.,* 27, 790, 1999.
70. Desagher, S. et al., Bid-induced conformational change of Bax is responsible for mitochondrial cytochrome c release during apoptosis, *J. Cell Biol.,* 144, 891, 1999.
71. Gogvadze, V. et al., Cytochrome c release occurs Via Ca^{2+}-dependent and Ca^{2+}-independent mechanisms that are regulated by Bax, *J. Biol. Chem.,* 276, 19066, 2001.
72. Woo, M. et al., Essential contribution of caspase 3/CPP32 to apoptosis and its associated nuclear changes, *Genes Dev.,* 12, 806, 1998.
73. Ghibelli, L. et al., Glutathione depletion causes cytochrome c release even in the absence of cell commitment to apoptosis, *FASEB J.,* 13, 2031, 1999.
74. Debatin, K. M. et al., Regulation of apoptosis through CD95 (APO-I/Fas) receptor-ligand interaction, *Biochem. Soc. Trans.,* 25, 405, 1997.
75. Peter, M. E. and Krammer, P. H., Mechanisms of CD95 (APO-1/Fas)-mediated apoptosis, *Curr. Opin. Immunol.,* 10, 545, 1998.
76. Martin, D. A. et al., Membrane oligomerization and cleavage activates the caspase-8 (FLICE/MACHalpha1) death signal, *J. Biol. Chem.,* 273, 4345, 1998.
77. Scaffidi, C. et al., FLICE is predominantly expressed as two functionally active isoforms, caspase-8/a and caspase-8/b, *J. Biol. Chem.,* 272, 26953, 1997.
78. Schmidt, M. et al., IL-10 induces apoptosis in human monocytes involving the CD95 receptor/ligand pathway, *Eur. J. Immunol.,* 30, 1769, 2000.
79. Stennicke, H. R. et al., Pro-caspase-3 is a major physiologic target of caspase-8, *J. Biol. Chem.,* 273, 27084, 1998.
80. Harvey, N. L., Butt, A. J., and Kumar, S., Functional activation of Nedd2/ICH-1 (caspase-2) is an early process in apoptosis, *J. Biol. Chem.,* 272, 13134, 1997.
81. Droin, N. et al., Involvement of caspase-2 long isoform in Fas-mediated cell death of human leukemic cells, *Blood,* 97, 1835, 2001.
82. Moss, R. B., Mercandetti, A., and Vojdani, A., TNF-alpha and chronic fatigue syndrome, *J. Clin. Immunol.,* 19, 314, 1999.
83. Dreisbach, A. W. et al., Elevated levels of tumor necrosis factor alpha in postdialysis fatigue, *Int. J. Artif. Organs,* 21, 83, 1998.
84. Lloyd, A. et al., Cell-mediated immunity in patients with chronic fatigue syndrome, healthy control subjects, and patients with major depression, *Clin. Exp. Immunol.,* 87, 76, 1992.
85. Rasmussen, A. K. et al., Chronic fatigue syndrome: a controlled cross sectional study, *J. Rheumatol.,* 21, 1527, 1994.
86. Cohen, G. M., Caspases: the executioners of apoptosis, *Biochem J.,* 326 (Pt 1), 1, 1997.
87. Schild, L. et al., Distinct Ca^{2+} thresholds determine cytochrome c release or permeability transition pore opening in brain mitochondria, *FASEB J.,* 15, 565, 2001.

88. Kharbanda, S. et al., Role for Bcl-xL as an inhibitor of cytosolic cytochrome c accumulation in DNA damage-induced apoptosis, *Proc. Natl. Acad. Sci. U.S.A.*, 94, 6939, 1997.
89. Chinnaiyan, A. M. et al., Molecular ordering of the cell death pathway. Bcl-2 and Bcl-xL function upstream of the CED-3-like apoptotic proteases, *J. Biol. Chem.*, 271, 4573, 1996.
90. Pan, G., O'Rourke, K., and Dixit, V. M., Caspase-9, Bcl-XL, and Apaf-1 form a ternary complex, *J. Biol. Chem.*, 273, 5841, 1998.
91. Jiang, X. and Wang, X., Cytochrome c promotes caspase-9 activation by inducing nucleotide binding to Apaf-1, *J. Biol. Chem.*, 275, 31199, 2000.
92. Saleh, A. et al., Cytochrome c and dATP-mediated oligomerization of Apaf-1 is a prerequisite for procaspase-9 activation, *J. Biol. Chem.*, 274, 17941, 1999.
93. Chua, B. T., Guo, K., and Li, P., Direct cleavage by the calcium-activated protease calpain can lead to inactivation of caspases, *J. Biol. Chem.*, 275, 5131, 2000.
94. Yoneda, T. et al., Activation of caspase-12, an endoplastic reticulum (ER) resident caspase, through tumor necrosis factor receptor-associated factor 2-dependent mechanism in response to the ER stress, *J. Biol. Chem.*, 276, 13935, 2001.
95. Nakagawa, T. and Yuan, J., Cross-talk between two cysteine protease families. Activation of caspase-12 by calpain in apoptosis, *J. Cell Biol.*, 150, 887, 2000.
96. Slee, E. A., Adrain, C., and Martin, S. J., Executioner caspase-3, -6, and -7 perform distinct, non-redundant roles during the demolition phase of apoptosis, *J. Biol. Chem.*, 276, 7320, 2001.
97. Blomgren, K. et al., Synergistic activation of caspase-3 by m-calpain after neonatal hypoxia- ischemia: a mechanism of "pathological apoptosis"? *J. Biol. Chem.*, 276, 10191, 2001.
98. Kaufmann, S. H. et al., Specific proteolytic cleavage of poly(ADP-ribose) polymerase: an early marker of chemotherapy-induced apoptosis, *Cancer Res.*, 53, 3976, 1993.
99. Nicholson, D. W. et al., Identification and inhibition of the ICE/CED-3 protease necessary for mammalian apoptosis, *Nature*, 376, 37, 1995.
100. Tewari, M. et al., Yama/CPP32 beta, a mammalian homolog of CED-3, is a CrmA-inhibitable protease that cleaves the death substrate poly(ADP-ribose) polymerase, *Cell*, 81, 801, 1995.
101. Mashima, T. et al., Actin cleavage by CPP-32/apopain during the development of apoptosis, *Oncogene*, 14, 1007, 1997.
102. Mashima, T., Naito, M., and Tsuruo, T., Caspase-mediated cleavage of cytoskeletal actin plays a positive role in the process of morphological apoptosis, *Oncogene*, 18, 2423, 1999.
103. Oliver, F. J. et al., Importance of poly(ADP-ribose) polymerase and its cleavage in apoptosis. Lesson from an uncleavable mutant, *J. Biol. Chem.*, 273, 33533, 1998.
104. Peitsch, M. C. et al., Characterization of the endogenous deoxyribonuclease involved in nuclear DNA degradation during apoptosis (programmed cell death), *EMBO J.*, 12, 371, 1993.
105. Suzuki, K. and Sorimachi, H., A novel aspect of calpain activation, *FEBS Lett.*, 433, 1, 1998.
106. Wang, K. K., Calpain and caspase: can you tell the difference? *Trends Neurosci.*, 23, 59, 2000.
107. Ohno, S. et al., Four genes for the calpain family locate on four distinct human chromosomes, *Cytogenet. Cell Genet.*, 53, 225, 1990.

108. Ono, Y., Sorimachi, H., and Suzuki, K., Structure and physiology of calpain, an enigmatic protease, *Biochem. Biophys. Res. Commun.*, 245, 289, 1998.
109. Takano, E. et al., Distribution of calpains and calpastatin in human blood cells, *Biochem. Int.*, 16, 391, 1988.
110. Ma, H. et al., Amino-terminal conserved region in proteinase inhibitor domain of calpastatin potentiates its calpain inhibitory activity by interacting with calmodulin-like domain of the proteinase, *J. Biol. Chem.*, 269, 24430, 1994.
111. Elce, J. S., Hegadorn, C., and Arthur, J. S., Autolysis, Ca2+ requirement, and heterodimer stability in m-calpain, *J. Biol. Chem.*, 272, 11268, 1997.
112. Carafoli, E. and Molinari, M., Calpain: a protease in search of a function? *Biochem. Biophys. Res. Commun.*, 247, 193, 1998.
113. Hosfield, C. M. et al., Crystal structure of calpain reveals the structural basis for Ca(2+)-dependent protease activity and a novel mode of enzyme activation, *EMBO J.*, 18, 6880, 1999.
114. Melloni, E. et al., Acyl-CoA-binding protein is a potent m-calpain activator, *J. Biol. Chem.*, 275, 82, 2000.
115. Saido, T. C. et al., Positive regulation of mu-calpain action by polyphosphoinositides, *J. Biol. Chem.*, 267, 24585, 1992.
116. Saido, T. C., Sorimachi, H., and Suzuki, K., Calpain: new perspectives in molecular diversity and physiological- pathological involvement, *FASEB J.*, 8, 814, 1994.
117. Schoenwaelder, S. M. et al., Distinct substrate specificities and functional roles for the 78- and 76-kDa forms of mu-calpain in human platelets, *J. Biol. Chem.*, 272, 24876, 1997.
118. Zhang, W., Lane, R. D., and Mellgren, R. L., The major calpain isozymes are long-lived proteins. Design of an antisense strategy for calpain depletion in cultured cells, *J. Biol. Chem.*, 271, 18825, 1996.
119. Vilei, E. M. et al., Functional properties of recombinant calpain I and of mutants lacking domains III and IV of the catalytic subunit, *J. Biol. Chem.*, 272, 25802, 1997.
120. Blomgren, K. et al., Calpastatin is up-regulated in response to hypoxia and is a suicide substrate to calpain after neonatal cerebral hypoxia-ischemia, *J. Biol. Chem.*, 274, 14046, 1999.
121. Shields, D. C. et al., A putative mechanism of demyelination in multiple sclerosis by a proteolytic enzyme, calpain, *Proc. Natl. Acad. Sci. U.S.A.*, 96, 11486, 1999.
122. Sorimachi, H., Ishiura, S., and Suzuki, K., Structure and physiological function of calpains, *Biochem. J.*, 328 (Pt 3), 721, 1997.
123. Squier, M. K. et al., Calpain and calpastatin regulate neutrophil apoptosis, *J. Cell Physiol.*, 178, 311, 1999.
124. Vanags, D. M. et al., Protease involvement in fodrin cleavage and phosphatidylserine exposure in apoptosis, *J. Biol. Chem.*, 271, 31075, 1996.
125. Aderem, A. A., How cytokines signal messages within cells, *J. Infect. Dis.*, 167, Suppl. 1, S2, 1993.
126. Satterwhite, L. L. and Pollard, T. D., Cytokinesis, *Curr. Opin. Cell Biol.*, 4, 43, 1992.
127. Heath, J. P. and Holifield, B. F., Cell locomotion: new research tests old ideas on membrane and cytoskeletal flow, *Cell Motil. Cytoskeleton*, 18, 245, 1991.
128. Mitchison, T. and Kirschner, M., Cytoskeletal dynamics and nerve growth, *Neuron*, 1, 761, 1988.
129. Dramsi, S. and Cossart, P., Intracellular pathogens and the actin cytoskeleton, *Annu. Rev. Cell Dev. Biol.*, 14, 137, 1998.
130. Okabe, S. and Hirokawa, N., Incorporation and turnover of biotin-labeled actin microinjected into fibroblastic cells: an immunoelectron microscopic study, *J. Cell Biol.*, 109, 1581, 1989.

131. Symons, M. H. and Mitchison, T. J., Control of actin polymerization in live and permeabilized fibroblasts, *J. Cell Biol.,* 114, 503, 1991.
132. Cao, L. G., Fishkind, D. J., and Wang, Y. L., Localization and dynamics of nonfilamentous actin in cultured cells, *J. Cell Biol.,* 123, 173, 1993.
133. Finlay, B. B. and Cossart, P., Exploitation of mammalian host cell functions by bacterial pathogens, *Science,* 276, 718, 1997.
134. Ireton, K. and Cossart, P., Interaction of invasive bacteria with host signaling pathways, *Curr. Opin. Cell Biol.,* 10, 276, 1998.
135. Pollard, T. D. and Cooper, J. A., Actin and actin-binding proteins. A critical evaluation of mechanisms and functions, *Annu. Rev. Biochem.,* 55, 987, 1986.
136. Theriot, J. A. and Mitchison, T. J., The three faces of profilin, *Cell,* 75, 835, 1993.
137. Safer, D., The interaction of actin with thymosin-beta 4, *J. Muscle Res. Cell Motil.,* 13, 269, 1992.
138. Matsudaira, P. and Janmey, P., Pieces in the actin-severing protein puzzle, *Cell,* 54, 139, 1988.
139. Hall, A., Small GTP-binding proteins and the regulation of the actin cytoskeleton, *Annu. Rev. Cell Biol.,* 10, 31, 1994.
140. Falkow, S., Isberg, R. R., and Portnoy, D. A., The interaction of bacteria with mammalian cells, *Annu. Rev. Cell Biol.,* 8, 333, 1992.
141. Shasby, D. M. et al., Role of endothelial cell cytoskeleton in control of endothelial permeability, *Circ. Res.,* 51, 657, 1982.
142. Alexander, J. S., Hechtman, H. B., and Shepro, D., Phalloidin enhances endothelial barrier function and reduces inflammatory permeability *in vitro, Microvasc. Res.,* 35, 308, 1988.
143. Suck, D., Lahm, A., and Oefner, C., Structure refined to 2Å of a nicked DNA octanucleotide complex with DNase I, *Nature,* 332, 464, 1988.
144. Kabsch, W. et al., Atomic structure of the actin:DNase I complex, *Nature,* 347, 37, 1990.
145. Yin, H. L., Gelsolin: calcium- and polyphosphoinositide-regulated actin-modulating protein, *Bioessays,* 7, 176, 1987.
146. Weeds, A. and Maciver, S., F-actin capping proteins, *Curr. Opin. Cell Biol.,* 5, 63, 1993.
147. Robinson, R. C. et al., Domain movement in gelsolin: a calcium-activated switch, *Science,* 286, 1939, 1999.
148. Kothakota, S. et al., Caspase-3-generated fragment of gelsolin: effector of morphological change in apoptosis, *Science,* 278, 294, 1997.
149. Koya, R. C. et al., Gelsolin inhibits apoptosis by blocking mitochondrial membrane potential loss and cytochrome c release, *J. Biol. Chem.,* 275, 15343, 2000.
150. Daiger, S. P., Schanfield, M. S., and Cavalli-Sforza, L. L., Group-specific component (Gc) proteins bind vitamin D and 25- hydroxyvitamin D, *Proc. Natl. Acad. Sci. U.S.A.,* 72, 2076, 1975.
151. Lee, W. M. et al., Diminished serum Gc (vitamin D-binding protein) levels and increased Gc:G-actin complexes in a hamster model of fulminant hepatic necrosis, *Hepatology,* 7, 825, 1987.
152. Goldschmidt-Clermont, P. J., Williams, M. H., and Galbraith, R. M., Altered conformation of Gc (vitamin D-binding protein) upon complexing with cellular actin, *Biochem. Biophys. Res. Commun.,* 146, 611, 1987.
153. Bouillon, R., Van Baelen, H., and De Moor, P., Physiology and pathophysiology of vitamin D-binding protein, *Colloquium Inserm,* 149, 333, 1986.
154. Petrini, M. et al., Gc (vitamin D-binding protein) binds to cytoplasm of all human lymphocytes and is expressed on B-cell membranes, *Clin. Immunol. Immunopathol.,* 31, 282, 1984.

155. Van Baelen, H., Bouillon, R., and De Moor, P., Vitamin D-binding protein (Gc-globulin) binds actin, *J. Biol. Chem.,* 255, 2270, 1980.
156. Haddad, J. G., Human serum binding protein for vitamin D and its metabolites (DBP): evidence that actin is the DBP binding component in human skeletal muscle, *Arch. Biochem. Biophys.,* 213, 538, 1982.
157. Haddad, J. G. et al., Angiopathic consequences of saturating the plasma scavenger system for actin, *Proc. Natl. Acad. Sci. U.S.A.,* 87, 1381, 1990.
158. Brancolini, C. et al., Dismantling cell–cell contacts during apoptosis is coupled to a caspase-dependent proteolytic cleavage of beta-catenin, *J. Cell Biol.,* 139, 759, 1997.
159. Kayalar, C. et al., Cleavage of actin by interleukin 1 beta-converting enzyme to reverse DNase I inhibition, *Proc. Natl. Acad. Sci. U.S.A.,* 93, 2234, 1996.
160. Martin, S. J. et al., Proteolysis of fodrin (nonerythroid spectrin) during apoptosis, *J. Biol. Chem.,* 270, 6425, 1995.
161. Mashima, T. et al., Aspartate-based inhibitor of interleukin-1 beta-converting enzyme prevents antitumor agent-induced apoptosis in human myeloid leukemia U937 cells, *Biochem. Biophys. Res. Commun.,* 209, 907, 1995.
162. Rudel, T. and Bokoch, G. M., Membrane and morphological changes in apoptotic cells regulated by caspase-mediated activation of PAK2, *Science,* 276, 1571, 1997.
163. Villa, P. G. et al., Calpain inhibitors, but not caspase inhibitors, prevent actin proteolysis and DNA fragmentation during apoptosis, *J. Cell Sci.,* 111, 713, 1998.
164. Croall, D. E. and DeMartino, G. N., Calcium-activated neutral protease (calpain) system: structure, function, and regulation, *Physiol Rev.,* 71, 813, 1991.
165. Roelens, S. et al., G-actin cleavage parallels 2-5A-dependent RNase L cleavage in peripheral blood mononuclear cells. Relevance to a possible serum-based screening test for dysregulations in the 2-5A pathway, *J. Chronic Fatigue Syndrome,* 8, 63, 2001.
166. Maruyama, W., Irie, S., and Sato, T. A., Morphological changes in the nucleus and actin cytoskeleton in the process of Fas-induced apoptosis in Jurkat T cells, *Histochem. J.,* 32, 495, 2000.
167. Spano, A. et al., Relationship between actin microfilaments and plasma membrane changes during apoptosis of neoplastic cell lines in different culture conditions, *Eur. J. Histochem.,* 44, 255, 2000.
168. Vukosavic, S. et al., Delaying caspase activation by Bcl-2: a clue to disease retardation in a transgenic mouse model of amyotrophic lateral sclerosis, *J. Neurosci.,* 20, 9119, 2000.
169. Yamazaki, Y. et al., Acceleration of DNA damage-induced apoptosis in leukemia cells by interfering with actin system, *Exp. Hematol.,* 28, 1491, 2000.
170. Iezzi, G., Karjalainen, K., and Lanzavecchia, A., The duration of antigenic stimulation determines the fate of naive and effector T cells, *Immunity,* 8, 89, 1998.
171. Wulfing, C. and Davis, M. M., A receptor/cytoskeletal movement triggered by costimulation during T cell activation, *Science,* 282, 2266, 1998.
172. Aderem, A. and Underhill, D. M., Mechanisms of phagocytosis in macrophages, *Annu. Rev. Immunol.,* 17, 593, 1999.
173. Ridley, A. J. and Hall, A., The small GTP-binding protein rho regulates the assembly of focal adhesions and actin stress fibers in response to growth factors, *Cell,* 70, 389, 1992.
174. Nobes, C. D. and Hall, A., Rho, rac, and cdc42 GTPases regulate the assembly of multimolecular focal complexes associated with actin stress fibers, lamellipodia, and filopodia, *Cell,* 81, 53, 1995.
175. Diakonova, M. et al., Localization of five annexins in J774 macrophages and on isolated phagosomes, *J. Cell Sci.,* 110, 1199, 1997.

176. Tanaka, H. and Samuel, C. E., Mechanism of interferon action: structure of the mouse PKR gene encoding the interferon-inducible RNA-dependent protein kinase, *Proc. Natl. Acad. Sci. U.S.A.,* 91, 7995, 1994.
177. Galabru, J. et al., The binding of double-stranded RNA and adenovirus VAI RNA to the interferon-induced protein kinase, *Eur. J. Biochem.,* 178, 581, 1989.
178. Roy, S. et al., The integrity of the stem structure of human immunodeficiency virus type 1 Tat-responsive sequence of RNA is required for interaction with the interferon-induced 68,000-Mr protein kinase, *J. Virol.,* 65, 632, 1991.
179. Wu, S. and Kaufman, R. J., A model for the double-stranded RNA (dsRNA)-dependent dimerization and activation of the dsRNA-activated protein kinase PKR, *J. Biol. Chem.,* 272, 1291, 1997.
180. Patel, R. C. and Sen, G. C., Requirement of PKR dimerization mediated by specific hydrophobic residues for its activation by double-stranded RNA and antigrowth effects in yeast, *Mol. Cell Biol.,* 18, 7009, 1998.
181. Lee, S. B. et al., The interferon-induced double-stranded RNA-activated human p68 protein kinase potently inhibits protein synthesis in cultured cells, *Virology,* 192, 380, 1993.
182. Ramaiah, K. V. et al., Expression of mutant eukaryotic initiation factor 2 alpha subunit (eIF-2 alpha) reduces inhibition of guanine nucleotide exchange activity of eIF-2B mediated by eIF-2 alpha phosphorylation, *Mol. Cell Biol.,* 14, 4546, 1994.
183. Williams, B. R. G., Role of the double-stranded RNA-activated protein kinase (PKR) in cell regulation, *Biochem. Soc. Trans.,* 25, 509, 1997.
184. Chong, K. L. et al., Human p68 kinase exhibits growth suppression in yeast and homology to the translational regulator GCN2, *EMBO J.,* 11, 1553, 1992.
185. Koromilas, A. E. et al., Malignant transformation by a mutant of the IFN-inducible dsRNA-dependent protein kinase, *Science,* 257, 1685, 1992.
186. Meurs, E. F. et al., Tumor suppressor function of the interferon-induced double-stranded RNA- activated protein kinase, *Proc. Natl. Acad. Sci. U.S.A.,* 90, 232, 1993.
187. Donze, O. et al., Abrogation of translation initiation factor eIF-2 phosphorylation causes malignant transformation of NIH 3T3 cells, *EMBO J.,* 14, 3828, 1995.
188. Kirchhoff, S. et al., IRF-1 induced cell growth inhibition and interferon induction requires the activity of the protein kinase PKR, *Oncogene,* 11, 439, 1995.
189. Raveh, T. et al., Double-stranded RNA-dependent protein kinase mediates c-Myc suppression induced by type I interferons, *J. Biol. Chem.,* 271, 25479, 1996.
190. Lee, S. B. and Esteban, M., The interferon-induced double-stranded RNA-activated protein kinase induces apoptosis, *Virology,* 199, 491, 1994.
191. Yeung, M. C., Liu, J., and Lau, A. S., An essential role for the interferon-inducible, double-stranded RNA-activated protein kinase PKR in the tumor necrosis factor-induced apoptosis in U937 cells, *Proc. Natl. Acad. Sci. U.S.A.,* 93, 12451, 1996.
192. Srivastava, S. P., Kumar, K. U., and Kaufman, R. J., Phosphorylation of eukaryotic translation initiation factor 2 mediates apoptosis in response to activation of the double-stranded RNA-dependent protein kinase, *J. Biol. Chem.,* 273, 2416, 1998.
193. Takizawa, T., Tatematsu, C., and Nakanishi, Y., Double-stranded RNA-activated protein kinase (PKR) fused to green fluorescent protein induces apoptosis of human embryonic kidney cells: possible role in the Fas signaling pathway, *J. Biochem. (Tokyo),* 125, 391, 1999.
194. Balachandran, S. et al., Activation of the dsRNA-dependent protein kinase, PKR, induces apoptosis through FADD-mediated death signaling, *EMBO J.,* 17, 6888, 1998.
195. Balachandran, S. et al., Alpha/beta interferons potentiate virus-induced apoptosis through activation of the FADD/Caspase-8 death signaling pathway, *J. Virol.,* 74, 1513, 2000.

196. Takizawa, T., Ohashi, K., and Nakanishi, Y., Possible involvement of double-stranded RNA-activated protein kinase in cell death by influenza virus infection, *J. Virol.,* 70, 8128, 1996.
197. Der, S. D. et al., A double-stranded RNA-activated protein kinase-dependent pathway mediating stress-induced apoptosis, *Proc. Natl. Acad. Sci. U.S.A.,* 94, 3279, 1997.
198. Vojdani, A. et al., Elevated apoptotic cell population in patients with chronic fatigue syndrome: the pivotal role of protein kinase RNA, *J. Intern. Med.,* 242, 465, 1997.
199. Black, T. L. et al., The cellular 68,000-Mr protein kinase is highly autophosphorylated and activated yet significantly degraded during poliovirus infection: implications for translational regulation, *J. Virol.,* 63, 2244, 1989.
200. Gil, J., Alcami, J., and Esteban, M., Induction of apoptosis by double-stranded-RNA-dependent protein kinase (PKR) involves the alpha subunit of eukaryotic translation initiation factor 2 and NF-kappaB, *Mol. Cell Biol.,* 19, 4653, 1999.
201. Lee, S. B. et al., The apoptosis pathway triggered by the interferon-induced protein kinase PKR requires the third basic domain, initiates upstream of Bcl-2, and involves ICE-like proteases, *Virology,* 231, 81, 1997.
202. Marissen, W. E. et al., Identification of caspase 3-mediated cleavage and functional alteration of eukaryotic initiation factor 2alpha in apoptosis, *J. Biol. Chem.,* 275, 9314, 2000.
203. Satoh, S. et al., Caspase-mediated cleavage of eukaryotic translation initiation factor subunit 2alpha, *Biochem. J.,* 342 (Pt 1), 65, 1999.
204. Donze, O., Dostie, J., and Sonenberg, N., Regulatable expression of the interferon-induced double-stranded RNA dependent protein kinase PKR induces apoptosis and Fas receptor expression, *Virology,* 256, 322, 1999.
205. de Martin, R., Schmid, J. A., and Hofer-Warbinek, R., The NF-kappaB/Rel family of transcription factors in oncogenic transformation and apoptosis, *Mutat. Res.,* 437, 231, 1999.
206. Kumar, A. et al., Double-stranded RNA-dependent protein kinase activates transcription factor NF-kappa B by phosphorylating I kappa B, *Proc. Natl. Acad. Sci. U.S.A.,* 91, 6288, 1994.
207. Zamanian-Daryoush, M. et al., NF-kappaB activation by double-stranded-RNA-activated protein kinase (PKR) is mediated through NF-kappaB-inducing kinase and IkappaB kinase, *Mol. Cell Biol.,* 20, 1278, 2000.
208. Bonnet, M. C. et al., PKR stimulates NF-kappaB irrespective of its kinase function by interacting with the I-kappaB kinase complex, *Mol. Cell Biol.,* 20, 4532, 2000.
209. Beg, A. A. and Baltimore, D., An essential role for NF-kappaB in preventing TNF-alpha-induced cell death, *Science,* 274, 782, 1996.
210. Yang, C. H. et al., IFNalpha/beta promotes cell survival by activating NF-kappa B, *Proc. Natl. Acad. Sci. U.S.A.,* 97, 13631, 2000.
211. Li, M. et al., The Rela(p65) subunit of NF-kappaB is essential for inhibiting double-stranded RNA-induced cytotoxicity, *J. Biol. Chem.,* 276, 1185, 2001.
212. Abbadie, C. et al., High levels of c-rel expression are associated with programmed cell death in the developing avian embryo and in bone marrow cells *in vitro, Cell,* 75, 899, 1993.
213. Lin, K. I. et al., Thiol agents and Bcl-2 identify an alphavirus-induced apoptotic pathway that requires activation of the transcription factor NF-kappa B, *J. Cell Biol.,* 131, 1149, 1995.
214. Kasibhatla, S. et al., DNA damaging agents induce expression of Fas ligand and subsequent apoptosis in T lymphocytes via the activation of NF-kappa B and AP-1, *Mol. Cell,* 1, 543, 1998.

215. Miyashita, T. et al., Identification of a p53-dependent negative response element in the bcl- 2 gene, *Cancer Res.,* 54, 3131, 1994.
216. Sionov, R. V. and Haupt, Y., The cellular response to p53: the decision between life and death, *Oncogene,* 18, 6145, 1999.
217. Vogelstein, B., Lane, D., and Levine, A. J., Surfing the p53 network, *Nature,* 408, 307, 2000.
218. Ding, H. F. et al., Essential role for caspase-8 in transcription-independent apoptosis triggered by p53, *J. Biol. Chem.,* 275, 38905, 2000.
219. Yeung, M. C. and Lau, A. S., Tumor suppressor p53 as a component of the tumor necrosis factor-induced, protein kinase PKR-mediated apoptotic pathway in human promonocytic U937 cells, *J. Biol. Chem.,* 273, 25198, 1998.
220. Kubbutat, M. H., Jones, S. N., and Vousden, K. H., Regulation of p53 stability by Mdm2, *Nature,* 387, 299, 1997.
221. Zhang, W. et al., Inhibition of the growth of WI-38 fibroblasts by benzyloxycarbonyl-Leu-Leu-Tyr diazomethyl ketone: evidence that cleavage of p53 by a calpain-like protease is necessary for G1 to S-phase transition, *Oncogene,* 14, 255, 1997.
222. Kubbutat, M. H. and Vousden, K. H., Proteolytic cleavage of human p53 by calpain: a potential regulator of protein stability, *Mol. Cell Biol.,* 17, 460, 1997.
223. Atencio, I. A. et al., Calpain inhibitor 1 activates p53-dependent apoptosis in tumor cell lines, *Cell Growth Differ.,* 11, 247, 2000.
224. Piechaczyk, M., Proteolysis of p53 protein by ubiquitous calpains, *Methods Mol. Biol.,* 144, 297, 2000.
225. Harada, H., Taniguchi, T., and Tanaka, N., The role of interferon regulatory factors in the interferon system and cell growth control, *Biochimie,* 80, 641, 1998.
226. Chapman, R. S. et al., A novel role for IRF-1 as a suppressor of apoptosis, *Oncogene,* 19, 6386, 2000.
227. Tamura, T. et al., Interferon-gamma induces ICE gene expression and enhances cellular susceptibility to apoptosis in the U937 leukemia cell line, *Biochem. Biophys. Res. Commun.,* 229, 21, 1996.
228. Tamura, T. et al., DNA damage-induced apoptosis and Ice gene induction in mitogenically activated T lymphocytes require IRF-1, *Leukemia,* 11 Suppl 3, 439, 1997.
229. Nguyen, H., Lin, R., and Hiscott, J., Activation of multiple growth regulatory genes following inducible expression of IRF-1 or IRF/RelA fusion proteins, *Oncogene,* 15, 1425, 1997.
230. Horiuchi, M. et al., Interferon regulatory factors regulate interleukin-1beta-converting enzyme expression and apoptosis in vascular smooth muscle cells, *Hypertension,* 33, 162, 1999.
231. Kano, A. et al., IRF-1 is an essential mediator in IFN-gamma-induced cell cycle arrest and apoptosis of primary cultured hepatocytes, *Biochem. Biophys. Res. Commun.,* 257, 672, 1999.
232. Sanceau, J. et al., IFN-beta induces serine phosphorylation of Stat-1 in Ewing's sarcoma cells and mediates apoptosis via induction of IRF-1 and activation of caspase-7, *Oncogene,* 19, 3372, 2000.
233. Zamanian-Daryoush, M., Der, S. D., and Williams, B. R., Cell cycle regulation of the double stranded RNA activated protein kinase, PKR, *Oncogene,* 18, 315, 1999.
234. Chow, W. A., Fang, J. J., and Yee, J. K., The IFN regulatory factor family participates in regulation of Fas ligand gene expression in T cells, *J. Immunol.,* 164, 3512, 2000.
235. Rager, K. J. et al., Activation of antiviral protein kinase leads to immunoglobulin E class switching in human B cells, *J. Virol.,* 72, 1171, 1998.

236. Komaroff, A. L. and Buchwald, D. S., Chronic fatigue syndrome: an update, *Annu. Rev. Med.,* 49, 1, 1998.
237. Everett, H. and McFadden, G., Apoptosis: an innate immune response to virus infection, *Trends Microbiol.,* 7, 160, 1999.
238. Kalvakolanu, D. V., Virus interception of cytokine-regulated pathways, *Trends Microbiol.,* 7, 166, 1999.
239. Gao, L. Y. and Kwaik, Y. A., The modulation of host cell apoptosis by intracellular bacterial pathogens, *Trends Microbiol.,* 8, 306, 2000.

7 RNase-L, Symptoms, Biochemistry of Fatigue and Pain, and Co-Morbid Disease

Neil R. McGregor, Pascale De Becker, and Kenny De Meirleir

CONTENTS

7.1 INTRODUCTION

Earlier chapters in this book have described in detail the structure and biochemical mechanisms leading to the activation and actions of the 2,5A-synthetase RNase-L system. This chapter will concentrate on the results or effects of activation or deregulation of this system within the chronic fatigue syndrome (CFS) patient and also try to assess some of the co-morbidity factors involved in the disease complexity. The major actions of the 2,5A-synthetase/RNase-L system are for antiviral defense

0-8493-1046-6/02/$0.00+$1.50

and controlled cellular degeneration or apoptosis. Deregulation of these actions via the removal or reduction of messenger RNA and the inhibition of protein synthesis, along with effects of the RNase-L fragments and other enzyme or receptor systems initiated by the deregulators of the RNase-L system, will have very large influences upon host homeostatic mechanisms and hence symptom expression. The reduced ability to produce proteins such as receptors, membrane pumps, and many other intracellular proteins and enzymes will alter control of cellular homeostasis. These biochemical changes are also influenced by the biochemical deregulating effects of the sporadic reactivation of persistent viruses or influences of bacterial pathogens. These alterations will significantly influence brain, immune, and other functions and hence symptom expression.

To establish the association between RNase-L activity and symptom expression was therefore a major priority in patients with CFS. In HIV-infected patients, the infection of cells which express the CD4 receptor results in reductions in both immune and neural function; these alterations make the patient susceptible to a series of opportunistic infections and alterations in central nervous system activity that leads to a very complex disease. In patients with a disturbed 2,5A-synthetase/RNase-L system, the deregulation of protein production and apoptosis will also have very significant consequences, which will lead to a complex disease process influenced by other co-morbid or secondary disease processes. This chapter will attempt to address some of these issues and bring some clarity to the understanding of these complex disease processes.

7.2 SYMPTOM CLUSTERS IN CFS

To understand the influence of the RNase-L system upon the biochemistry of the host in CFS, we need to understand the symptom groupings or clusters in CFS patients. To assess these symptom associations, De Becker et al.[1] used factor analysis of symptom variation in a large population of defined CFS patients and fatigued patients (n = 2073) who were excluded by the CFS definition criteria. The factor analysis revealed the presence of four-factor symptom groupings termed: 1) general CFS symptoms; 2) neurocognitive symptoms; 3) musculoskeletal symptoms; and 4) mood change and psychiatric symptoms (Table 7.1).

The primary symptom factor grouping that differentiated the CFS patients from the excluded fatigued patients was a group of general CFS definition-related symptoms (fever, sore throat, flu-like symptoms), gastrointestinal symptoms, viral reactivation problems (aphthous ulceration, shingles, cold sores), and a group of general symptoms including urinary frequency, nonrestorative sleep, and alterations in taste, smell, and hearing. Interestingly, the neurocognitive and musculoskeletal symptoms which are part of the CFS definitions were separated from the major defining CFS symptoms (fever, sore throat, etc.). This suggests that the neurocognitive and musculoskeletal symptoms may represent or be due to separate influences or host responses. This can also be stated for the mood change and psychiatric grouping, which was separated from the other symptom groupings.

The general CFS, neurocognitive, and musculoskeletal symptom groupings were important in differentiating the defined CFS patients from the excluded fatigued

TABLE 7.1
Symptom Factor Clusters and Loadings (L) for the Four-Factor Groupings in Chronically Fatigued Patients

Symptom Factor and Score		Symptom Factor and Score	
General CFS Symptoms	**L**	**Neurocognitive Symptoms**	**L**
Sore throat	61	Memory disturbance	75
Recurrent flu-like symptoms	60	Difficulty with calculations	71
Hot flushes	56	Attention deficit	70
Low-grade fever	54	Spatial dysfunction	68
Lymph nodes	50	Difficulty with words	66
Cold sores and shingles	46	Blackouts	55
Recurrent aphthous ulceration	43	Disequilibria	38
Gastrointestinal disturbance	41	Light-headedness	35
New sensitivities to food, drugs, etc.	39	Visual acuity	35
Urinary frequency	38	Speech difficulties	35
Allergies	37	**Musculoskeletal Symptoms**	
Rashes	37	Myalgia	69
Cough	36	Arthralgia	65
Cold extremities	35	Muscle fasciculation	60
Gingivitis/Periodontitis	35	Numbness or paraesthesia	59
Nonrefreshed sleep	35	Muscle weakness	50
Dry eyes	34	Chest pain	34
Night sweats	34	Tinnitus	31
Diarrhea	33	Paralysis	31
Symptom exacerbation in extremes of temperature	31	**Mood Change/Psychiatric Symptoms**	
		Depression	73
Headache	30	Anxiety	69
Altered taste, smell, and hearing	29	Emotional labiality	65
Post-exertional fatigue	28	Personality change	63
Hair loss	27	Nightmares	42
Weight loss	24	Psychosis	37

Source: From De Becker, P.J., et al., Ph.D. thesis, 2000, submitted for publication.

patients, while the mood change and psychiatric factor grouping was unrelated to defined CFS or the other symptom factor groupings within this fatigued group of patients. Illness duration was associated with increases in all factor symptom scores while a sudden onset of fatigue was associated with an increase in the CFS general symptoms and reduction of the mood change and psychiatric symptoms. Exercise capacity, as assessed by VO_2Max/Kg and maximum watts/Kg, was strongly inversely related to the musculoskeletal symptom factor grouping. The general CFS, neurocognitive, and mood change and psychiatric symptom factor groupings were not strongly associated with the reduction in exercise parameters. Thus, a sudden onset of CFS was associated with an increase in general CFS factor symptoms, while the duration of CFS was associated with increased severity of all four factor groupings. The reduction in exercise capacity was principally associated with the musculoskeletal symptom factor score. Therefore, using the factor analysis modeling system, it was found that symptom expression within CFS patients is a complex interaction of four major symptom groupings or clusters.

These same indices were calculated in three other study data sets which contained control subjects for comparison. This was done because we had previously reported that infectious symptoms were associated with greatest symptom variability within CFS patients, but the host response symptoms had the highest variability when the controls were included.[2] Multiple regression analysis using the factor indices in the other three study groups revealed that the musculoskeletal factor grouping was the primary determinant between CFS patients and controls, and the mood change and psychiatric factor score was once again not a significant variable in the group differences.[3] These data are consistent with the findings of De Becker et al.[1] and a disease model of an underlying host-based change with symptom variability being associated with co-morbid infectious events.

7.3 RNase-L PROTEINS, sIL-2r, IL-6, CFS, AND SYMPTOMS

Suhadolnik et al.[4] first described the RNase-L anomaly in CFS patients and showed an increase in total RNase-L activity that was associated with reduced cellular protein levels. In a second article, Suhadolnik et al.[5] reported that there was a complex group of different forms of RNase-L in CFS patients and that the interactions were not a simple increase in activity. The study by De Meirleir et al.[6] revealed that, in CFS patients there was an increased prevalence of the fragmented 37- and the 40-kDa RNase-L proteins but not the monomeric 80-kDa protein; however, there was an increase in the quantities of all three RNase-L proteins in the CFS patients. The ratio of the 37- to the 80-kDa RNase-L monomer protein levels (×10) was found to be a good predictor of defined CFS (odds ratio >100). These studies suggest that a defect in the 2,5A synthetase/RNase-L system may be the core host-based change, not unlike HIV in the AIDS disease model. These findings are supported by other investigators[7,8] who showed that there were increases in RNase-L proteins as well as a reduction in RNase-L inhibitory protein in CFS patients. They also showed that these mechanisms could be activated by both viral and chemical mechanisms and were mediated by the IFN-β receptor. Therefore, a complex interaction between the effects of the different RNase-L proteins and their degradation fractions may exist and these variations may be associated with differences in symptom expression.

Metcalf et al.[9] assessed the total RNase-L activity, irrespective of the actions of any individual RNase-L protein, in CFS patients and control subjects.[10] The total RNase-L activity was found to be a good selective biochemical measure for CFS, which is consistent with the findings of the Suhadolnik and De Meirleir studies. The total RNase-L activity was strongly associated with the reporting of fatigue, low-grade fever, neurocognitive dysfunction, paraesthesia, unrefreshed sleep, and headaches — primary defining CFS symptoms. Table 7.2 shows the association between De Becker's[1] factor scores, total RNase-L, sIL-2r, and IL-6 from Metcalf's study. The total RNase-L activity was most strongly associated with the general CFS and

TABLE 7.2
Associations between the De Becker Factor Scores,[a] RNase-L, sIL-2r, and IL-6 (n = 66; 33 CFS and 33 age-sex-matched controls)

Grouping (Multiple Regression)	Parameter	*P*-Value
General CFS	RNase-L	<0.0005
$R^2 = 0.38$, F = 18.3, $P<0.000001$	sIL-2r	<0.001
Musculoskeletal	RNase-L	<0.0004
$R^2 = 0.40$, F = 19.4, $P<0.000001$	sIL-2r	<0.0007
Neurocognitive	sIL-2r	<0.002
$R^2 = 0.31$, F = 13.1, $P<0.00002$	RNase-L	<0.006
Mood Change/Psychiatric	sIL-2r	<0.0004
$R^2 = 0.28$, F = 11.4, $P<0.00006$		

Correlations	Gen.CFS	Neurocog.	Musculo.	Mood
RNase-L	0.48 <0.001	0.39 <0.005	0.46 <0.001	0.29 <0.050
sIL-2r	0.53 <0.001	0.50 <0.001	0.57 <0.001	0.45 <0.001
IL-6	0.31 <0.030	0.21	0.27	0.17

[a] Statistical methods = Spearman rank correlations and multiple regression.
* = $P<0.05$.

Source: From Metcalf, L.N., et al., Honors Thesis, University of Newcastle, Australia.

musculoskeletal symptom factors. These data suggest that the higher the total RNase-L levels, the higher the general CFS and musculoskeletal symptom scores.

As the total RNase-L activity was associated with the primary CFS defining symptoms, we needed to assess if this also applied to the fragmentation of RNase-L. Within De Becker's study group,[1] 535 subjects had an RNase-L test and the 37-kDa:80-kDa RNase-L ratio was significantly associated with only the general CFS symptom factor score. This suggests that the higher the level of fragmentation of RNase-L, the higher the level of infectious symptoms and the higher the prevalence and severity of the symptoms of reactivation of persistent viruses. Interestingly, Metcalf's[9] data show increases in IL-6 with the general CFS symptom factor and the Richards[11] study data showed an increase in C-reactive protein with the general CFS symptom factor, which is consistent with evidence of an increased acute phase reaction and viral reactivation. Thus, during the periods of infectious symptoms and viral reactivation there is increased fragmentation of the RNase-L proteins and an increase in the total RNase-L activity. The increase in total RNase-L activity is prominently associated with the development of the general CFS and musculoskeletal symptoms and to a lesser degree the levels of neurocognitive symptoms, while the highest level of fragmentation occurs during the periods when pathogen-associated events appear to be occurring. This would also suggest that the musculoskeletal, neurocognitive, and mood change and psychiatric symptoms may be host-based consequences of the pathogen-associated increased RNase-L activity.

Analysis of the Metcalf data[3,9] for associations between RNase-L, sIL-2r, and IL-6 and the factor symptom groupings revealed that the general CFS and the musculoskeletal symptom factors were associated predominately with increases in RNase-L, the neurocognitive factor score was predominately associated with sIL-2r and RNase-L, while the mood change and psychiatric factor score was solely associated with increases in sIL-2r levels.[3] Thus, increases in RNase-L activity were prominently associated with the general CFS symptom and musculoskeletal symptom factor groupings. Combined increases in RNase-L and sIL-2r were associated with the neurocognitive factor symptoms, and sIL-2r alone was strongly associated with the mood change and psychiatric factor score. This is consistent with the strong associations between RNase-L and the general CFS, musculoskeletal, and neurocognitive factor scores and defined CFS in the Suhadolnik,[4,5] De Meirleir,[6] and De Becker[1] studies, as well as the failure to find a strong association between the mood change and psychiatric factor and defined CFS. While mood change is triggered by the onset of CFS[12] as well as the therapeutic use of cytokines,[13,14] the evidence suggests that it is a host response and environmental-influenced symptom cluster relatively independent of the other symptom factors. This finding has significant ramifications for the understanding of CFS and also the development of depression and anxiety disorders within both CFS and normal populations.

To assess whether RNase-L activity and cytokines may interact to lead to symptom variation, we used weighted statistical analysis. A combined increase in RNase-L activity and sIL-2r was strongly associated with increases in neurocognitive and mood disturbance scores, while a combined increase in RNase-L and IL-6 was strongly associated with increased muscle pain and fatigue scores. The RNase-L activity and the immune activation markers were not only associated with different symptoms, but they also appeared to act in a combined manner to selectively increase expression of specific symptoms. Thus, the interactions between effects of RNase-L and the various cytokines appear to introduce significant heterogeneity into symptom expression and may therefore be the basis of the symptom heterogeneity and the development of the host-based changes which initiate the musculoskeletal, neurocognitive, and mood change and psychiatric symptoms.

In conclusion, these experimental data show that variation in prevalence of the different RNase-L proteins and immune-related cytokines are associated with alterations in symptom expression. These findings are strongly supported by observations of the therapeutic use of cytokines for the treatment of various other diseases and the variation of cytokines within other illnesses.[13-15] With therapeutic use of interferon alpha in patients with hepatitis C, different patients developed fatigue, myalgia, and mood alterations; however, not all subjects developed the same symptom complexes.[13,14] This strongly suggests that variation in the host response to the challenge or interactions with other co-morbid disease or environmental influences results in variation in symptom expression. This introduces the first significant complexity in the understanding of the association between RNase-L activity and symptom expression, the roles of the host response, environmental influences, and co-morbid disease.

7.4 THE COMPLEX DISEASE PROCESS

In the De Becker study[1] the general CFS factor grouping was the major discriminate factor grouping for separating CFS patients from the excluded fatigued patients; however, we needed to know which factor was more important for separating the CFS patients from control subjects. We therefore assessed the Metcalf data to compare the different factor scores between CFS patients and controls.[3] Discriminate function analysis revealed that the musculoskeletal index was the primary factor grouping that differentiated the CFS patients from the control subjects (+ve $P<0.00001$). This was followed by the general CFS factor symptoms and the neurocognitive factor symptoms. Once again the mood change and psychiatric factor symptoms were not significant. This is supported by the observation that the host-based symptoms (fatigue or myalgia) were good predictors of CFS while the infectious symptoms were the best determinates of symptom variation within the CFS patients.[2] This comparison was also performed in the second data set used to assess oxidative damage markers in CFS patients.[3,11] This comparison showed the same result as that seen in the Metcalf study where the musculoskeletal symptom score was the prime determinant when comparing CFS and control patients.[3] Thus, the RNase-L associated symptom factor groupings (general CFS, musculoskeletal) were the best predictors of CFS when compared with controls. From these data and the clinical histories of patients, a model can be developed which may explain this complex set of data.

The literature shows that CFS patients report frequent recurring bouts of low-grade fever, lymphodynia, and reactivated viruses and that the severity of the fatigue, neurocognitive, and musculoskeletal symptoms increases for a period after these infectious events. In the De Becker[1] study, low-grade fever, lymphodynia, and recurrent oral aphthous ulceration were reported by 58, 46, and 30%, respectively, while in the Metcalf[9] study these symptoms were reported in the previous 7 days by 25, 38, and 34%, respectively. These data would suggest that the general CFS symptoms occured between 25 to 50% of CFS patients at any one point in time. This is consistent with the prevalence of reactivation of viruses in CFS patients reported in the literature.[17-20] These reactivation and infection events were associated with increases in total RNase-L activity and RNase-L fragmentation in CFS patients. In non-CFS subjects reactivation of viruses is associated with an increase in RNase-L activity (reviewed in Reference 21); this would suggest that the general CFS symptoms are the normal reactions associated with a reactivated virus or infection and most likely represent an increase in total RNase-L activity and the fragmentation of the RNase-L enzyme system proteins. Table 7.3 shows that the general CFS symptoms were also associated with increases in oxidative markers and C-reactive protein, which suggests an increase in an acute phase process.

The musculoskeletal and neurocognitive symptom groupings were the best predictors of the difference between CFS patients and controls, and in many patients the increase in severity over the first 2 to 6 months of their illness. Table 7.3 also shows that increases in oxidative markers are associated with each of the factor symptom groupings. However, the variation in the symptom factor scores was associated with variation in oxidative markers. Interestingly, increases in oxidative

TABLE 7.3
Associations between the De Becker Factor Scores,[a] Blood Cell, Oxidative Damage and Serum Biochemical Parameters (n = 60; 33 CFS and 27 age-sex-matched controls)

	General CFS	Neurocog.	Musculoskel.	Mood Change
Methaemoglobin	0.39 <0.009	0.31 <0.05	0.42 <0.005	0.35 <0.03
2,3-biphosphoglycerate	0.33 <0.03	0.42 <0.004	0.35 <0.03	0.28
Malondialdehyde	0.25	0.35 <0.03	0.37 <0.02	0.30 <0.05
C-reactive Protein	0.32 <0.04	0.13	0.18	0.16

Sources: De Becker, P.J., et al., Ph.D. thesis, 2000, submitted for publication; McGregor, N.R., et al., AACFS Conf., 2001, Abstract 59; Richards, R.S., et al., submitted for publication.

radicals are able to initiate apoptosis via release of cytochrome C and subsequent activation of caspase 9. Those patients who develop mood change and psychiatric symptoms also develop these symptoms after the onset of the disease.[12] The therapeutic use of cytokines clearly establishes these proteins as host-based factors that can initiate musculoskeletal, neurocognitive, and mood change and psychiatric symptoms,[13,14] which supports the contention that these host-based symptoms are a result of the disease process and are the persistent symptoms noted in CFS patients.

When examining the four-symptom factor groupings for changes in total urinary amino and organic acid excretion and serum free fatty acids, we found that a fall in urinary amino and organic acids was negatively associated with all four factor scores, with the most prominent association the musculoskeletal and mood change scores. Conversely, total serum free fatty acid levels were positively associated with musculoskeletal symptom score and unrelated to the other factor scores.[22] However, changes in each factor score were associated with changes in different patterns of urinary amino and organic acids and serum free fatty acids. These changes support the association between host response changes and the cognitive, musculoskeletal, and mood change factor scores.

Figure 7.1 shows the variation in factor scores for a CFS patient and a control subject over eight sequential assessments. In the control subject the scores are all low and the various factor scores are independent of each other. In the CFS patient the scores are all high and there is a relatively strong association between the general CFS factor scores and the musculoskeletal factor scores — both seem to rise and fall together. The neurocognitive scores seem to be relatively even across the duration of measurement while the mood and psychiatric scores seem to fluctuate to a greater degree. From this we can hypothesize that the general CFS factor symptoms are associated with the intermittent or sporadic periods of fevers and viral reactivation which are a prominent feature of CFS. These periods of fever and viral reactivation are most likely associated with increases in total RNase-L quantity and activity, and represent a common feature of a cytokine-driven set of changes in the host. In the CFS patients, as distinct from non-CFS patients, the general CFS symptoms are also associated with increases in the fragmentation of RNase-L (the 37-kDa RNase-L).

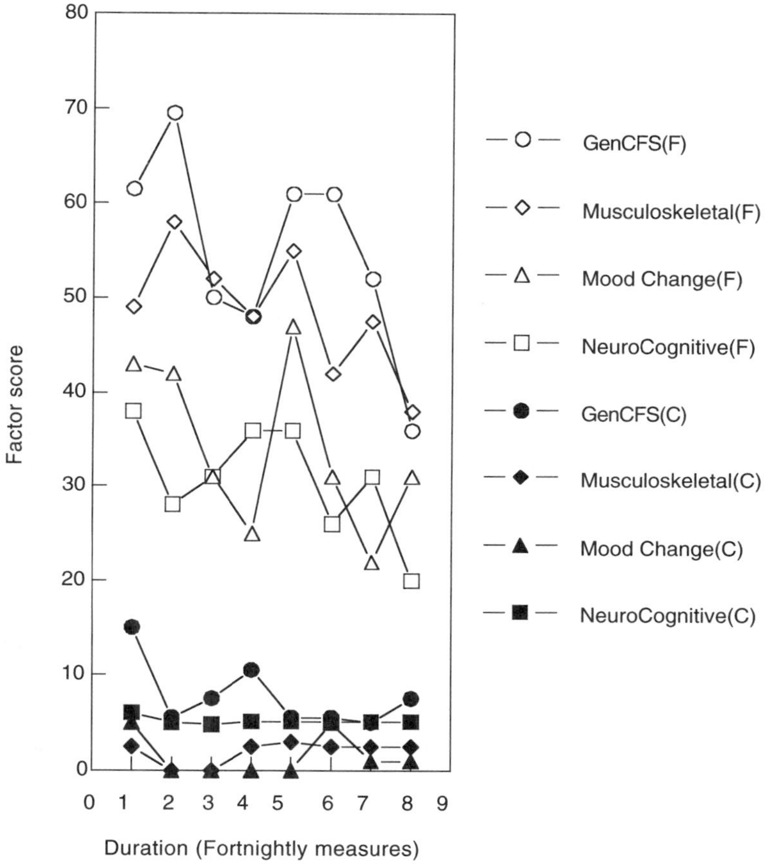

FIGURE 7.1 The association between the four factor scores in a CFS patient and control subject with time — subjects completed questionnaires every 2 weeks for 16 weeks. There is a close relationship between the general CFS factor score and the musculoskeletal factor score. The mood change factor score did show a closer relationship to the general CFS and musculoskeletal scores than the cognitive score in this CFS patient. The general CFS factor symptoms show two sporadic increases in this 16-week period. We suggest that each general CFS symptom bout is associated with an increase in the total and fragmented RNase-L levels and the alteration in urinary output, muscle protein, and RNA content. Similarly, activation of t-lymphocytes with the production of sIL-2r and, possibly, other cytokines influences the degree of cognitive disturbance and alteration in mood.

These features are associated with the musculoskeletal factor score, which is associated with falls in muscle total protein and RNA levels[23,24] and increases in urinary 3-methylhistidine excretion, as well as the reduction in urinary total amino and organic acids and increases in total serum lipid levels. The musculoskeletal, cognitive, and mood change factors appear to represent the symptoms which are host derived and result from the disease process.

7.5 CHANGES IN BIOCHEMISTRY ASSOCIATED WITH MUSCLE PAIN AND FATIGUE

A number of amino acid excretion anomalies were found in the CFS patients compared with the control subjects.[25,26] Increases in excretion of tyrosine and 3-methylhistidine were observed in the CFS patients.[10] The increase in tyrosine is associated with a fall in leucine excretion in the majority of polysymptomatic patients including myofascial pain and fibromyalgia syndrome patients. The tyrosine:leucine ratio was found to be correlated to pain intensity. The increase in 3-methylhistidine is associated with the degradation of actin, as it is a unique methylated amino acid found only in cytoskeletal and muscle actin.[27] Thus, increases in 3-methylhistidine indicate the degradation of the contractile or fibrillar proteins. This process of degradation of fibrillar proteins is distinct from the degradation of the other proteins inside the cell — non-fibrillar proteins.[28] The tyrosine:leucine ratio is more indicative of this process. The non-fibrillar pattern is associated with alterations in muscle pain, while the increased excretion of 3-methylhistidine is prominently associated with arthralgia and arthritic symptoms. Interestingly, the release of 3-methylhistidine requires IL-1 and IL-6,[29] which are also increased in patients with arthritis such as rheumatoid arthritis.[30]

Myalgia and lethargy are common symptoms following interferon injection.[13,14] Immune activation resulting in increases in cytokines is common in CFS patients[31-37] and likely to play a significant role in symptom development. Importantly, upregulation of nitric oxide synthesis as a result of interferon-gamma (IFN-γ) and tumor necrosis factor (TNF) that occurs via the glucocorticoid-regulated enzyme argininosuccinate synthetase[38] will result in an alteration in the redox potential (NADPH:NADP$^+$ ratio) and effect mitochondrial citric acid cycle regulation. IFN-γ and TNF increase glycolysis and inhibit complex 1 (NADH:ubiquinone oxidoreductase) of the mitochondrial respiratory chain.[39] Similarly, short-term up-regulation of nitric oxide production stimulated by IFN-γ and bacterial lipopolysaccharide (LPS) initially inhibits mitochondrial cytochrome c oxidase, while prolonged upregulation of nitric oxide production not only inhibits mitochondrial respiratory chain cytochrome-c oxidase but also succinate-cytochrome-c reductase activity.[40] This difference in the IFN-γ/LPS-associated nitric oxide response with time has been shown to be associated with alterations in glycolytic activity.[41] During the initial phase of exposure to IFN-γ and LPS, there was a threefold increase in glycolytic activity in fibroblasts, and inhibition of glycolysis (glucocorticoids, iodoacetate, NAD$^+$) during this phase was associated with cell survival. Following prolonged exposure, withdrawal of glucose or the use of glycolytic inhibitors did not protect the cells from IFN-γ/LPS-induced cell death. Mitochondrial function, specifically malate- and succinate-associated respiration, was depressed with combined IFN-γ and LPS and the cells had reduced ATP levels. We need to examine CFS patients to ascertain if any of these types of changes are present.

These possibilities were assessed by examining the alterations in the patterns of amino and organic acid excretion and serum changes. There is a negative correlation between the fatigue severity scores and the levels of excretion of succinic acid and asparagine, and the serum reductions in tyrosine and phenylalanine. Similarly, patients with CFS have been found to have a reduction in excretion of both asparagine

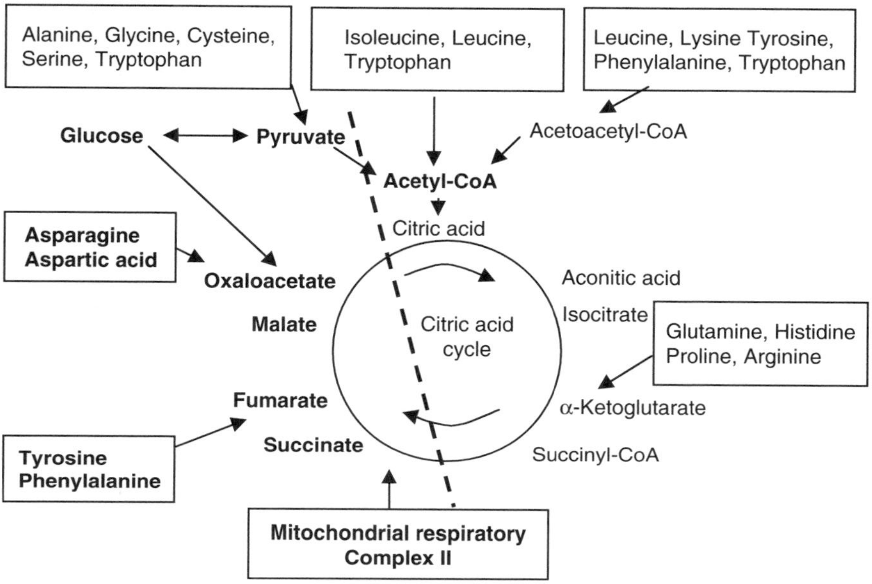

FIGURE 7.2 Summary of the metabolism of the oxidative positions of glycolysis and the various amino acids for provision of energy to the citric acid cycle. The highlighted components on the left side of the dashed line are likely to have their normal homeostasis disturbed by the chronic cytokine events seen in CFS patients. Any co-morbid disease which prolongs the cytokine mediated response or alters the metabolism of these components is likely to enhance the morbidity of CFS patients.

and succinic acid and falls in serum tyrosine and phenylalanine concentrations compared with controls.[42] The excretion of succinic acid in CFS patients was positively correlated with asparagine and phenylalanine (as well as the phenylalanine metabolite phenylacetic acid), and it is noted that both of these amino acids are catabolized into the citric acid cycle as oxaloacetate and fumarate, respectively (Figure 7.2). Both oxaloacetate and fumarate have a negative feedback on succinate oxidation but also pyruvate metabolism.[43,44]

These data suggest that there is increased catabolism of asparagine, phenylalanine, and possibly tyrosine in the latter part of the citric acid cycle and an increase in glycolysis in CFS patients. In support there is a positive correlation between serum glucose and succinate excretion, suggesting an increase in glycolysis and an inhibition of oxidative phosphorylation as has been reported by other investigations of CFS patients.[45-47] Wong et al.[45] found that in CFS patients at exercise-associated exhaustion, the level of adenosine triphosphate (ATP) was significantly reduced and concluded that there was an acceleration of glycolysis in the working skeletal muscles of CFS patients and that this may be associated with a defect of oxidative metabolism. Conversely, Barnes et al.[48] found no consistent difference in glycolysis or mitochondrial activity in CFS patients but did find heterogeneity in the biochemical responses. Similarly, Lane et al.[46] reported a heterogeneous lactate response suggestive of a deregulation of glycolysis that was associated with a reduction in the number of type I (oxidative muscle fibers) in the quadriceps muscle. Interestingly, fatigue is commonly reported by patients with uremia and has been associated with an increase in the rate of glycolysis and intracellular phosphate levels and a reduction in maximal oxidative capacity,[49] suggesting a similar mechanism is involved. As described above, the events associated with enhanced glycolysis and inhibition of oxidative phosphorylation are usually initiated in cells by a nitric oxide-mediated cytokine response, but are normally an acute response. The chronic nature of the low-grade cytokine response appears to have resulted in depletion of the precursors and the development of symptoms related with these changes, which is also consistent with the previously reported changes.

This has immense ramifications for many cytokine-related conditions as the catecholamine precursors appear to be catabolized by this mechanism, which may lead to the development of many of the sympathetic nervous system- and catecholamine-associated symptoms seen in CFS. The general CFS symptom factor scores are therefore likely to represent the increases in the acute cytokine-mediated responses while the neurocognitive, musculoskeletal, and mood change and psychiatric factor scores are likely to represent the variations in the resultant depletion or accumulation of the components altered by the different cytokine responses or the co-morbid conditions that influence the homeostatic processes.

Chronic pain consists of alterations in nociception and central nervous system processing of the signals. While the processes involved in the acute pain processes are becoming known and therapeutic interventions are quite advanced, the processes involved in chronic pain are little understood. Coderre and Yashpal[50,51] have proposed a mechanism for the development of spinal hyperalgesia in the acute pain response. The acute pain response consists of two major components: 1) the peripheral nerve-based actions associated with tissue injury and nociception and the brain's response to those signals; and 2) the development of spinal column hyperalgesia, which results in the increased sensitivity of the pain response. This hyperalgesia response is important in the development of the protective responses following tissue injury. However, the chronic pain response lacks the specific peripheral traumas associated with these processes. We know that cytokines can initiate the chronic pain response and that many of these substances are also involved in the acute pain response. Therefore, we propose that many different factors that can initiate inflammatory substance increases can also act to initiate the chronic pain response. Where the chronic pain response differs is that the chronic nature of the driving forces that initiate the response result in changes in homeostasis, which result in deficiencies or excesses of various metabolic products which potentiate the whole process. Figure 7.3 shows the hypothetical model of the mechanisms involved in the development of the chronic pain hyperalgesia response. Instead of tissue damage being the driving force, as in the acute response, upregulation of immune or cytokine-mediated products and the co-morbid pathogen-related toxins appear to drive the response.

There are common biochemical features in the renal diuresis response and the spinal hyperalgesia response. Both can be initiated by nitric oxide and arachidonic acid and inflammatory products and both are inhibited by the catecholamines. Analysis of the data from over 1500 patients shows that the more severe the pain and fatigue symptoms in a patient, the higher the level of serum lipids and lower the

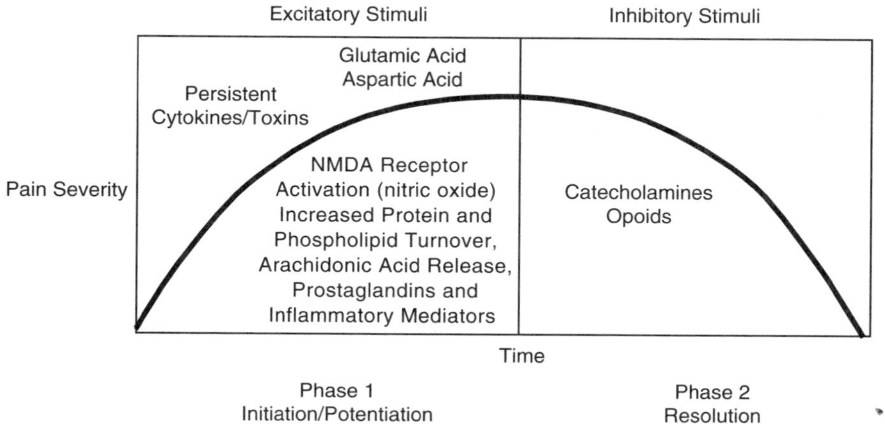

FIGURE 7.3 The proposed phases of the chronic spinal hyperalgesia pain response. Phase 1. Initiation/Potentiation. Cytokine/toxin mediated/potentiated activation of nitric oxide response with initiation and potentiation of the hyperalgesia response in the spinal column. The same biochemical changes initiate diuresis, low-grade aminoaciduria, and sodium loss from the kidney[52] and an alteration in the hypothalamic pituitary adrenal and gonadal axes.[53] Increased excitatory amino acid levels activate the NMDA receptor and initiate a nitric oxide-mediated response within the spinal column. Increased phospholipase activity and release of lipid-associated pain mediators, such as arachidonic acid, facilitate increases in prostaglandins and leukotrienes. Na^+K^+ATPase activity is inhibited and the hyperalgesia response is initiated as a result of alteration in membrane potential (a nitric oxide-mediated Na^+K^+ATPase channelopathy-like situation). The reduction in serum tyrosine will result in a reduction in catecholamines and failure to inhibit the pain response and a reduction of dopamine that will facilitate the kidney diuresis and loss of electrolyte and amino acids. Progressive loss of amino and organic acid components in urine results in additional loss of neurotransmitter precursors and deregulation of cognitive and neurological functions that prevent the initiation of the resolution phase. Phase 2. Resolution. A reduction in the cytokine/toxin stimuli and potentiation phase components, along with restoration in amino and organic acid loss and increases in catecholamines and opoids, inhibits the hyperalgesia response. (With permission from McGregor, N.R., et al., *J. CFS,* 2000, 7:3–21; McGregor, N.R., Ph.D. thesis, University of Sydney, 2000.)

excretion of amino acids.[22,42] Figure 7.4 shows the hypothetical model of this situation.[44] Table 7.4 shows the increase in the De Becker factor scores[1] in relationship to these changes; increases in the n-6 fatty acids were strongly associated with all factor scores and the saturated fatty acids levels were negatively associated with the general CFS, cognitive, and mood change and psychiatric factor scores.[22] These changes are consistent with an increase in the proinflammatory or prohyperalgesia response. These data therefore strongly support proposed mechanisms behind development of the chronic pain response.

Table 7.5 shows the alterations in the various amino and organic acids in a group of patients with myofascial pain syndrome as well as the possible roles that each plays in normal homeostasis. Variation in the different amino and organic acids is associated with differences in homeostasis, which in turn influence symptom expression in various tissues. The variation in each of these metabolites and the patient's

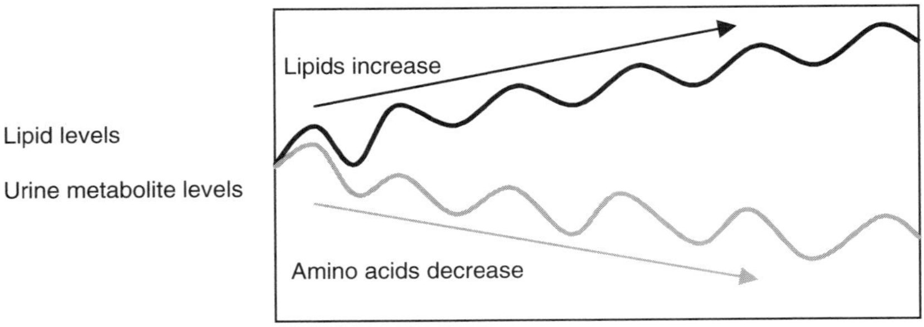

FIGURE 7.4 The hypothetical model for the loss of amino acids and increase in proinflammatory fatty acid precursors with duration of the symptoms. Each general CFS symptom factor score increase is associated with a cytokine-driven aminoacidaemia and diuresis-associated loss of sodium and amino acids. The alterations in the lipid and amino acids fractions results in increased hyperalgesia, alterations in tissue precursor, and byproduct availability, which in turn influences symptom development. The symptoms developed fall predominately into the musculoskeletal, neurocognitive, and mood change and psychiatric factor groupings.

genetics or co-morbid disease will also influence the changes seen in these metabolites and hence symptom expression within the patient.

Thus, a complex relationship exists between the variation in the cytokine-driven changes and the body's response to these stimuli. The analysis of these relationships is therefore of primary importance not only to understand the disease process, but also to develop the therapeutic interventions required by clinicians.

TABLE 7.4
Factor Scores, Lipid Totals, and Ratios — Multivariate Analysis (n = 1573)

Totals	Gen CFS T, *P*	Cognitive T, *P*	Musculo. T, *P*	Mood T, *P*
PUFA	4.7 < 0.001	4.0 < 0.001	4.9 < 0.001	5.7 < 0.001
DGLA: Linoleic acid	3.4 < 0.001	2.6 < 0.01	4.1 < 0.001	3.4 < 0.001
Palmitic: Stearic	4.0 < 0.001	2.9 < 0.01	—	3.5 < 0.001
Oleic: Linoleic acid	2.3 < 0.05	—	—	2.9 < 0.01
Palmitoleic: Palmitic	—	—	2.7 < 0.01	—
Saturated FA	–3.1 < 0.01	–2.6 < 0.01	—	–3.0 < 0.01
Oleic: Stearic	–2.9 < 0.01	–2.1 < 0.05	—	–2.2 < 0.05
Sterols	–2.5 < 0.05	—	–2.9 < 0.01	—

Note: Statistical method = Multiple regression analysis; T = t-score; p = p-value of statistical significance; PUFA = polyunsaturated fatty acids; DLGA = di-homo-gamma-linoleic acid.

TABLE 7.5
Summary of the Potential Metabolic Associations Which May Be Effected by the Various Changes in Urine Excretion Which Were Correlated with Illness Duration and Pain Severity

Metabolite	Metabolic Functions	Symptom
Total metabolites	Kidney resorption — nitric oxide diuresis, catabolism	Pain intensity Duration
Aconitic acid	Citric acid cycle intermediate	Pain intensity
Alanine	Nitrogen metabolism, transamination, gluconeogenesis	Duration
Aspartate	Excitatory amino acid, nitric oxide precursors via AST and arginino-succinate synthetase	Pain intensity Duration
Ethanolamine	Present in the polar head group of important complex lipids as components of cell membranes	Pain intensity
Glutamate	Excitatory amino acid, nitrogen metabolism, transamination, gluconeogenesis	Pain intensity Duration
Glycine	Neurotransmitter; used in liver as a conjugate to form hippuric acid; formation of bile salts	Pain intensity
Hippuric acid	Impaired urea cycle function, liver detoxification	Duration
Hydroxyproline, Proline	Connective tissue turnover	Pain intensity Duration
Leucine	Indicator of protein synthesis, regulator of protein catabolism	Pain intensity
3-Methylhistidine	Marker of actin catabolism	
Phenylalanine	An essential amino acid; precursor for catecholamine synthesis	Pain intensity
Serine	A precursor to glycine and ethanolamine; required for the formation of tetrahydrofolate derivatives; present in the polar head group of important complex lipids as components of cell membranes	Pain intensity
Succinic acid	The citric acid cycle and mitochondrial oxidative phosphorylation	Duration
Threonine	An essential amino acid	Pain intensity
Valine	An essential branched chain amino acid	Pain intensity

7.6 CO-MORBID DISEASE IN CFS PATIENTS

The understanding of co-morbid disease in CFS and chronic pain patients is complex, but the available evidence would suggest that a model based upon HIV or AIDS would be applicable. In CFS patients the fall in energy availability appears to be the major underlying disease process, while in HIV it is the disturbance of CD4 receptor-expressing cells. Therefore, discussion of the various factors involved in the CFS disease processes is required. These can be divided into two major categories:

1. those that predominately increase cytokine production and 2. those that predominately alter energy availability.

7.6.1 Factors Influencing Cytokine Production

7.6.1.1 Viruses

The reactivation of viruses is the one major factor seen in the majority of CFS patients and it is logical that variation in the reactivation of these viruses will influence symptom expression. While there is no evidence of an increase in the prevalence of the various viruses, in most studies there is a significant increase in the number of subjects showing evidence of reactivation. This is typical of a co-morbid condition and not a causative condition. Viral reactivation is associated with De Becker's general CFS factor grouping[1] and with an increase in RNase-L fragmentation. This leads to one simple question: do viruses influence the symptom expression in CFS patients as they do in patients with HIV?

Figure 7.5 shows a summary of the areas of influence of some members of the herpes family of viruses (HSV-1, EBV, CMV, HHV-6) and the adenoviruses on the interferon/2-5A synthetase/RNase-L enzyme system. Each of the viruses influences different parts of the interferon/2-5A synthetase/RNase-L enzyme system and is likely to influence differences in chemistry, which in turn are likely to influence symptom expression.

From the De Meirleir data, a small subset of 94 CFS patients were assessed for CMV antibodies. Forty-five (47.9%) had a positive CMV antibody titer and there was a positive correlation between the IgM and IgG titers ($r = +0.31$ — $P<0.001$). Sixty (68.2%) of 88 CFS patients had a positive HSV-1 antibody titer. Table 7.6 shows the symptoms that were positively correlated with the CMV and HSV-1 IgG titers within the CFS patients. The general CFS factor score was positively correlated with the CMV IgG ($r = 0.39$ — $P<0.03$), while the musculoskeletal factor score was positively correlated with the CMV IgG ($r = 0.32$ — $P<0.03$) and the HSV-1 IgG ($r = 0.27$ — $P<0.05$) titers. Both CMV and HSV-1 were associated with differences in symptom expression in CFS patients and were predominantly associated with the general CFS and musculoskeletal symptom scores. Interestingly, the symptoms associated with CMV antibody titers in these CFS patients are very similar to those seen in patients with HIV.[54,55] Lehner et al.[56] have suggested that CMV infection of cardiac muscle is a cause of CFS and is associated with flattening or inversion in the EEG t-wave. HSV-1 was also associated with increased symptom expression that had a distinct symptom pattern. The increase in CMV IgG was positively associated with increases in CD3+HLADR+ levels, suggesting that increases in lymphocyte activation are associated with increases in viral reactivation; yet this was not increased compared with the controls.

Figure 7.6 shows the canonical plot of the different serum lipid profiles of control patients and CFS patients divided based on the presence of Epstein–Barr virus early antigen (EBEA). The presence of EBEA is associated with a very distinct change

EBEA

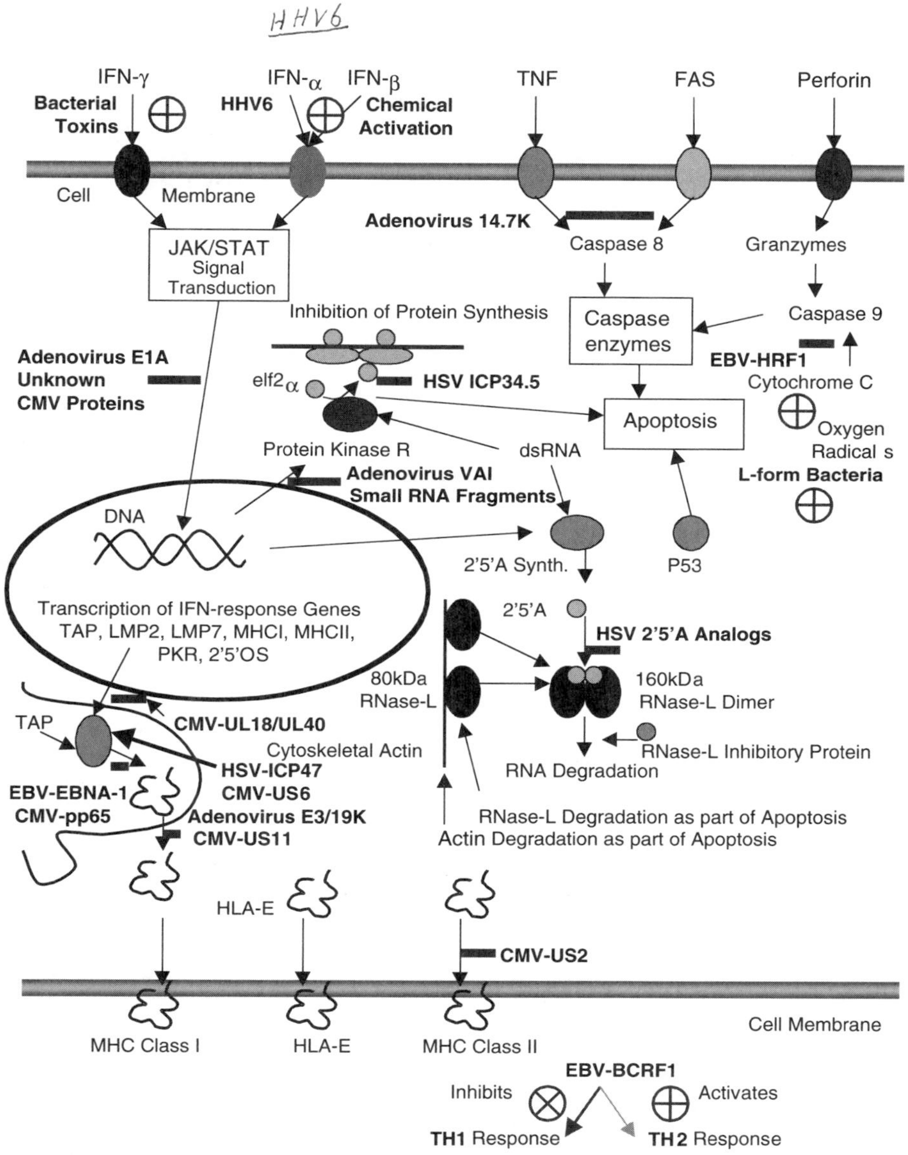

Abreviations: IFN=Interferon; **TNF**=Tumour necrosis factor; **CMV**=Cytomegalovirus; **HSV**=Herpes Simplex virus; **dsRNA**=double stranded RNA; MHC=major histocompatibility complex; **EBV**=Epstein-Barr virus; ▬ =Virus inhibition; **elf2α**=eukaryotic translation initiation factor.

FIGURE 7.5 Viral mechanisms of interference with apoptosis/interferon/2'5'A synthetase.

TABLE 7.6
Correlations between Symptoms and CMV IgG and HSV-1 IgG Titers

Symptom	CMV r-Value *P*-Value	HSV-1 r-Value *P*-Value
Anxiety	0.22 <0.03	0.22 <0.04
Loss of libido	0.35 <0.002	
Shingles	0.34 <0.002	
Cardiac palpitations	0.28 <0.006	
Arthralgia	0.27 <0.01	
Hair loss	0.25 <0.02	
Sleep disturbance	0.25 <0.02	
Weight loss	0.23 <0.03	
Diarrhea	0.22 <0.04	
Flu-like symptoms	0.22 <0.04	
Myalgia	0.20 <0.05	
Spatial dysfunction		0.29 <0.006
Tinnitus		0.29 <0.006
Blackouts		0.29 <0.007
Muscle fasciculation		0.26 <0.02
New symptoms of sensitivity		0.26 <0.03
Cough		0.25 <0.03
Chest pain		0.24 <0.04
Numbness/Paraesthesia		0.22 <0.04

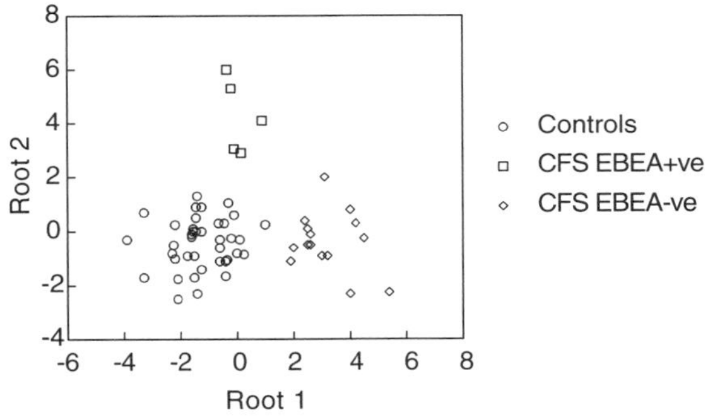

FIGURE 7.6 Canonical plot of the differences in serum lipid profiles in relationship to the presence and absence of EBV early antigen in CFS patients. This scatter plot shows the high degree of separation of the lipid profiles indicating that the virus early antigen is associated with altering the patients' lipid chemistry.

in the lipid profiles of the CFS patients who are positive for this antigen.[57] Thus, the presence of a virus is likely to be associated not only with alterations in symptom expression but also with changes in the chemistry of the patient that are involved in the symptom presentation.

Thus, reactivation of the different viruses or even their simple presence is associated with variation in lymphocyte activation, most likely cytokine production, changes in the biochemistry, and increases in symptom expression. Symptom and lymphocyte activation heterogeneity is likely to be a result of the different patterns of viral reactivation, which may help to explain the heterogeneous results obtained in many CFS studies. Studies are required to appropriately investigate these significant confounding factors in the study of CFS, as they are consistent with those found in HIV.

7.6.1.2 L-Form Bacteria

Both HIV and CFS patients have increased carriage of L-form bacteria such as *Mycoplasma spp.*[58-60] In both CFS and HIV, between 40 and 60% of subjects have positive PCR or serology for *Mycoplasma spp.* Nicholson has also suggested that the number of detectable *Mycoplasma spp.* increases with duration of the illness. This complies with a typical co-morbid pathogen for both CFS and HIV. Interestingly, *Mycoplasma fermentans* has been found to promote apoptosis by several different mechanisms,[61,62] and in De Meirleir's study group a positive PCR test for *Mycoplasma spp.* was associated with an increase in RNase-L levels, while a positive *Mycoplasma spp.* antibody test was unrelated to RNase-L levels. The positive response was most highly associated with the *Mycoplasma fermentans* positive PCR result. While treatment with antibiotics such as Azithromycin has been associated with a reduction in symptoms and RNase-L levels, several questions remain unresolved as to whether the *Mycoplasma spp.* are inducing an alteration in RNase-L levels, are present as a result of the elevated RNase-L fragmentation, if the PCR tests are actually measuring DNA fragments that have homology with *Mycoplasma spp.*, or if the antibiotics are inhibiting the apoptotic mechanisms.

7.6.1.3 Bacterial Toxins

Bacterial toxins can initiate and also potentiate cytokine activity. Our group has found a significant relationship between lipid soluble membrane-damaging toxin production by skin coagulase-negative staphylococci (CoNS) and myofascial pain syndrome (MFPS).[42,63-65] MFPS is defined as palpable muscle tenderness that occurs predominantly in a facio–scapulo–humeral distribution and occurs in 48% of the population.[42] Twenty percent of the population will report MFPS while only 5% will actively seek treatment. As with CFS, MFPS has a high female to male ratio (3:1). This picture is quite different in CFS patients: 29% reported MFPS to clinicians, with 69% of patients reporting the symptoms of MFPS by questionnaire.[42] MFPS patients were more likely to report that their partners have similar problems, which suggested that a transmissible pathogen was involved.[42,63] The same associations between toxin-producing staphylococci and symptoms and biochemical changes were seen in the CFS patients and the control subjects. Thus the condition fulfills the criteria for a co-morbid condition.

Interestingly, the prevalence and severity of symptoms were higher in the CFS patients with myofascial pain compared with the remaining CFS patients.[42] In CFS patients the number of δ-toxin producing strains of CoNS was positively correlated with the general CFS symptom scores, but not the other factor scores.[1] Similarly, the levels of each of the toxins were found to be associated with exacerbation of specific symptoms that correlated with RNase-L activity, suggesting that toxin-associated potentiation of the cytokine activity was occurring.[24,42] These data are consistent with a co-morbid condition which exacerbates the condition in CFS patients. This association also exists between CFS and fibromyalgia where the patients with both conditions had a greater degree of disability compared with the other patients.[66,67] This also suggests that fibromyalgia may be a co-morbid condition and may have a similar origin. These interesting interactions in CFS patients need to be investigated further.

De Becker[1] found that CFS patients have a much higher level of reporting of gastrointestinal symptoms than previously reported, and these were found to be predominately associated with the general CFS symptom factor (Table 7.1). Our group has developed an assay for the assessment of the bowel microbial flora to ascertain the changes that may be present in these situations. CFS patients were found to have significant disturbances in the bowel microflora compared with controls.

Table 7.7 shows the comparison of 831 patients with severe fatigue compared with 61 no fatigue controls. The severe-fatigue patients were more likely to have *Enterococcus spp.* and *Staphylococcus spp.* isolated from their bowel and less likely to have *Bifidobacterium spp.* or *Lactobaccillus spp.* isolated from their bowel. Examination of the colony counts shows similar relationships except that the Candida colony counts were lower in the severe-fatigue patients. Forward stepwise multiple regression on the *Enterococcus spp.* count and the CPRU questionnaire (86 symptoms) revealed that fatigue was the prime discriminate variable (n = 1466 subjects). Multiple regression analysis was also applied to assess possible relationships with the De Becker factor scores in the 1466 subjects. *Enterococcus spp.* were positively associated with the musculoskeletal factor score, while the *Bifidobacterium spp.* were negatively associated. Both the *Lactobaccillus spp.* and *Clostridia spp.* counts were negatively associated with the mood change and psychiatric factor score. None of the other species counts were associated with changes in the factor scores. These data show that patients with severe fatigue do have significant gut dysbiosis and that variation in the colony counts of certain microorganisms is associated with symptom presentation. The changes in symptom presentation were not associated with changes in the infective events in the general CFS factor score, but related to the musculoskeletal and mood change and psychiatric factor scores. These findings do not support the suggestion that candid species are significant pathogens in CFS patients. Further studies are required to more fully assess these relationships and any possible therapeutic interventions.

TABLE 7.6
Multi- and Univariate Analysis of the Changes in the Bowel Microflora in 831 Patients with Severe Fatigue Compared with 61 Subjects with No Fatigue

Organism	Severe Fatigue	Control	Multivariate *P*-value	Odds Ratio (95% CI)
Increased				
Enterococcus	$33.5E^6$	$17.7E^6$	<0.00005	3.3(1.9–5.6) <0.00002
Staphylococcus	95,784	934	<0.03	3.7(1.1–11.8) <0.009
Decreased				
Lactobaccillus	$229E^6$	$546E^6$	<0.02	0.22(0.05–0.91) <0.008
Bifidobacterium	$52.5E^6$	$105.4E^6$	<0.008	0.47(0.26–0.85) <0.009
Candida	$0.6E^5$	$2.2E^5$	<0.02	NS
No Change				
Bacteriodes	$233.7\ E^9$	$27.1E^9$	NS	NS
E.coli	$62.4E^6$	$41.9E^6$	NS	NS
Klebsiella	$21.9E^6$	$1.7E^6$	NS	NS
Clostridia	$19.3E^6$	$0.5E^6$	NS	NS
Eubacterium	$3.6E^6$	$1.5E^6$	NS	NS

Note: Wilks' $\lambda = 0.937$, $F(12,879) = 4.95$ — $P<0.0000$; the microbial flora is measured as colony-forming units/liter.

7.6.2 Factors Influencing Energy Supply

The depletion of amino acids is a consistent finding in CFS patients and this usually results in an increased dependence upon glycolytic metabolism as is evident from a number of studies, including the cytokine-based studies.[38-46] Therefore, any condition associated with glucose intolerance is likely to have a significant effect upon energy availability, as the uptake of glucose will result in reductions in ability of cells to survive the cytokine-associated changes. No study has been conducted to ascertain the associations with these conditions in CFS patients; however, the influences upon the altered homeostatic mechanisms are likely to be profound as indicated by our pilot study data. These conditions range from disaccharidase deficiencies (primary and secondary) and insulin disorders to inflammatory bowel disease and malabsorption disorders. Both hyper- and hypoglycemic conditions that alter glucose utilization will have an influence upon the host metabolism during these periods. Insulin problems are also likely to alter amino acid transport, which is an insulin-mediated process for certain amino acids.

We have also noted a significant increase in prevalence of CFS patients with elevated urinary lysine excretion. Multiple regression analysis on 3625 subjects showed a positive association between urinary lysine excretion and the musculoskeletal factor score, and a negative association with the mood change and psychiatric score. Thus, genetic disturbances in amino acid transport are likely to affect the RNase-L system as increases in lysine are likely to reduce cellular arginine transport. Reduced arginine availability has been shown to initiate apoptosis[61] and to have profound effects upon nitric oxide production.

Emms[66] has found that food intolerances in CFS patients, when treated, result in reduction in symptom severity. Many of the gastrointestinal symptoms, such as irritable bowel, completely resolve while the fatigue and myalgia scores reduce in intensity. This is typical of a co-morbid condition and strongly suggests that treatment of these co-morbid problems does assist in reducing CFS patient morbidity even though the etiological problems have not been treated.

The induction of the cytokine-mediated aminoacidemia and aminoaciduria is also associated with changes in renal function, which results in reductions in urinary volume and amino and organic acid excretion, which in turn contributes to the exacerbation of symptoms.[42] This cytokine-induced change in renal function is also likely to result in increased sodium excretion, which in turn is likely to contribute to the development of orthostatic hypotension seen in up to 40% of CFS patients.[69-71] Once again the susceptibility to this disease response may be predisposed in many patients by the reduction in collagen cross-linking seen in patients with joint hypermobility as suggested by Rowe's group, e.g., Ehlers Danlos syndrome. Current therapies for the correction of the orthostatic hypotension in CFS patients involve the use of dietary modification to facilitate salt and fluid replacement, and the use of drugs such as Florinef. These therapies appear to benefit that segment of CFS patients with hypotension.

7.7 CONCLUSIONS

The biochemistry of CFS patients is very complex, as the disease appears to be associated with not only the underlying viral- or pathogen-induced changes, such as the anomalies in the 2-5A synthetase RNase-L system, but also changes induced by co-morbid disease. The understanding of the variation in symptoms achieved by De Becker's study[1] has, for the first time, allowed researchers to begin to understand this complex disease process. Identification of the underlying changes and their related chemistry will in the long term allow researchers and clinicians to develop appropriate therapies.

ACKNOWLEDGMENTS

The data in this chapter are a reflection of the collective knowledge gained through interactions with members of: 1. the CPRU (Henry L. Butt, R. Hugh Dunstan, Iven J Klineberg, Tim K. Roberts, Lee Metcalf, Tania Emms, Katrina King, Suzy Niblet, Mariann Zerbes, and the staff); and 2. the Royal North Shore Hospital CFS Research Unit (Phillip Clifton Bligh, Leigh Hoskin, Greg Fulcher, and Julie Dunsmore).

REFERENCES

1. De Becker, P.J., McGregor, N.R., and De Meirleir, K.L., A factor analysis study of symptoms in 1573 patients with chronic fatigue syndrome, Ph.D. thesis, 2000, paper submitted for publication.
2. McGregor, N.R. et al., Heterogeneity of symptom, onset, and biochemical profiles in "defined" CFS patients. AACFS conference, Boston, 1998, Abstract No. 59.
3. McGregor, N.R. et al., Analysis of biochemical changes associated with the different chronic fatigue syndrome factor analysis symptom clusters, AACFS Conf., Seattle, Jan. 2001, Abstract 59.
4. Suhadolnik, R.J. et al., Up-regulation of the 2-5A synthetase/RNase L pathway associated with chronic fatigue syndrome, *Clin. Infect. Dis.*, 1994; 18:S96–S104.
5. Suhadolnik, R.J. et al., Biochemical evidence for a novel low molecular weight 2-5A-dependent RNase L in chronic fatigue syndrome, *J. Interferon Cytokine Res.*, 1997; 17:377–385.
6. De Meirleir, K. et al., A 37-kDa 2-5A binding protein as a potential biochemical marker for chronic fatigue syndrome, *Am. J. Med.*, 2000; 108:99–105.
7. Vojdani, A., Choppa, P.C., and Lapp, C.W., Down-regulation of RNase L inhibitor correlates with up-regulation of interferon-induced proteins (2-5A synthetase and RNase L) in patients with chronic fatigue immune dysfunction syndrome, *J. Clin. Lab. Immunol.*, 1998; 50:1–16.
8. Vojdani, A. and Lapp, C.W., Interferon-induced proteins are elevated in blood samples of patients with chemically or virally induced chronic fatigue syndrome, *Immunopharm. Immunotox.*, 1999; 21:175–202.
9. Metcalf, L.N. et al., RNase-L, soluble IL-2 receptor, and IL-6 levels in patients with chronic fatigue syndrome, Honours Thesis, University of Newcastle, Australia.
10. McGregor, N.R. et al., The biochemistry of chronic pain and fatigue, *J. CFS,* 2000; 7:3–21.
11. Richards, R.S. et al., Blood parameters indicative of oxidative stress are associated with symptom expression in chronic fatigue syndrome (submitted for publication).
12. White, P.D. et al., Incidence, risk, and prognosis of acute and chronic fatigue syndromes and psychiatric disorders after glandular fever, *Brit. J. Psychiatry,* 1998; 173:475–481.
13. Dusheiko, G., Side effects of alpha interferon in chronic hepatitis C, *Hepatology,* 1997; 26:112S–121S.
14. Valentine, A.D. et al., Mood and cognitive side effects of interferon-alpha therapy, *Semin. Oncol.*, 1998; 25:39–47.
15. Allen-Mersh, T.G. et al., Relation between depression and circulating immune products in patients with advanced colorectal cancer, *J. R. Soc. Med.*, 1998; 91:408–413.
16. Richards, R.S. et al., Investigation of erythrocyte oxidative damage in rheumatoid arthritis and chronic fatigue syndrome, *J. CFS,* 2000; 6:37–46.
17. Nairn, C., Galbraith, D.N., and Clements, G.B., Comparison of coxsackie B neutralisation and enteroviral PCR in chronic fatigue patients, *J. Med. Virol.*, 1995; 46:310–313.
18. Patnaik, M. et al., Prevalence of IgM antibodies to human herpesvirus 6 early antigen (p41/38) in patients with chronic fatigue syndrome, *J. Infect. Dis.*, 1995; 172:1364–1367.
19. Sairenji, T. et al., Antibody responses to Epstein–Barr virus, human herpesvirus 6 and human herpesvirus 7 in patients with chronic fatigue syndrome, *Intervirology,* 1995; 38:269–273.

20. Nakaya, T. et al., Demonstration of Borna disease virus RNA in peripheral blood mononuclear cells derived from Japanese patients with chronic fatigue syndrome, *FEBS Lett.,* 1996; 378:145–149.
21. Player, M.R. and Torrence, P.F., The 2-5A system: modulation of viral and cellular processes through acceleration of RNA degradation, *Pharmacol. Ther.,* 1998; 78:55–113.
22. Dunstan, R.H. et al., Analysis of serum lipid changes associated with self-reported fatigue, muscle pain, and the different chronic fatigue syndrome factor analysis symptom clusters, AACFS Conf. Seattle, Jan. 2001, Abstract 62.
23. Young, V.R., The role of skeletal and cardiac muscle in the regulation of protein metabolism, in *Mammalian Protein Metabolism,* Munro, H.M., Ed., 1970, vol. 4, pp. 586–674, Academic Press, New York.
24. Pacy, P.J. et al., Post-absorptive whole body leucine kinetics and quadreceps muscle protein synthetic rate (MPSR) in the post-viral syndrome, *Clin. Sci.,* 1988; 75:36–37.
25. McGregor, N.R. et al., Preliminary determination of a molecular basis to chronic fatigue syndrome, *Biochem. Mol. Med.,* 1996; 57:73–80.
26. McGregor, N.R. et al., Preliminary determination of the association between symptom expression and urinary metabolites in subjects with chronic fatigue syndrome, *Biochem. Mol. Med.,* 1996; 58:85–92.
27. Thompson, M.G. et al., Measurement of protein degradation by release of labeled 3-methylhistidine from skeletal muscle and non-muscle cells, *J. Cell. Physiol.,* 1996; 166:506–511.
28. Mortimore, G.E. and Poso, A.R., Intracellular protein catabolism and its control during nutrient deprivation and supply, *Annu. Rev. Nutr.,* 1987; 7:539–564.
29. Goodman, M.N., Interleukin-6 induces skeletal muscle protein breakdown in rats, *Proc. Soc. Exp. Biol. Med.,* 1994; 205:182–185.
30. Choy, E.H.S. and Panayi, G.S., Cytokine pathways and joint inflammation in rheumatoid arthritis, *N. Engl. J. Med.,* 2001; 344:907–916.
31. Visser, J. et al., CD4 T lymphocytes from patients with chronic fatigue syndrome have decreased interferon-gamma production and increased sensitivity to dexamethasone, *J. Infect. Dis.,* 1998; 177:451–454.
32. Buchwald, D. et al., Markers of inflammation and immune activation in chronic fatigue and chronic fatigue syndrome, *J. Rheumatol.,* 1997; 24:372–376.
33. Gupta, S. et al., Cytokine production by adherent and nonadherent mononuclear cells in chronic fatigue syndrome, *J. Psych. Res.,* 1997; 31:149–156.
34. Gupta, S. and Vayuvegula, B., A comprehensive immunological analysis in chronic fatigue syndrome, *Scand. J. Immunol.,* 1991; 33:319–327.
35. Chao, C.C. et al., Altered cytokine release in peripheral blood mononuclear cell cultures from patients with the chronic fatigue syndrome, *Cytokine,* 1991; 3:292–298.
36. Patarca, R. et al., Dysregulated expression of tumor necrosis factor in chronic fatigue syndrome: interrelations with cellular sources and patterns of soluble immune mediator expression, *Clin. Infect. Dis.,* 1994; 18:S147–S153.
37. Buchwald, D. et al., Markers of inflammation and immune activation in chronic fatigue and chronic fatigue syndrome, *J. Rheumatol.,* 1997; 24:372–376.
38. Hattori, Y., Campbell, E.B., and Gross, S.S., Argininosuccinate synthetase mRNA and activity are induced by immunostimulants in vascular smooth muscle. Role in the regeneration or arginine for nitric oxide synthesis, *J. Biol. Chem.,* 1994; 269:9405–9408.

39. Geng, Y., Hansson, G.K., and Holme, E., Interferon-gamma and tumor necrosis factor synergize to induce nitric oxide production and inhibit mitochondrial respiration in vascular smooth muscle cells, *Circ. Res.,* 1992; 71:1268–1276.
40. Bolanos, J.P. et al., Nitric oxide-mediated inhibition of the mitochondrial respiratory chain in cultured astrocytes, *J. Neurochem.,* 1994; 63:910–916.
41. Dijkmans, R. and Billiau, A., Interferon-gamma/lipopolysaccharide-treated mouse embryonic fibroblasts are killed by a glycolysis/L-arginine-dependent process accompanied by depression of mitochondrial respiration, *Eur. J. Biochem.,* 1991; 202:151–159.
42. McGregor, N.R., An investigation of the association between toxin-producing staphylococcus, biochemical changes, and jaw muscle pain, Ph.D. thesis, University of Sydney, 2000.
43. Moraga-Amador, D.A. et al., Asparagine catabolism in rat liver mitochondria, *Arch. Biochem. Biophys.,* 1989; 268:314–326.
44. Bahl, J.J. et al., *In vitro* and *in vivo* suppression of gluconeogenesis by inhibition of pyruvate carboxylase, *Biochem. Pharmacol.,* 1997; 53:67–74.
45. Wong, R. et al., Skeletal muscle metabolism in the chronic fatigue syndrome. *In vivo* assessment by 31P nuclear magnetic resonance spectroscopy, *Chest,* 1992; 102:1716–1722.
46. Lane, R.J. et al., Muscle fibre characteristics and lactate responses to exercise in chronic fatigue syndrome, *J. Neurol. Neurosurg. Psych.,* 1998; 64:362–367.
47. Plioplys, A.V. and Plioplys, S., Serum levels of carnitine in chronic fatigue syndrome: clinical correlates, *Neuropsychobiology,* 1995; 32:132–138.
48. Barnes, P.R. et al., Skeletal muscle bioenergetics in the chronic fatigue syndrome, *J. Neurol. Neurosurg. Psych.,* 1993; 56:679–83.
49. Thompson, C.H. et al., Effect of chronic uraemia on skeletal muscle metabolism in man, *Nephrol. Dial. Transplant.,* 1993; 8:218–222.
50. Coderre, T.J., The role of excitatory amino acid receptors and intracellular messengers in persistent nociception after tissue injury in rats, *Mol. Neurobiol.,* 1993; 7:229–246.
51. Coderre, T.J. and Yashpal, K., Intracellular messengers contributing to persistent nociception and hyperalgesia induced by L-glutamate and substance P in the rat formalin pain model, *Eur. J. Neurosci.,* 1994; 6:1328–1334.
52. Haynes, W.G. et al., Physiological role of nitric oxide in regulation of renal function in humans, *Am. J. Physiol.,* 1997; 272:F364–F371.
53. Wilder, R.L., Neuroendocrine-immune interactions and autoimmunity, *Annu. Rev. Immunol.,* 1995; 13:307–338.
54. Bowles, N.E. et al., The detection of viral genomes by polymerase chain reaction in the myocardium of pediatric patients with advanced HIV disease, *J. Am. Coll. Cardiol.,* 1999; 34:857–865.
55. Monkemuller, K.E. and Wilcox, C.M., Diagnosis and treatment of colonic disease in AIDS, *Gastro. Endo. Clin. N. Am.,* 1998; 8:889–911.
56. Lerner, A.M., Cardiac involvement in patients with CFS, *Proc. Intl. Conf. CFS,* Sydney, Australia, 1998, pp. 21–25.
57. McGregor, N.R. et al., Assessment of the plasma lipid homeostasis in relationship to Epstein–Barr virus titres in patients with chronic fatigue syndrome, *Proc. Intl. Conf. CFS,* Sydney, Australia, 1998, pp. 267.
58. Choppa, P.C. et al., Multiplex PCR for the detection of *Mycoplasma fermentans, M. hominis,* and *M. penetrans* in cell cultures and blood samples of patients with chronic fatigue syndrome, *Mol. Cell Probes,* 1998; 12:301–308.

59. Vojdani, A. et al., Detection of *Mycoplasma* genus and *Mycoplasma fermentans* by PCR in patients with chronic fatigue syndrome, *FEMS Immunol. Med. Micro.,* 1998; 22:355–365.
60. Horowitz, S. et al., Antibodies to *Mycoplasma fermentans* in HIV-positive heterosexual patients: seroprevalence and association with AIDS, *J. Infect.,* 1998; 36:79–84.
61. Gong, H. et al., Arginine deaminase inhibits cell proliferation by arresting cell cycle and inducing apoptosis, *Biochem. Biophys. Res. Comm.,* 1999; 261:10–14.
62. Paddenberg, R. et al., Mycoplasma nucleases able to induce internucleosomal DNA degradation in cultured cells possess many characteristics of eukaryotic apoptotic nucleases, *Cell Death Different.,* 1998; 5:517–528.
63. McGregor, N.R. et al., Assessment of pain (distribution and onset), symptoms, SCL-90-R inventory responses and the association with infectious events in patients with chronic orofacial pain, *J. Orofacial Pain,* 1996; 10:339–350.
64. Butt, H.L. et al., An association of membrane damaging toxins from coagulase-negative staphylococcus and chronic orofacial muscle pain, *J. Med. Microbiol.,* 1998; 47:577–584.
65. McGregor, N.R. et al., Coagulase-negative staphylococcal membrane damaging toxins are associated with pain-related metabolic changes in chronic research diagnostic criteria type 1a muscle pain patients, submitted for publication.
66. Bombardier, C.H. and Buchwald, D., Chronic fatigue, chronic fatigue syndrome, and fibromyalgia. Disability and health-care use, *Med. Care,* 1996; 34:924–930.
67. Buchwald, D. et al., Functional status in patients with chronic fatigue syndrome, other fatiguing illnesses, and healthy individuals, *Am. J. Med.,* 1996; 101:364–370.
68. Emms, T.M. et al., Food intolerance in chronic fatigue syndrome, AACFS Conf., Seattle, Jan. 2001, Abstract 15.
69. Rowe, P.C. and Calkins, H., Neurally mediated hypotension and chronic fatigue syndrome, *Am. J. Med.,* 1998; 105:15S–21S.
70. De Lorenzo, F., Hargreaves, J., and Kakkar, V.V., Pathogenesis and management of delayed orthostatic hypotension in patients with chronic fatigue syndrome, *Clin. Autonomic Res.,* 1997; 7:185–90.
71. Bou-Holaigah, I. et al., The relationship between neurally mediated hypotension and the chronic fatigue syndrome, *JAMA,* 1995; 274:961–967.

8 CFS Etiology, the Immune System, and Infection

Kenny De Meirleir, Pascale De Becker, Jo Nijs, Daniel L. Peterson, Garth Nicolson, Roberto Patarca-Montero, and Patrick Englebienne

CONTENTS

0-8493-1046-6/02/$0.00+$1.50

8.1 ETIOLOGY AND TRIGGERING FACTORS

Over the years, a large number of studies have been conducted to unravel the pathogenesis of chronic fatigue syndrome (CFS). CFS has been attributed to a variety of infectious agents, including Epstein–Barr virus (EBV), human herpesvirus 6 (HHV-6), cytomegalovirus, enteroviruses, retroviruses, stealth viruses, Borna virus, and Ross River virus.[1-11] However, the results of these studies are not consistent.[12-17] Many of these viruses (EBV and HHV-6) are endemic within human beings, with infections occurring early in life.[2] Therefore, a new primary infection by these viruses is unlikely. Most evidence points to a reactivation of these endemic viruses, rather than primary infection.[12,14,18] The potential role of several infectious agents will be outlined in detail in this chapter.

Not being able to identify a viral agent does not mean *per se* that a new viral infection did not take place. Levy proposed a "hit and run" effect, whereby a virus might infect the host, cause immune abnormalities leading to CFS, and then be eliminated, leaving the immune system in an activated state.[18] Alternate hypotheses suggest that bacterial stealth infections, including Brucella species,[10] Mycoplasma species,[19,20] and *Chlamydia pneumoniae*[21] may be important in patient morbidity. Thus, although it is clear that no single etiologic agent can be unequivocally associated with most cases of CFS, a number of infections do seem to precede its development.

Studies investigating the precipitating factors for CFS reveal a high percentage of patients who attribute their disorder to some kind of infectious agent. Salit observes that 72% of CFS patients report an apparently infectious illness associated with the development of CFS.[22] Earlier studies also report a similar percentage of viral illnesses preceding the onset of CFS.[11,23-26] Alternatively, other investigators suggest that noninfectious factors may also play a role in the etiopathogenesis of CFS. In the 3 months to 1 year preceding CFS, stressful and negative life events take place very commonly in patients who later develop CFS.[22,27] It is clear that there is no consensus among researchers and clinicians regarding the onset of CFS and there is a high degree of heterogeneity in the results.

Thus, a number of infectious agents and environmental factors may serve as triggers for immunoregulatory abnormalities known to persist in these patients. In a large retrospective study by De Becker et al. including 1546 patients,[28] infectious agents seem to play an important role in the onset of CFS (Table 8.1). Upper respiratory tract infection was the most common preceding illness before the development of CFS. There were ten distinct groupings of factors involved in the onset and most involved infectious events. Blood transfusion and hepatitis B vaccination were also important in two onset clusters. It was concluded that the hypothesis of immune dysregulation after a (viral) infection or persistent stealth infections remains eligible and that the simultaneous occurrence of infectious and noninfectious factors seem to be important onset-associated events in CFS.

8.2 THE IMMUNE SYSTEM IN CFS

Immunological function in CFS has been the focus of intense studies by numerous investigators. These studies always aimed at detecting an underlying immunological

TABLE 8.1
Prevalence of Various Onset Factors in CFS

Onset Parameter	Total CFS N (%)	Holmes Criteria N (%)
Upper respiratory tract infection	376 (24.3)	297 (48.0)*
Unknown gradual	237 (15.3)	208 (33.6)*
Flu-like illness	213 (13.8)	47 (7.6)*
Viral infection	170 (11.0)	35 (5.7)*
Other infection (bacterial)	163 (10.5)	23 (3.7)*
Stress	152 (9.8)	33 (5.3)*
Mononucleosis	142 (9.2)	31 (5.0)**
Motor vehicle accident	103 (6.7)	27 (4.4)***
Pneumonia	103 (6.7)	16 (2.6)*
Blood transfusion	98 (6.4)	31 (5.0)
Insect bites	79 (5.1)	14 (2.3)**
Hepatitis B vaccine	78 (5.0)	17 (2.7)***
Unknown sudden	75 (4.8)	20 (3.2)
Gastrointestinal infections	59 (3.8)	22 (3.6)
Postpartum	54 (3.5)	29 (4.7)
Surgery	47 (3.0)	18 (2.9)
Sinusitis	43 (2.8)	21 (3.4)
Pain	22 (1.4)	14 (2.3)
Hepatitis	21 (1.4)	5 (0.8)
Sleep disturbance	20 (1.3)	2 (0.3)***
CMV	18 (1.2)	6 (1.0)
Mental trauma	12 (0.8)	2 (0.3)
Toxoplasmosis	11 (0.7)	3 (0.5)
Intoxication	10 (0.6)	5 (0.8)
Meningitis	5 (0.3)	1 (0.2)
Neurological problems	4 (0.3)	2 (0.3)

Note: Holmes were compared with the Fukuda criteria. Statistical method = μ^2 analysis; *P*-values — * <0.001; ** <0.01; *** <0.05.

abnormality that could be a cause for CFS. Another aim was to discover and characterize secondary immune abnormalities that are in reality responses to an underlying infection and that cause the manifestation of the syndrome as an unintended side effect. When we review the literature on the immunology of CFS, two problems seem to be consistently present in these patients: poor cellular function and immune activation. Furthermore, there is increased incidence of allergy and autoimmunity. Several publications suggest a long-term systemic shift toward a T_H2-cytokine balance in CFS patients.[29,30]

8.2.1 Poor Cellular Function

A number of controlled studies have examined both the NK number and function (natural killer cell cytotoxicity) in patients with CFS. Estimations of NK cell numbers in the peripheral blood were variable: low,[31,32] normal,[12,33-35] or elevated.[36-39] Although the CFS cases were carefully characterized, none of these studies controlled for factors such as medication, alcohol ingestion, smoking, or state of anxiety and mood. It is therefore important that future studies of NK cells in patients with CFS include analyses of these confounding factors.[40] Despite the contrasting results of natural killer cell numbers, several studies have found consistent evidence of impaired NK cell function in CFS patients.[34,36,41-45] The changes in NK cell cytolytic activity are probably related to different factors:

1. CD56+CD3– cells are the lymphoid subset with the highest NK activity and their decrease has a high impact on NK cell activity.
2. Reduction in CD4+CD45+T cells may also result in decreased induction of suppressor/cytotoxic T cells.
3. Changes in the ability of the NK cells to respond to IL-2 and interferon-γ.

The exact mechanism for this remains unclear. Among the many possible explanations put forward, the study of Ogawa et al.[46] fits best with the model we present in this book. They reveal a possible dysfunction in the nitric oxide (NO)-mediated NK cell activation in CFS patients, as arginine enhances NK cell activity in controls but not in CFS patients. See et al.[47] in an *in vitro* study showed that addition of a glyconutrient compound to peripheral cells of CFS patients enhanced NK cell activity. To this we can add that NO itself is found to be toxic to NK cells. As more NO is released in the peripheral circulation of CFS patients, this may be a direct explanation for low NK cell function. Vitamin B12, an NO· scavenger, has a favorable effect on NK cell function.[48] Furthermore, the phagocytic function of NK cells is probably decreased because their cytoskeleton is damaged. G-actin is cleaved by calpain and caspase 3 (Chapter 6).

Another abnormality in cellular function is the poor lymphocyte response to mitogens in culture. Controlled studies of T-cells in CFS patients have included the enumeration of the peripheral blood T-cell subsets by flow cytometry, an assessment of their proliferative capacity after stimulation *in vitro* by mitogens, and the *in vivo* evaluation of their capacity to respond to previously encountered antigens via delayed-type hypersensitivity (DTH) skin testing.[12,33,36,41,49-51] Depressed responses to phytohemagglutinin (PHA) and pokeweed mitogen (PWM), an indication of dysfunction in cellular immunity, were found in the CFS patients studied by most teams,[33,36,41,49-55] while Mawle and coworkers[56] found no change. In terms of B-cell function, spontaneous and mitogen-induced immunoglobulin synthesis is also affected; in 10% of patients the immunoglobulin synthesis is depressed.[57,58] CFS patients have decreased amounts of immunoglobulins of A, G, M, or D classes.[4,41,42,50,57,59-61] IgG subclass deficiency (particularly of IgG1 and IgG3) can be demonstrated in a substantial percentage of CFS patients.[36,44,51,61,62] A defective humoral immune response to the infections thought to precipitate CFS may allow antigens to evade immune mechanisms.[40] Spontaneous and mitogen-stimulated immunoglobulin synthesis is depressed in 10% of patients with CFS.[36,58]

Studies of the humoral immune response to specific agents thought to initiate CFS have focused particularly on the herpes and enterovirus groups. The pattern of development and persistence of IgG antibodies to these viral antigens was previously proposed as a laboratory marker for CFS. However, neither the presence nor the titers of IgG antibodies to these viruses has any diagnostic usefulness in CFS,[63,64] although polyclonal activation of antibodies to herpesviruses (EBV, herpes simplex, CMV, and HHV-6) has been demonstrated.[12,63] This antibody production may occur as a result of reactivation of latent viral infection[64] or of altered T-cell regulation of immunoglobulin production.[40]

8.2.2 Immune Activation, Cytokines, Allergy, and Autoimmunity

The literature of CFS reports elevated numbers of activated CD3+, CD4+, and CD8+ T lymphocytes (HLADR+, CD25+, CD26+, CD38+). These changes, however, have been insufficiently documented to allow drawing a possible association with specific symptoms or subgroups of patients. They may be the result of an ongoing illness-related process, persistent opportunistic infection, or intercurrent, transient common condition. Stimulated lymphoid cells either express or induce the expression in other cells of a heterogeneous group of soluble mediators that exhibit either effector or regulatory functions. The regulation of a cell-mediated immune response to antigens is critically dependent upon the activity of cytokines, which function as intercellular messengers. Although essential for effective immunity, cytokines may also directly produce clinical symptoms including central nervous system and muscle symptoms.[40]

The decreased NK cell cytotoxic and lymphoproliferative activities and the increased allergic and T_H2 autoimmune manifestations in CFS would be compatible with the hypothesis that the immune system of affected individuals is biased toward a T_H2 type or humoral immunity-oriented cytokine pattern.[29] Vaccines and stressful stimuli known to be able to trigger CFS[22,27] have been shown to lead to long-term, nonspecific shifts in cytokine balance.[30]

Several groups have noted alterations in cytokine levels in patients with CFS. Increased levels of interleukin-1,[65,66] interleukin-2,[67] interleukin-6,[68-70] neopterin,[70-71] interferon-α,[72,73] tumor necrosis factor-α and -β,[65] and tumor growth factor-β[56,65,69,74] have been reported in patients with CFS compared to normal controls. However, there are just as many studies which report normal levels of these cytokines. Normal levels have been reported in CFS patients for interleukin-1,[33,39] interleukin-2,[65,66] interleukin-6,[25,56,65,66,69,71] neopterin,[65,66] interferon-α,[66,75] tumor necrosis factor-α and -β,[25,39,69,75,76] and tumor growth factor-β.[25]

Cytokine production by cultured lymphocytes following mitogenic stimulation has also yielded conflicting results. *In vitro* production by mitogen-stimulated lymphocytes of interleukin-1β,[76] interleukin-2,[77] interferon-γ,[36] transforming growth factor-β,[69] and tumor necrosis factor-α[76] were decreased. By contrast, increased production of interleukin-1β,[69] interleukin-2,[17] interleukin-6,[69] and tumor necrosis factor-α,[69] as well as normal production of interleukin-1β,[56,69] interleukin-2,[56,69] interleukin-6,[56] interferon-γ,[56] transforming growth factor-β,[69] and tumor necrosis factor-α[69] by cultured lymphocytes following stimulation with various mitogenic agents have been observed in patients with CFS. Spontaneous production of several proinflammatory cytokines including interleukin-1α, interleukin-1β, interleukin-6, and tumor necrosis factor-α and -β by cultured lymphocytes has consistently been shown to be normal in patients with CFS compared to healthy controls.[56,69,76] Interestingly, some of these proinflammatory cytokines are known to induce symptoms similar to those reported in patients with CFS.[74] In particular, data from clinical trials of interferon have provided indirect evidence which supports the possible involvement of interferon in CFS.[78,79]

Investigators have reported clusters of cytokine abnormalities in patients with CFS. One cluster of abnormalities includes tumor necrosis factor receptor type 1, soluble interleukin-6 receptor, and β2-microglobulin. The other cluster of abnormalities includes tumor necrosis factor-α, interleukin-1α, interleukin-4, soluble interleukin-2 receptor, and interleukin-1 receptor antagonist.[80] A recent study showed a significant increase in serum tumor necrosis factor-α in patients with CFS.[81] Whether these abnormalities have any relationship to the symptoms reported by patients with CFS remains unclear. In the only study to examine this question, clinical improvement in CFS was not associated with changes in lymphocyte subsets or activation.[39]

8.2.2.1 Circulating Immune Complexes

Elevated levels of immune complexes have been reported in a number of studies,[41,49,58] in contrast to two studies which could not find differences in the level of circulating immune complexes when CFS patients were compared with controls.[35,56] Depressed levels of complement have also been reported in 0 to 25% of patients.[35,41,49,56,58] Buchwald and colleagues found elevated levels of C-reactive protein among CFS patients.[71]

8.2.2.2 Allergies

There is an apparent association between atopy and CFS. A history of atopy (inhalant, food, or drug allergy) and skin reactivity to food and inhalant antigens were found in a high percentage of CFS patients.[82-84] Also, nickel allergy is a common finding in CFS patients.[85] The atopy could be a manifestation of a disturbed cell-mediated immunity which can also predispose to an aberrant response to viral or intracellular infection, perhaps via altered cytokine production.[40]

Respectively, 90 and 83% of CFS patients have positive histories of allergy or positive immediate skin tests. The immediate cutaneous reactivity rate to selected allergens of minimally 50%[84,86] is much higher than the 20 to 30% observed in unselected Caucasian adults of similar age reported in the literature.[86] Straus and associates[84] have suggested that the increased reactivity of patients to allergens that results in atopic disease may also result in CFS because of the heightened response to the putative infection causing CFS. Immunological studies have demonstrated a positive correlation between allergy severity and the magnitude of the Epstein–Barr virus serological response. In atopic patients, higher titers of EBV antibodies have been demonstrated when compared to controls.[87] CFS patients with a positive EBV serology show an enhanced allergen-induced responsiveness compared with patients suffering from allergies in the absence of CFS.[88] Recently, a remarkably high percentage of bronchial hyperreactivity has been demonstrated in CFS patients.[89]

8.2.2.3 Autoantibodies

Konstantinov et al.[90] found that approximately 52% of sera from CFS patients react with nuclear envelope antigens. The presence of rheumatoid factor,[50,52,55,91-93] antinuclear antibodies,[17,50,52,55,91,94-97] antithyroid antibodies,[49,55,98] antismooth-muscle antibodies,[49] antigliadin, cold agglutinins, cryoglobulins, and false serological positivity for syphilis[49] have also been reported.

8.3 INFECTIOUS AGENTS

The sudden onset of CFS, often after an infectious-like period[23,24] and recurrent flu-like symptoms,[99,100] has prompted several researchers to study the role of infectious agents in CFS.

8.3.1 VIRUSES

In the search for an etiologic agent, three families of viruses have been primarily discussed: herpesviruses, enteroviruses, and retroviruses.[18] More particular attention has been focused on the following DNA and RNA viruses in connection with CFS: Epstein–Barr virus (EBV), human herpesvirus 6 (HHV-6), group B Coxsackie viruses, HTLV (human T-lymphotropic virus)-II-like virus, spumavirus, hepatitis C virus, and human herpesvirus 7 (HHV 7). Also a novel type of CMV-related stealth viruses[8,101] has been mentioned. These viruses are thought to have significant DNA homologies to CMV and have been isolated repeatedly from one patient with CFS.[102]

A possible link between active infection with Borna disease virus, a virus that can produce central nervous system disease in animals, has also been suggested.[9,103,104] Chronic parvovirus infection can also result in CFS,[105] and several studies suggest that CFS may follow Ross River virus infection.[11,51]

8.3.1.1 EBV

Due to the finding that CFS occasionally develops after an episode of infectious mononucleosis and evidence of high titers of antibodies to Epstein–Barr virus cited in the early reported cases of CFS,[1,17,50,55,63,106] an infection of the B lymphocytes by EBV was first considered the cause of CFS.[18] Atypical profiles of antibody responses to EBV were found in many cases,[50,63] but could not be confirmed in other studies[36,35,55,107,108] and there was considerable overlap between patients and controls, as well as those who had recovered from infectious mononucleosis.[1,106,107,109] Controlled studies of seroepidemiology[1,17,63] and antiviral therapy[110] have shown that EBV infection cannot be the sole explanation for most cases of CFS. Moreover, no correlation has been found between serologic parameters of EBV activity and the incidence of CFS.[17] In any case, a causal link between EBV and CFS could not be established,[1,66,96,111] and detection of antibodies to EBV is not considered helpful in the clinical evaluation of CFS patients.[63,107] The relationship between EBV and illness illustrates the complexity of interaction between pathogen and host. It is more likely an epiphenomenon of the subtle immunologic abnormalities that characterize the syndrome. Either the syndrome permits more frequent reactivation of latent EBV, which in turn stimulates antibody responses, or a nonspecific polyclonal activation augments the titers of antibodies to many viruses.[112] Because antibody titers to other viruses such as CMV, HSV-1 and 2, and measles are also elevated in CFS, the latter scenario is the more appealing.[63] The evidence is thus growing that the serologic findings of an enhanced EBV state in CFS patients, as well as the subsequent reports of increased antibody titers to other viruses, reflect a generalized underlying immunologic dysfunction in those patients.[109]

8.3.1.2 HHV-6

Human herpesvirus type 6 (HHV-6) is another herpesvirus that infects T lymphocytes.[18] Antibodies to HHV-6 have also been found to be increased in CFS patients, implying an etiologic role.[3-5,112-114] The induction of certain cytokines and the *in vitro* infection of natural killer cells provides indirect evidence of the involvement of HHV-6 in CFS.[18] When reactivated or during reinfection, HHV-6 may contribute to the symptoms of CFS.[18] There is much more evidence of involvement of HHV-6 in CFS compared to other human herpesviruses such as EBV, cytomegalovirus, herpes simplex virus types 1 and 2, varicella-zoster virus, and HHV-7.[5,12-14,18,31,59] Thus, the report that recurrent or persistent HHV-6 infection is present in 73% of CFS patients[59] may not have much etiological significance.[115] HHV-6, like EBV, may be a ubiquitous virus reactivated in the host as the result of an altered immune system, but not necessarily a causative factor.[12-14,18,63] Both these herpesviruses are most likely reactivated as the consequence of an altered immune system.[5,12,14,18] Therefore, although perhaps contributing to some of the symptoms in CFS, neither of these herpesviruses nor cytomegalovirus can be linked etiologically with CFS.[2,12,14,18,63,107] Recently, Herst et al. reported a strong correlation between HHV-6 infection and upregulation of 2'-5'A synthetase/RNase L enzyme activity in CFS and MS. In this study the evidence of HHV-6 infection was determined by nested PCR, the presence of IgM antibody, an elevated IgG antibody, and short-term culture of PBMC's.[116]

8.3.1.3 Enteroviruses

In approximately 20% of a selected patient population, a persistent enterovirus infection was identified.[117,118] Antibodies to the virus have been detected, and enterovirus antigens and RNA have been found in the muscle and in other tissues of CFS patients.[49,117,119] Antibodies to Coxsackie B virus are commonly found in the blood of CFS patients.[117,120,121] This is only indirect evidence since these viruses are common in the community[118] and the percentage positive rates of antibodies against Coxsackie B virus reflects the background in the population and therefore does not help diagnose CFS.[64] Also, Coxsackie virus-specific RNA has been reported in skeletal muscle biopsies,[119,122] but again the evidence is not always consistent.[123] Gow and colleagues detected enterovirus sequences and virus isolation from patients with CFS, but he did not find any specific enterovirus type that could be linked to CFS.[6] However, the serological observations have not been confirmed by other laboratories.[64,76] A recent study indicates a persistent enteroviral infection in a small group of CFS patients.[124]

8.3.1.4 HTLV

Among the retroviruses, two major subgroups have been cited as probable agents of CFS: human T-cell leukemia virus type 2 (HTLV-II) and the spumaviruses.[7,18] There is evidence for HTLV-II-like infection of white blood cells from CFS patients and the presence of antibodies to the major core protein of this retrovirus has also been reported.[7] Interestingly, not only the samples from CFS patients were associated with HTLV-II-like genes and HTLV-II reactive antibodies, but also samples from a significant proportion of their nonsexual contacts were positive, though to a lesser extent.[7] This may not be unimportant, since in some studies horizontal, casual transmission is suggested, as in the epidemics of Lake Tahoe[63] and Lyndonville.[120] Rather than an etiologic agent, it may be a benign secondary infection to which immunologically compromised patients are susceptible. Or it may be one of two viruses that, when coinfecting the same hematopoetic cells, induce immune dysfunction.[7] Despite the initial positive results,[7] these data could not be confirmed in other laboratories.[15,16] One laboratory reported the recovery of large numbers of strains of spumaviruses from CFS patients.[125] Other investigators did not find evidence of retroviral activity in CFS patients.[12,13,15,16,64]

8.3.1.5 Other Viruses

HHV-7 was isolated from the peripheral blood mononuclear cells of a CFS patient, but the antibody titers were not significantly increased in CFS compared to controls.[126] Varicella-zoster virus and hepatitis B virus have also been implicated in CFS.[127] Patients developing CFS after recombinant hepatitis B immunization have anecdotally been reported. Given the fact that hepatitis B can trigger the onset of multiple sclerosis (MS) in some cases, the authors suggest that CFS can be an autoimmune reaction without the consequences of demyelinization seen in MS.[128]

test for hep B

8.3.2 Bacteria and Parasites

Microorganisms such as Mycoplasma, Chlamydia, Brucella, Rickettsia, Coxiella, and possibly others appear to be important in causing patient morbidity or exacerbating the major signs and symptoms observed in patients with chronic illness.[20] Recently a bacterium which received the provisional name of Human Blood Bacterium (Lindner and MacPhee — U.S. Patent 6,255,467 B1 — issued July 3, 2001) has also been associated with morbidity in CFS.

8.3.2.1 Mycoplasma

Several research groups have been interested in the association of these chronic infectious agents with CFS because these microorganisms of the class Mollicutes can potentially cause a lot of the symptoms found in these patients.[129] This a class of bacteria lacking cell walls, and some species are capable of invading several types of human cells and tissues and are associated with a wide variety of human diseases.[130]

ref

Using highly sensitive and specific forensic polymerase chain reaction (FPCR) or PCR with single or multiple sets of primers, several groups have analyzed the leukocytes of CFS patients for the presence of mycoplasmas.[131-135] In 50 to 70% of these samples one or more mycoplasmal infections were detected. When an FPCR with genetracking for *M. fermentans, M. pneumoniae, M.penetrans,* and *M. hominis* was performed, multiple infections were found in 17 to 50% of the CFS patients.[132,135] CFS patients with multiple infections generally had a longer history of illness, suggesting that they may have contracted additional infections during their illness.[132] Studies that reported on a sufficiently large number of controls show that less than 17% (range 9.9 to 16.8%) of them carry *mycoplasma spp.* Mycoplasmas have been proposed to interact nonspecifically with B-lymphocytes, resulting in modulation of immunity, autoimmune reactions, and promotion of rheumatic diseases.[136] They have also been associated with the progression of autoimmune and immunosuppressive diseases such as HIV-AIDS. Blanchard and Montangier[137] have proposed that HIV-AIDS progression can be enhanced by mycoplasmas like *M. fermentans* which could in part account for the increased susceptibility of AIDS patients to additional opportunistic infections.

There is further evidence that mycoplasmal infections are associated with various autoimmune disorders, such as multiple sclerosis, amyotrophic lateral sclerosis, lupus, autoimmune thyroid disease, and possibly others.[129] The autoimmune signs and symptoms could be the result of intracellular pathogens such as mycoplasmas in these disorders escaping from cellular components and incorporating into their own structures pieces of host cell membranes that contain important host antigens. This may trigger autoimmune responses; mycoplasma surface components may also stimulate directly autoimmune responses, possibly due to molecular mimicry of host antigens.[138]

Chronic airway disorders such as asthma, airway inflammation, and bronchial hyperreactivity (known to be present in CFS patients), are also known to be associated with mycoplasmal infections.[89,139-141] In a group of 206 CFS patients, using PCR with nucleoprotein gene tracking, we found that approximately 70% showed mycoplasmal infection;[135] compared to the patients who were *mycoplasma spp.* negative, the *mycoplasma spp.* positive CFS patients had a significantly higher LMW $\times$ 10/HMW RNase L ratio. Mycoplasmas could cleave the RNase L using the following mechanism: they have been shown to overexpress an apoptotic-like endonuclease,[142] which acts on the nuclear fraction of the host cells. Small base-pair fragments (< 25 base pairs) induce bad activation of the 2'-5' A synthetase and leave the monomeric RNase L unprotected to cleavage by calpain, elastase, and Cathepsin G. Mycoplasmas could also induce a direct proteolytic cleavage of the RNase L (Figure 8.1).

8.3.2.2 Chlamydia

Chlamydiae are intracellular prokaryotic cells, frequently associated with fatigue in infected patients.[21] Currently, the following species are recognized to be associated with human disease: *C. pneumoniae, C. trachomatis,* and *C. psittaci*. *C. pneumoniae* is transmitted by the respiratory route[143] and is a widely recognized pathogen in atherosclerosis,[143-147] sinusitis,[144] bronchitis,[144,146] pneumonia,[144] and adult-onset asthma[146,147] as well as myocarditis, pericarditis, and endocarditis.[143,145] *C. pneumoniae* has been suggested to contribute to patients' morbidity in subsets of chronic

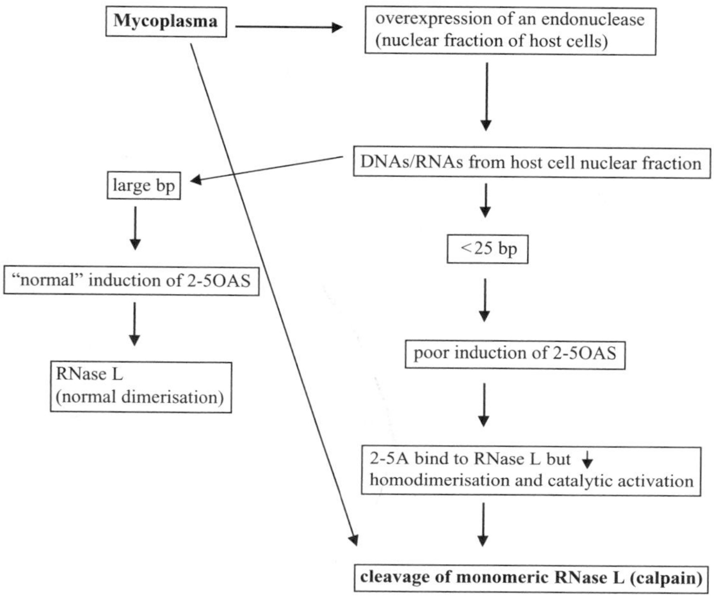

FIGURE 8.1 Proposed mechanism by which mycoplasmas could contribute to the cleavage of RNase L. (Abbreviations: DNA = desoxyribonucleic acid; RNA = ribonucleic acid; bp = base pairs.)

fatigue syndrome[21] and fibromyalgia.[148] Indeed, Chia and Chia[21] found elevated titers of antibody to *C. pneumoniae* in 10 out of 171 patients with symptoms of chronic fatigue long after initial respiratory infection. Most patients responded well to a 1- to 2-month course of azithromycin. Elevated antibody titers of *C. pneumoniae* are present in ±25% of our patients (De Meirleir, personal communication). Machtey[148] suggested that the possible etiopathogenic connection of *C. pneumoniae* and fibromyalgia should be considered after he established a prevalence rate of 18/23 (78.3%) (*C. pneumoniae* antibodies). In addition, *Chlamydia trachomatis* antibodies were found in 12 out of 21 fibromyalgia patients (57.1%).[148] These epidemiological data should be interpreted with caution due to the high prevalence in asymptomatic humans (more than 50% of the overall population, with increasing prevalence rates with age).[143-145,149] Adult women appear to have antibodies less often, while virtually everyone will be infected at some time in his or her life (and reinfection is common). Many infected asymptomatic persons function as transmitters of *C. pneumoniae* infection, probably by respiratory tract secretions.

Intracellular infection with *C. pneumoniae* is initiated by a metabolically inactive form of the pathogen (an "elementary body").[150] These elementary bodies differentiate into large metabolic active reticulate bodies, which in turn divide to form intracellular inclusion bodies. Infection is spread by redifferentiation of reticulate bodies into elementary bodies, which are released by exocytosis or host cell lysis.[147,150] Body defenses against *C. pneumoniae* consist of numerous pathophysiological responses. First, despite the intracellular nature of these bacterial infections, neutralizing antibodies are produced.[150] However, these effector molecules of the humoral immune system provide a defensive action of limited efficacy.[145] Interestingly, Rottenberg and colleagues[144] found no evidence for a shift to a T_H2 cytokine profile in *C. pneumoniae*-infected mice. Generally, two different stages characterize immunity to *C. pneumoniae*: First, IFN-γ inhibits bacteria growth during the early stages,[144,146,149,151] and later there is an adaptive immune response consisting of CD4+ and CD8+ T-cells.[144] IFN-γ stimulates the expression of a specific enzyme (indoleamine 2.3-dioxygenase) which degrades L-tryptophan.[144,150] Consequently, *Chlamydia* replication is inhibited. IFN-γ augments nitric oxide production as well, providing an additional pathway to inhibit chlamydial growth; IL-12 is suggested to enhance IFN-γ mRNA.

These early stages of body defenses against *C. pneumoniae* do not require natural killer cell cytotoxicity and associated perforin-mediated exocytosis,[144] nor CD8+-mediated perforin-release.[149] Nevertheless, recent data indicate that other unidentified factors (partially from IFN-γ) are involved in the mechanisms of growth limitation in *C. pneumoniae*-infected human monocytes.[146] Concerning cell-mediated immunity, some studies have revealed a stronger T-helper 1 (T_H1)-type local immune response in the lungs of *C. pneumoniae*-infected mice.[145] As in *C. trachomatis* and *C. psittaci* infections, T-cell responses are essential in body defenses against *C. pneumoniae*. INF-γ is believed to be of great importance for cell-mediated immunity against these pathogens.[145,151] Therefore, CFS patients with a deregulated 2-5A synthetase/RNase L antiviral pathway and associated TNF-α converting enzyme (TACE)-degradation by elastase[152] appear to be more vulnerable to chronic *C. pneumoniae*-infection.

C. pneumoniae-infected monocytes induce the proliferation of CD4+ T-lymphocytes. CD4+ cells promote bacterial growth and disease early after infection, while they play a protective role (control of growth and protection against reinfection) later.[149] CD8+ T cells inhibit this early deleterious activity of CD4+ cells. Acquired immunity against *C. pneumoniae* was shown to be CD8+-dependent (CD8+ cells are dominant over CD4+). Nevertheless, the exact role of cytotoxic T-cells in chlamydial infections remains to be established, since there is no evidence available pointing to cytotoxic T-lymphocyte activity. CD8+ T-cell protection is mediated, at least partly, through IL-4, IL-10, and IFN-γ.[149] Research results from Halme et al.,[151] however, suggest that overproduction of IL-10 at some stages may inhibit the immune response and IFN-γ secretion upon *C. pneumoniae* infection, increasing in turn the susceptibility to reinfection. Indeed, IL-10 augments T_H2-type immunity, while inhibiting T_H1-responses. Additionally, *C. pneumoniae*-infected smooth muscle cells produce significant amounts of bFGF (basic fibroblast growth factor) and IL-6.[147] IL-6 activates B-cells, while bFGF is a multifunctional cytokine involved in proliferation and differentiation of many cell types. (It enhances collagenase expression.)

Survival of *C. pneumoniae* relies on host cell integrity. Indeed, intracellular chlamydial growth is established by nutrients supplied by the host cells.[153] Of course, host cells protect intracellular pathogens from phagocytosis.[153] Therefore, developing strategies to inhibit host cell apoptosis may be of great value for these intracellular pathogens. Recently, researchers have provided evidence for such strategies in *C. pneumoniae*-infected human blood peripheral mononuclear cells.[153] Resistance of *C. pneumoniae* to apoptosis appears to be mediated through IL-10, possibly downregulating IL-12 and

INF-γ. (INF-γ activates caspases and synthesis of nitric oxide (NO), the latter causing damage to DNA.) In addition, Rödel and colleagues[147] found that chlamydiae are slow inducers of cellular cytokine responses in contrast to other bacterial pathogens. The outlined mechanisms are essential for *C. pneumoniae* to generate chronic, systemic infections. Monocytes and macrophages have recently been shown to be responsible for dissemination of the infection from the respiratory tract to other organs,[154] providing a theoretical basis for multiorgan involvement, as seen in CFS.

8.3.2.3 Chronic Borrelia, Brucella, and Rickettsial Infections

These chronic infections often present with very similar symptoms as in CFS patients. They often remain undiagnosed, but represent a treatable subset of CFS patients. Their diagnosis can still be problematic because the physician does not associate the clinical manifestations of the disorder with these hidden infections. Furthermore, laboratory tests needed to diagnose them lack sensitivity and false positive results create doubt about the role of these microorganisms in the pathophysiology of the patient's medical problem.

Chronic borreliosis is a recognized entity treated by administration of antibiotics for a long period of time. It remains difficult to diagnose because of the lack of widespread availability of a specific PCR. Using an immunofluorescence test, which gave similar results to an ELISA test, Dr. C. Jadin diagnosed many chronic Rickettsial infections in patients with persistent fatigue. She has successfully (recovery rate 84 to 96%) treated them with tetracyclines and other antibiotics.[155]

8.4 ONSET, IMMUNE CHANGES, AND INFECTION — AN INTEGRATED MODEL THAT EXPLAINS THE SYMPTOMS OF CFS

Figure 8.2 gives a model of how different onset mechanisms and changes in the immune system can elicit a number of events and symptoms that will not reverse spontaneously; it explains, for example, why the symptoms wax and wane, why women are more affected than men, and why an increased incidence of certain cancers is likely to be expected after a few years of natural course of the disorder.

The proposed immune alterations (poor cellular immunity and T_H2-dominated immunity) can result from at least seven groups of onset and predisposing factors:

1. Cellular stress can result from blood transfusion, fetal cells, or radiation exposure. We have reported[28] that, in a significant number of cases and within days of the onset of the CFS-related symptoms, our patients had received a blood transfusion. Furthermore, we have observed that many female CFS sufferers developed their first symptoms shortly after pregnancy.[28] Inhalation of radioactive material or uraniumoxide will also lead to increased cellular stress.
2. Several toxins such as heavy metals, organophosphates, and pentachlorphenol (PCP) can all exert immune dysfunctional effects; PCP impaired *in vitro* lymphocyte stimulation responses in 65% of 188 patients.[156]

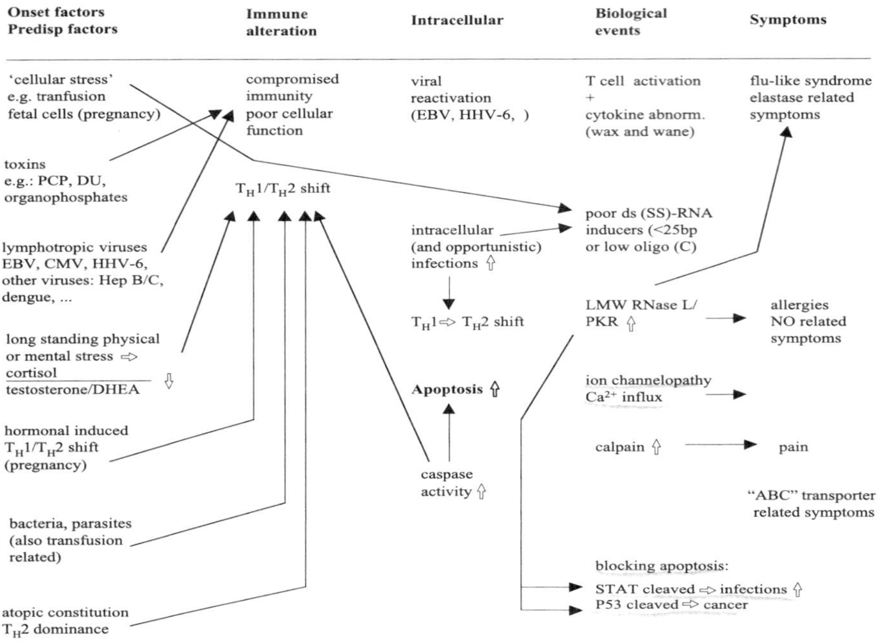

FIGURE 8.2 Proposed pathophysiologic mechanisms in CFS: integrated model with onset and predisposing factors, immune alterations, intracellular and biological events, and direct relationship with symptoms. (Abbreviations: PCP = pentachlorophenol; DU = depleted uranium; EBV = Epstein–Barr virus; CMV = cytomegalovirus; HHV-6 = human herpes virus 6; Hep B/C = hepatitis B/C; DHEA = dehydroepiandrosterone; T_H = T helper; ds-RNA = double-stranded ribonucleic acid; LMW = low molecular weight (37-kDa); RNase L = ribonuclease L; bp = base pairs; C = cytosine; PKR = protein kinase R; Ca^{2+} = calcium^{2+}; NO = nitric oxide; STAT = signal transducers and activators of transcription (gene-regulatory proteins); P53 = a tumor suppressor; ABC = ATP binding cassette (ATP = adenosine triphosphate).)

Others showed that it could also induce activation of T-cells[157] and cause severe T-lymphocyte dysfunction,[156] and is immunotoxic to human immunocompetent cells.[158] In a few families several siblings developed CFS after an acute PCP intoxication (De Meirleir, personal observation). Heavy metals are generally supposed to stimulate the immune system in low concentrations,[159-161] but not in high concentrations, which are toxic.[159] Zinc, in low concentrations, is crucial for normal development and function of the immune system. In high concentrations, however, zinc appears to be immunotoxic *in vitro* to phagocytic function,[162,163] to interfere with the ability of monocytes to respond to activation signals,[164] and to inhibit spontaneous cell-mediated cytotoxity and antibody-dependent cell-mediated cytotoxicity.[165] Zinc and other metal cations are released from dental alloys and they produce dose-dependent cytopathogenic effects in distinct cell types, including human gingival fibroblasts and human tissue mast cells.[166] Cadmium appears to be immunotoxic as well, even in low concentrations.[159] Indeed, this heavy metal was found to be even more immunotoxic than zinc *in vitro* to phagocytic function.[162,163] The latter is consistent with the results of Leibbrandt and Koropatnick,[164] who reported that cadmium exposure interferes with the ability of monocytes to respond to activation signals. Additionally, cadmium suppresses lymphocyte-proliferation in mice,[167] and inhibits spontaneous cell-mediated cytotoxity and antibody-dependent cell-mediated cytotoxicity.[165] These and other harmful effects of zinc, cadmium, lead, chromium, nickel, mercury, and arsenic are presented in Table 8.2.

3. Earlier in this chapter we discussed the role of viruses. Some acute viral infections such as EBV and CMV induce a suppression of the immune system for long periods of time. This creates opportunities for endogenous viruses to reactivate and for opportunistic infections to develop. The role of hepatitis B vaccination as an initiating event in CFS is still under debate.
4. Longstanding physical or mental stress has a negative impact on both cellular immunity and cytokine balance. In this situation (Figure 8.1) we observe a shift in T_H1/T_H2 balance; high plasma cortisol levels decrease macrophage function and create a favorable situation for viral reactivation and development of other stealth infections. Females are more likely to develop a T_H1 response after challenge with an infectious agent or antigen. Although the proinflammatory cytokines are stimulated, the function of the T_H1 response is abnormal.[168] We have observed increased IL 12 and Interferon-γ serum levels in 71 CFS patients compared to controls.[152] IL 12 was originally described as a natural killer cell stimulatory factor, but in CFS we see low NK function in the presence of high IL12 serum levels. Other factors, including NO toxicity, are probably responsible for this low NK function. Gamma interferon cannot exert its effects at cellular level because STAT1 is cleaved. So although it seems that a shift toward T_H1 occurs, functionally there is a shift toward T_H2.
5. Pregnancy and physiological and pathological situations with high estrogen levels tend to shift the T_H1/T_H2 balance toward T_H2. Estrogen has biphasic dose effects, with lower levels enhancing and higher levels (such as those found in pregnancy) inhibiting specific immune activities.[168] Sex hormones such as estrogen can, therefore, even in the presence of high IL 12 and gamma interferon levels, elicit a further functional shift toward T_H2.
6. Infections that are eliminated slowly can lead to a shift in T_H1/T_H2 balance. Opportunistic infections can enter the body via blood transfusion; certainly immediately after pregnancy a T_H2 environment prevails which favors the chances of intracellular microorganisms like chlamydiae and mycoplasmas to survive. Viral reactivation and reactivation of toxoplasma will also be enhanced.
7. Having an atopic constitution (T_H2 dominance) will also favor the development of stealth infections.

The immune alterations have been extensively described in Section 8.2. With intermittent viral reactivation, T-cell activation and cytokine abnormalities wax and

TABLE 8.2
Heavy Metals and Immunopathology

Heavy Metal	Immunopathologic Effect	References
Zinc *(Zn)*	1. immunotoxic to phagocytic function 2. reduced responsiveness of monocytes to activation signals 3. inhibition of spontaneous cell-mediated cytotoxicity and antibody-dependent cell-mediated cytotoxicity 4. significant reduction of macrophage endocytosis in mice 5. inhibition of hematopoiesis in animal and human marrow	1. Brousseau et al., 2000,[162] Fugère et al., 1996[163] 2. Leibbrandt and Koropatnick, 1994[164] 3. Stacey, 1986[165] 4. Skornik and Brain, 1983[175] 5. Lutton et al., 1997[176]
Cadmium *(Cd)*	1. *in vitro* immunotoxic to phagocytic function 2. reduced responsiveness of monocytes to activation signals 3. suppression of lymphocyte-proliferation in mice 4. inhibition of spontaneous cell-mediated cytotoxicity and antibody-dependent cell-mediated cytotoxicity 5. affects both humoral and cell-mediated immune response in animals	1. Brousseau et al., 2000[162] Fugère et al., 1996[163] 2. Leibbrandt and Koropatnick, 1994[164] 3. Ohsawa et al., 1986[167] 4. Stacey, 1986[165] 5. Nath et al., 1984[177]
Lead *(Pb)*	1. positive correlation between [blood lead] and virgin and activated lymphocytes, due to vehicular traffic 2. immune dysfunctions: reduced numbers of CD3+ T-cells, CD4+ cells (helper/inducer T-cells) and impaired functional integrity of T-cells	1. Boscolo et al., 2000[178] 2. Fischbein et al., 1993[179]
Chromium *(Cr)*	1. positive correlation between [urine chromium] and memory lymphocytes, due to vehicular traffic 2. dose-dependent cytopathogenic effects on human gingival fibroblasts and human tissue mast cells	1. Boscolo et al., 2000[178] 2. Schedle et al., 1995[179]
Nickel *(Ni)*	1. positive correlation between [urine nickel] and memory lymphocytes, due to vehicular traffic 2. dose-dependent cytopathogenic effects on human gingival fibroblasts and human tissue mast cells	1. Boscolo et al., 2000[178] 2. Schedle et al., 1995[166]
Mercury *(Hg)*	1. *in vitro* highly toxic to both cell viability and phagocytic activity 2. *in vitro* immunotoxic to phagocytic function, even more than zinc and cadmium	1. Fugère et al., 1996[163] 2. Brousseau et al., 2000[162]

TABLE 8.2 *(Continued)*
Heavy Metals and Immunopathology

Heavy Metal	Immunopathologic Effect	References
Arsenic *(As)*	1. multi-organ tumor promotor 2. inhibition of IL-2 secretion and reduced activated T-cell proliferation in human peripheral blood mononuclear cells 3. defective IL-2 receptor expression in lymphocytes of patients with arsenic-induced Bowen's disease 4. impairs human lymphocyte stimulation and proliferation	1. Hughes et al., 2000[179] 2. Vega et al., 1999[180] 3. Yu et al., 1998[181] 4. Gonsebat et al., 1992[182]

wane. The intermittent increases in lymphotoxins and T_H1 cytokines are probably responsible for the flu-like symptoms and tender lymph nodes. At the same time (Figure 8.2), poor double-stranded (or single-stranded) RNA inducers are responsible for making the RNase L vulnerable to cleavage (Chapter 2), and there is a downregulation of RLI at both mRNA and protein expression.[169] These events are responsible for an increased RNase L activity; because it is likely that, as a result of interaction of ABC transporters with the ankyrin fragment of RNase L upon its release by proteolytic cleavage, improper ion channel function will develop, leading toward an acquired channelopathy of the ABC transporters. This abnormality is clinically very important as it explains many symptoms we observe in CFS. Table 4.6 (Chapter 4) gives a list of the ABC transporters and the relationship of their dysfunction and the symptoms that we observe in CFS: transient hypoglycemia, periodic night sweats, downregulation of pain threshold, hypersensitivity to toxic chemicals and drugs, macrophage dysfunction, depression due to deficient tryptophan uptake, other CNS abnormalities due to dysfunction in transport of monoamine neurotransmitters, loss of potassium for cells leading to decreased pain threshold, and a number of other symptoms, defects in processing and presentation of antigens by MHC class I leading to immunosuppression and T_H2 switch, defects in heme transport leading to anemia, defective elimination of apoptotic cells by macrophages, and all-transretinaldehyde accumulation leading to visual problems.

The acquired channelopathy with loss of intracellular potassium will lead to other metabolic and intracellular abnormalities (Figure 8.3): decreased secretion of anti-diuretic hormone with the subsequent loss of capability to concentrate urine during the night, extracellular sodium retention, central fatigue and sleep disturbances, secondary intracellular hypomagnesemia leading to muscle weakness, weakness of respiratory muscles thus affecting respiration, drop-in intracellular pH with many metabolic and cellular consequences, metabolic alkalosis and hyperventilation, bronchial hyperreactivity, abnormal response to exercise, low blood volume, cardiac manifestations such as ectopic beats and prominent U-wave on the ECG, arterial hypotension due to decreased angiotensin II activity, increased prostaglandin production leading to bladder problems, premenstrual syndrome, and increased HCl

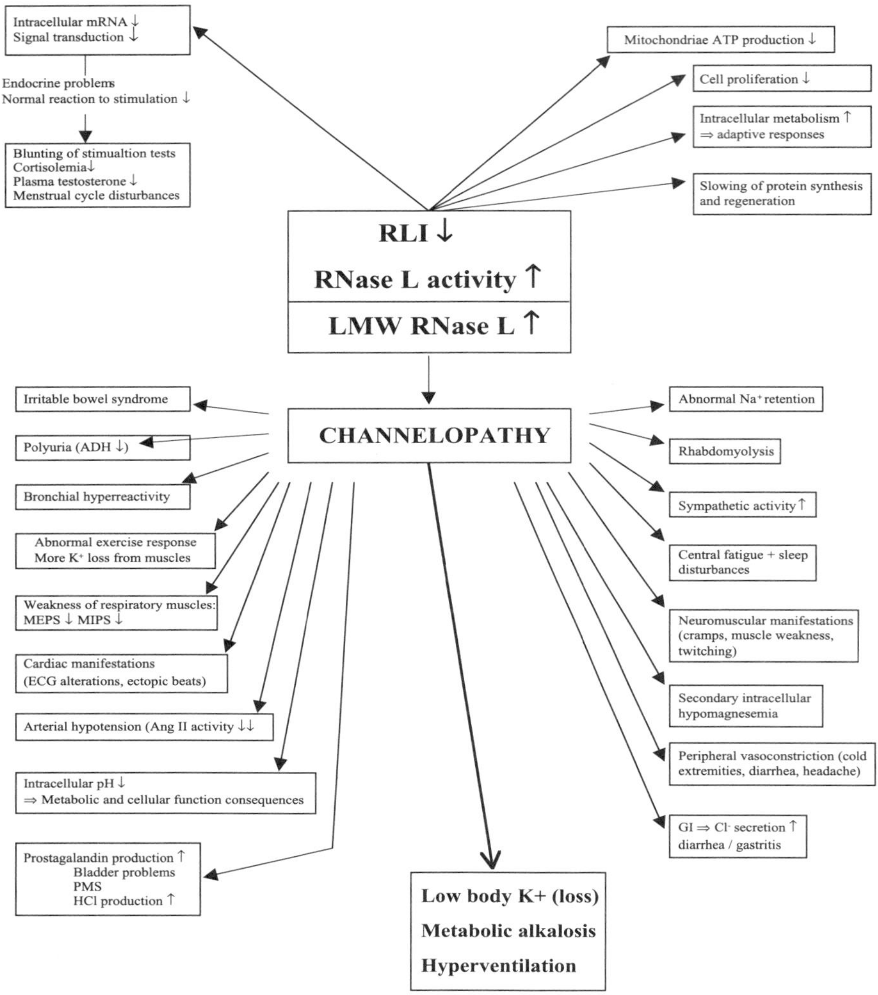

FIGURE 8.3 Symptoms and metabolic and intracellular abnormalities that are associated with dysregulation of the 2'-5'A synthetase/RNase L pathway. (Abbreviations: mRNA = messenger ribonucleic acid; RLI = RNase L inhibitor; RNase = ribonuclease; ATP = adenosine triphosphate; LMW = low molecular weight = 37-kDa; ADH = antidiuretic hormone; K^+ = potassium cation; Na^+ = sodium cation; Cl^+ = chloride anion; ECG = electrocardiogram; Ang II = angiotensine II; PMS = premenstrual syndrome; HCl = hydrogen chloride; GI = gastro-intestinal.)

production in the stomach, further leading to diarrhea and chronic gastritis; there will be an increased risk for rhabdomyolysis. Increased intracellular calcium will elicit increased calpain activation, which will contribute to increased pain sensation (Figure 8.2).

Due to the downregulation of RLI and increased RNase L activity, other elements will be added to the pathophysiology of CFS: slowing of protein resynthesis (slow recovery after exercise, slow healing of wounds), slower production of hormones, peripheral resistance to hormones (decreased signal transduction due to destruction of mRNA), slowing of cell proliferation, and decreased mitochondrial ATP production (Figure 8.3).

Elastase is a proteolytic enzyme contained in the azurophylic granules of polymorphonuclear leukocytes, and has been reported to cause epithelial damage, vascular hyperpermeability, mucus hypersecretion, mucus gland metaplasia, and a reduction in mucociliary clearance. Increased release from leukocytes will also contribute to the symptoms in CFS. Elastase releases tissue kallikrein and cleaves both high- and low-molecular weight kininogen to yield lysyl-bradykinin. Bradykinin causes bronchoconstriction.[169] It is tempting to speculate that increased elastase release also induces widespread changes in elasticity of connective tissue in CFS patients. Cervical and lumbal hernias, diaphragmatic hernias with reflux esophagitis, and mitral valve prolapse are found at a high incidence in patients with longstanding CFS. Loss of elastin fibers due to abnormally high elastase activity on the connective tissue could at least in part explain these changes. As we have observed that many CFS patients have mild pulmonary diffusion abnormalities (decreased CO-transfer-factor and CO-transfercoefficient), elastase might also play a role here. We speculate that some CFS patients also suffer from mild pulmonary hypertension; this has not been reported or studied in detail. Exertional dyspnea, tachypnea, chest pain, and lightheadedness are common complaints in CFS patients. Progression of pulmonary hypertension is associated with increased serine elastase activity,[171] and it is therefore not unlikely that CFS patients have a mild form of pulmonary hypertension that could be treated with an elastase inhibitor.

In Chapter 6, the role in CFS of another ds RNA-activated enzyme called protein kinase (PKR) was thoroughly discussed. The elevated activity in CFS is not only responsible for induction of apoptosis, but also through stimulation of NFκB will lead to increased tissue NO production. Although nitric oxide has a potent antimicrobial activity, by itself it can also induce a number of symptoms in these patients. NO acts in the brain as a neurotransmitter and neuromodulator. It causes smooth muscle cells to relax and dilate the blood vessels.[172] Together with low blood volume and increased sympathetic activity, it explains some of the vascular abnormalities observed in CFS patients: vasodilation in the large blood vessels and vasoconstriction in the peripheral vessels. Together with low angiotensin II activity and decreased mineralocorticoid activity, increased nitric oxide contributes to low blood pressure in CFS patients.

Finally, one report on increased incidence of cancer in CFS patients was published.[173] As to its possible mechanisms — if confirmed — this does not come as a surprise as NK activity is decreased and there is a subgroup of patients in whom apoptosis is blocked and p53 is cleaved (Chapter 6).

8.5 CONCLUSION

The etiology of CFS is complicated but finds its basis in acquired immune dysregulation and persistent infections. Small fragments of ds (ss)-RNA initiate a dysregulation of the 2'-5'A synthetase/RNase L pathway and cause upregulation of the PKR pathway. An acquired channelopathy involving homology of products of the dysfunctional RNase L pathway with the ABC transporters seems responsible for many of the symptoms experienced by CFS patients. Other symptoms are attributed to increased RNase L activity, the metabolic consequences of PKR activation, and cytokines.

In some CFS patients, a precancerous state develops, particularly when apoptotic activity is decreased and p53 is cleaved.

REFERENCES

1. Buchwald, D., Sullivan, J., and Komaroff, A., Frequency of "chronic active Epstein–Barr virus infection" in a general medical practice, *JAMA*, 257, 2303, 1987.
2. Ablashi, D.V., Josephs, S.F., and Buchbinder, A., Human B-lymphotropic virus (human herpesvirus-6), *J. Virol. Methods*, 21, 29, 1988.
3. Luka, J., Okano, M., and Thiele, G., Isolation of human herpesvirus-6 from clinical specimens using human fibroblasts cultures, *J. Clin. Lab. Anal.*, 4, 483, 1990.
4. Buchwald, D. and Komaroff, A., Review of laboratory findings for patients with Chronic Fatigue Syndrome, *Rev. Infect. Dis.*, 13 (suppl 1), S12, 1991.
5. Josephs, S.F. et al., HHV-6 reactivation in chronic fatigue syndrome, *Lancet*, 337, 1346, 1991.
6. Gow, J. et al., Studies on enterovirus in patients with Chronic Fatigue Syndrome, *Clin. Infect. Dis.*, 18 (suppl 1), S126, 1994.
7. Defreitas, E. et al., Retroviral sequences related to human T-lymphotropic virus type II in patients with chronic fatigue immune dysfunction syndrome, *Proc. Natl. Acad. Sci. U.S.A.*, 88, 2922, 1991.
8. Martin, W.J., Detection of RNA sequences in cultures of a stealth virus isolated from the cerebrospinal fluid of a health care worker with chronic fatigue syndrome, case report, *Pathobiology*, 65, 57, 1997.
9. Nakaya, T. et al., Borna disease virus infection in two family clusters of patients with chronic fatigue syndrome, *Microbiol. Immunol.*, 43, 679, 1999.
10. Klonoff, D.C., Chronic fatigue syndrome, *Clin. Infect. Dis.*, 15, 812, 1992.
11. Lloyd, A.R., Hickie, I., Boughton, C.R., et al., The prevalence of chronic fatigue syndrome in an Australian population, *Med. J. Aust.*, 153, 522, 1990.
12. Landay, A.L. et al., Chronic fatigue syndrome, clinical condition associated with immune activation, *Lancet*, 338, 707, 1991.
13. Levine, P.H. et al., Clinical, epidemiological, and virological studies in four clusters of the chronic fatigue syndrome, *Arch. Intern. Med.*, 152, 1611, 1992.
14. Buchwald, D. et al., A chronic illness characterized by fatigue, neurologic and immunologic disorders, and active human herpes type 6 infection, *Annu. Intern. Med.*, 116, 103, 1992.
15. Khan, A.S. et al., Assessment of a retrovirus sequence and other possible risk factors for the chronic fatigue syndrome in adults, *Annu. Intern. Med.*, 118, 241, 1993,
16. Heneine, W. et al., Lack of evidence for infection with known human and animal retroviruses in patients with chronic fatigue syndrome, *Clin. Infect. Dis.*, 18 (suppl. 1), S121, 1994.
17. Gold, D. et al., Chronic fatigue. A prospective clinical and virological study, *JAMA*, 302, 140, 1990.

18. Levy, J., Viral studies of chronic fatigue syndrome, introduction, *Clin. Infect. Dis.*, 18 (suppl. 1), S117, 1994.
19. Nicolson, G.L. et al., Diagnosis and treatment of mycoplasmal infections in fibromyalgia and chronic fatigue, *Biomed. Therapy*, 16, 266, 1998.
20. Nicolson, G.L. and Nicolson, N.L., Diagnosis and treatment of mycoplasmal infections in Persian Gulf War illness — CFIDS patients, *Intern. J. Occ. Med. Immunol. Toxicol.*, 5, 69, 1996.
21. Chia, J.K.S. and Chia, L.Y., Chronic *Chlamydia pneumoniae* infection: a treatable cause of chronic fatigue syndrome, *Clin. Infect. Dis.*, 29, 452, 1999.
22. Salit, I.E., Precipitating factors for the chronic fatigue syndrome, *J. Psychiatr. Res.*, 31, 59, 1997.
23. Komaroff, A.L. and Buchwald, D., Symptoms and signs of chronic fatigue syndrome, *Rev. Infect. Dis.*, 13 (suppl. 1), S8, 1991.
24. Schluederberg, A. et al., Chronic fatigue syndrome research; definition and medical outcome research, *Annu. Intern. Med.*, 117, 325, 11992.
25. MacDonald, K.L. et al., A case-control study to assess possible triggers and cofactors in chronic fatigue syndrome, *Am. J. Med.*, 100, 548, 1996.
26. Bock, G.R. and Whelan, J., eds., *Chronic Fatigue Syndrome,* Chichester, John Wiley & Sons, 1993, pp. 280–297.
27. Theorell, T. et al., Critical life events, infections, and symptoms during the year preceding chronic fatigue syndrome (CFS): an examination of CFS patients and subjects with a nonspecific life crisis, *Psychosom. Med.*, 61, 304, 1999.
28. De Becker, P., McGregor, N., and De Meirleir, K., Possible triggers and mode of onset of chronic fatigue syndrome, *J. Chronic Fatigue Syndrome*, in press.
29. Rook, G.A. and Zumla, A., Gulf War syndrome, is it due to a systemic shift in cytokine balance towards a Th2 profile? *Lancet*, 349, 1831,1997.
30. Patarca-Montero, R. et al., Immunology of chronic fatigue syndrome, *J. Chronic Fatigue Syndrome,* 6, 69, 2000.
31. Gupta, S. and Vayuvegula, B., A comprehensive immunological analysis in chronic fatigue syndrome, *Scand. J. Immunol.*, 33, 319,1991.
32. Matsuda, J., Gohchi, K., and Gotoh, N., Serum concentrations of 2',5'-oligo-adenylate synthetase, neopterin, and beta-glucan in patients with chronic fatigue syndrome and in patients with major depression, *J. Neurol. Neurosurg. Psychiatry*, 57, 1015, 1994.
33. Lloyd, A. et al., Cell-mediated immunity in patients with chronic fatigue syndrome, healthy control subjects, and patients with major depression, *Clin. Exp. Immunol.*, 87, 76, 1992.
34. Barker, E. et al., Immunologic abnormalities associated with chronic fatigue syndrome, *Clin. Infect. Dis.*, 18, S136, 1994.
35. Natelson, B.H. et al., Immunologic parameters in chronic fatigue syndrome, major depression, and multiple sclerosis, *Am. J. Med.*, 105, 43S, 1998.
36. Klimas, N.G. et al., Immunologic abnormalities in chronic fatigue syndrome, *J. Clin. Microbiol.*, 28, 1403, 1990.
37. Morrison, L.J.A., Behan, W.H.M., and Behan, P.O., Changes in natural killer cell phenotype in patients with post-viral fatigue syndrome, *Clin. Exp. Immunol.*, 83, 441, 1991.
38. Tirelli, U. et al., Immunological abnormalities in patients with chronic fatigue syndrome, *Scand. J. Immunol.*, 40, 601, 1994.
39. Peakman, M. et al., Clinical improvement in chronic fatigue syndrome is not associated with lymphocyte subsets of function or activation, *Clin. Immunol. Immunopathol.*, 82, 83, 1997.

40. Lloyd, A.R. et al., Immunologic and psychologic therapy for patients with the chronic fatigue syndrome, a double-blind, placebo-controlled trial, *Am. J. Med.*, 94, 197, 1993.
41. Straus, S.E. et al., Persisting illness and fatigue in adults with evidence of Epstein–Barr virus infection, *Annu. Intern. Med.*, 102, 7, 1985.
42. Dubois, R.E., Gamma globulin therapy for chronic mononucleosis syndrome, *AIDS Res.*, 2, S191, 1986.
43. Caligiuri, M. et al., Phenotypic and functional deficiency of natural killer cells in patients with chronic fatigue syndrome, *J. Immunol.*, 139, 3306, 1987.
44. Linde, A., Hammerstrom, L., and Smith, C.I., IgG subclass deficiency and chronic fatigue syndrome, *Lancet*, 1, 885, 1988.
45. Whiteside, T.L. and Friberg, D., Natural killer cells and natural killer cell activity in chronic fatigue syndrome, *Am. J. Med.*, 105, 27S, 1998.
46. Ogawa, M. et al., Decreased nitric oxide-mediated natural killer cell activation in chronic fatigue syndrome, *Eur. J. Clin. Invest.*, 28, 937, 1998.
47. See, D.M. et al., The *in vitro* immunomodulatory effects of glyconutrients on peripheral blood mononuclear cells of patients with chronic fatigue syndrome, *Int. Physiol. Behav. Sci.*, 33, 280, 1998.
48. Pall, M.L., Cobalamin used in chronic fatigue syndrome therapy is a nitric oxide scavenger, *J. Chronic Fatigue Syndrome,* 8, 39, 2001.
49. Behan, P.O., Behan, W.M., and Bell, E.J., The postviral fatigue syndrome — an analysis of the findings in 50 cases, *J. Infect.*, 10, 211, 1985.
50. Jones, J.F. et al., Evidence for active Epstein–Barr virus infection in patients with persistent, unexplained illnesses, and elevated anti-early antigen antibodies, *Annu. Intern. Med.*, 102, 1, 1985.
51. Lloyd, A.R. et al., Immunological abnormalities in the chronic fatigue syndrome, *Med. J. Aust.*, 151, 122, 1989.
52. Jones, J., Serologic and immunologic responses in chronic fatigue syndrome with emphasis on the Epstein–Barr virus, *Rev. Infect. Dis.*, 13, S26, 1991.
53. Aoki, T. et al., Low natural killer syndrome, clinical and immunological features, *Nat. Immun. Cell Growth Regul.*, 6, 116, 1987.
54. Hassan, I.S. et al., A study of the immunology of the chronic fatigue syndrome: correlation of immunologic parameters to health dysfunction, *Clin. Immunol. Immunopathol.*, 87, 60, 1998.
55. Tobi, M. et al. Prolonged atypical illness associated with serological evidence of persistent Epstein–Barr virus infection, *Lancet*, 1, 61, 1982.
56. Mawle, A.C., Nisenbaum, R., and Dobbins, J.G., Immune responses associated with chronic fatigue syndrome, a case-control study, *J. Infect. Dis.*, 175, 136, 1997.
57. Tosato, G. et al., Characteristic T-cell dysfunction in patients with chronic active Epstein–Barr virus infection (Chronic Infectious Mononucleosis), *J. Immunol.*, 5, 3082, 1985.
58. Borysiewicz, L.K. et al., Epstein–Barr virus-specific immune defects in patients with persistent symptoms following infectious mononucleosis, *Q. J. Med.*, 58, 111, 1986.
59. Hilgers, A. and Frank, J., Chronic fatigue syndrome, evaluation of a 30-criteria score and correlation with immune activation, *J. Chronic Fatigue Syndrome*, 2, 34, 1996.
60. Read, R. et al., IgG 1 subclass deficiency in patients with chronic fatigue syndrome, *Lancet*, 1, 241, 1988.
61. Peterson, P.K. et al., A controlled trial of intravenous immunoglobulin G in chronic fatigue syndrome, *Am. J. Med.*, 89, 554, 1990.
62. Komaroff, A.L., Geiger, A.M., and Wormsley, S., IgG subclass deficiencies in chronic fatigue syndrome, *Lancet*, 1, 1288, 1988.

63. Holmes, G.P. et al., A cluster of patients with a chronic mononucleosis-like syndrome. Is Epstein–Barr virus the cause? *JAMA,* 257, 2297, 1987.
64. Miller, N.A. et al., Antibody to Coxsackie B virus in diagnosing postviral fatigue syndrome, *BMJ*, 302, 140, 1991.
65. Patarca, R. et al., Dysregulated expression of tumor necrosis factor in chronic fatigue syndrome, interrelations with cellular sources and patterns of soluble immune mediator expression, *Clin. Infect. Dis.*, 18, S147, 1994.
66. Linde, A. et al., Serum levels of lymphokines and soluble receptors in primary EBV infection and in patients with chronic fatigue syndrome, *J. Infect. Dis.*, 165, 994, 1992.
67. Cheney, P.R., Dorman, S.E., and Bell, D.S., Interleukin-2 and the chronic fatigue syndrome, *Annu. Intern. Med.*, 110, 321, 1989.
68. Gupta, S. et al., Cytokine production by adherent and nonadherent mononuclear cells in chronic fatigue syndrome, *J. Psychiatr. Res.*, 31, 149, 1997.
69. Chao, C.C. et al., Altered cytokine release in peripheral blood mononuclear cell cultures from patients with the chronic fatigue syndrome, *Cytokine*, 3, 292, 1991.
70. Chao, C.C. et al., Serum neopterin and interleukin-6 levels in chronic fatigue syndrome, *J. Infect. Dis.*, 162, 1412, 1990.
71. Buchwald, D. et al., Markers of inflammation and immune activation in chronic fatigue and chronic fatigue syndrome, *J. Rheumatol.*, 24, 372, 1997.
72. Vojdani, A. et al., Elevated apoptotic cell population in patients with chronic fatigue syndrome, the pivotal role of protein kinase RNA, *J. Intern. Med.*, 242, 465, 1997.
73. Ho-Yen, D.O., Carrington, D., and Armstrong, A.A., Myalgic encephalomyelitis and alpha-interferon, *Lancet*, 1, 125, 1988.
74. Bennett, A. et al., Elevation of bioactive transforming growth factor-beta in serum from patients with chronic fatigue syndrome, *J. Clin. Immunol.*, 17, 160, 1997.
75. Lloyd, A., Hickie, I., and Wakefield, D., eds., *Chronic Fatigue Syndrome,* New York, Marcel Dekker, 1994, 353–386.
76. Swanink, C.M. et al., Lymphocyte subsets, apoptosis, and cytokines in patients with chronic fatigue syndrome, *J. Infect. Dis.*,173, 460, 1996.
77. Kibler, R. et al., Immune function in chronic active Epstein–Barr virus infection, *J. Clin. Immunol.*, 5, 46, 1985.
78. Brook, M., Bannister, B., and Weir, W., Interferon-alpha therapy for patients with chronic fatigue syndrome, *J. Infect. Dis.*, 168, 791, 1993.
79. See, D.M. and Tilles, J.G., Alpha interferon treatment of patients with chronic fatigue syndrome, *Immunol. Invest.*, 25, 153, 1996.
80. Plioplys, S. and Plioplys, A.V., Chronic fatigue syndrome (Myalgic encephalopathy), *S. Med. J.*, 88, 993, 1995.
81. Moss, R.B., Mercandetti, A., and Vojdani, A., TNF-alpha and chronic fatigue syndrome, *J. Clin. Immunol.*, 19, 314–316, 1999.
82. Bell, K.M. et al., Risk factors associated with chronic fatigue syndrome in a cluster of pediatric cases, *Rev. Infect. Dis.*, 13, S32, 1991.
83. Valesini, G., Priori, R., and Conti, F., Chronic fatigue syndrome, what factors trigger it off? *Clin. Exp. Rheumatol.*, 12, 473, 1994.
84. Sraus, S.E. et al., Allergy and the chronic fatigue syndrome, *J. Allergy Clin. Immunol.*, 1988c, 81, 791–795.
85. Sterzl, I. et al., Mercury and nickel allergy: risk factors in fatigue and autoimmunity, *Neuroendocrinol. Lett.,* 20, 221, 1999.
86. Steinberg, P. et al., Double-blind placebo-controlled study of the efficacy of oral terfenadine in the treatment of chronic fatigue syndrome, *J. Allergy Clin. Immunol.*, 97, 119, 1996.

87. Rystedt, I., Strannegard, I.L., and Strannegard, O., Increased serum levels of antibodies to Epstein–Barr virus in adults with history of atopic dermatitis, *Int. Arch. Allergy Appl. Immunol.*, 75, 179, 1984.
88. Olson, G.B. et al., Correlation between allergy and persistent Epstein-Barr virus infections in chronic-active Epstein–Barr virus-infected patients, *J. Allergy Clin. Immunol.*, 78, 308, 1986.
89. De Meirleir, K. et al., Pulmonary function testing in chronic fatigue syndrome (CFS), *Chest*, 116, 331S, 1999.
90. Konstantinov, K. et al., Autoantibodies to nuclear antigens in chronic fatigue syndrome, *J. Clin. Invest.*, 98, 1888, 1996.
91. Jones, J.F. and Straus, S.E., Chronic Epstein–Barr virus infection, *Annu. Rev. Med.*, 38, 195, 1987.
92. Prieto, J. et al., Naloxone-reversible monocyte dysfunction in patients with chronic fatigue syndrome, *Scand. J. Immunol.*, 30, 13; 1989.
93. Roubalova, K. et al., Antibody response to Epstein–Barr virus antigens in patients with chronic viral infection, *J. Med. Virol.*, 25, 115, 1988.
94. Salit, I.E., Sporadic postinfectious neuromyasthenia, *Can. Med. Assoc. J.*, 133, 659, 1985.
95. Von Mikecz, A. et al., High frequency of autoantibodies to insoluble cellular antigens in patients with chronic fatigue syndrome, *Arthritis Rheum.*, 40, 295, 1997.
96. Bates, D. et al., Clinical laboratory test findings in patients with chronic fatigue syndrome, *Arch. Intern. Med.*, 155, 97, 1995.
97. Nishikai, M. and Kosaka, S., Incidence of antinuclear antibodies in Japanese patients with chronic fatigue syndrome, *Arthritis Rheum.*, 40, 2095, 1997.
98. Weinstein, L., Thyroiditis and "chronic infectious mononucleosis" (letter), *N. Engl. J. Med.*, 317, 1225, 1987.
99. Fukuda, K. et al., The chronic fatigue syndrome; a comprehensive approach to its definition and study, *Annu. Intern. Med.*, 121, 953, 1994.
100. Buchwald, D. et al.,Viral serologies in patients with chronic fatigue and chronic fatigue syndrome, *J. Med. Virol.*, 50, 25, 1996.
101. Martin, J.W. and Glass, T.R., Acute encephalopathy induced in cats with a stealth virus isolated from a patient with chronic fatigue syndrome, *Pathobiology*, 63, 115, 1995.
102. Martin, J. et al., Cytomegalovirus-related sequence in an atypical cytopathic virus repeatedly isolated from a patient with chronic fatigue syndrome, *Am. J. Pathol.*, 145, 440, 1994.
103. Nakaya, T. et al., Demonstration of Borna disease virus RNA in peripheral blood mononuclear cells derived from Japanese patients with chronic fatigue syndrome, *FEBS Lett.*, 378, 145,1996.
104. Dobbins, J.G. et al., Borna disease virus in the Atlanta CFS case-control study, *J. Chronic Fatigue Syndrome*, 2, 82, 1996.
105. Jacobson, S.K. et al., Chronic parvovirus B19 infection resulting in chronic fatigue syndrome, case history and review, *Clin. Infect. Dis.*, 24, 1048, 1997.
106. Natelson, B.H. et al., High titers of anti-Epstein–Barr virus DNA polymerase are found in patients with severe fatiguing illness, *J. Med. Virol.*, 42, 42, 1994.
107. Hellinger, W.C. et al., Chronic fatigue syndrome and the diagnostic utility of antibody to Epstein–Barr virus early antigen, *JAMA*, 260, 971, 1988.
108. Wallace, H.L., II et al., Human herpesviruses in chronic fatigue syndrome, *Clin. Diagn. Lab. Immunol.*, 6, 216, 1999.

109. Sumaya, C., Serologic and virologic epidemiology of Epstein–Barr virus: relevance to chronic fatigue syndrome, *Rev. Infect. Dis.*, 13 (suppl. 1), S19, 1991.
110. Straus, S.E. et al., Acyclovir treatment of the chronic fatigue syndrome, *N. Engl. J. Med.*, 319, 1692, 1988.
111. Whelton, C.L., Salit, I., and Moldofsky, H., Sleep, Epstein–Barr virus infection, musculoskeletal pain, and depressive symptoms in chronic fatigue syndrome, *J. Rheumatol.*, 19, 939, 1992.
112. Straus, S.E., The chronic mononucleosis syndrome, *J. Infect. Dis.*, 157, 405, 1988.
113. Levine, P. and Komaroff, A., Human herpesvirus type 6 and chronic fatigue syndrome (letter), *Arch. Intern. Med.*, 153, 661, 1993.
114. Ablashi, D.V., Summary: viral studies of chronic fatigue syndrome, *Clin. Infect. Dis.*, 18, S130, 1994.
115. Dickinson, C.J., Chronic fatigue syndrome — aetiological aspects, *Eur. J. Clin. Invest.*, 27, 257, 1997.
116. Herst, C. et al., Fiedman 4th Int. Conf. Human Herpesviruses 6, 7, and 8, Paris, May, 2001.
117. Yousef, G.E. et al., Chronic enterovirus infection in patients with post-viral fatigue syndrome, *Lancet*, 1, 146, 1988.
118. Gow, J.W. et al., Enteroviral RNA sequences detected by polymerase chain reaction in muscle of patients with postviral fatigue syndrome, *BMJ*, 302, 692, 1991.
119. Archard, L.C. et al., Postviral fatigue syndrome, persistence of enterovirus RNA in muscle and elevated creatine kinase, *J. R. Soc. Med.*, 81, 326, 1988.
120. Bell, E.J., McCartney, R.A., and Riding, M.H., Coxsackie B viruses and myalgic encephalomyelitis, *J. R. Soc. Med.*, 81, 329, 1988.
121. Dowsett, E.G. et al., Myalgic encephalomyelitis — a persistent enteroviral infection? *Postgrad. Med. J.*, 66, 526, 1990.
122. Cunningham, L. et al., Persistence of enteroviral RNA in chronic fatigue syndrome is associated with the abnormal production of equal amounts of positive and negative strands of enteroviral RNA, *J. Gen. Virol.*, 71, 1399, 1990.
123. Dillon, M.J. et al., Epidemic neuromyasthenia; outbreak among nurses at a children's hospital, *BMJ*, 1, 301, 1974.
124. Galbraith, D.N., Clements, G.B., and Nairn, C., Evidence for enteroviral persistence in humans, *J. Gen. Virol.*, 78, 307, 1997.
125. Martin, J., Chronic fatigue syndrome (letter), *Science*, 255, 663, 1992.
126. Berneman, Z.N. et al., Human herpesvirus 7 is a T-lymphotropic virus and is related to, but significantly different from, human herpesvirus 6 and human cytomegalovirus, *Proc. Natl. Acad. Sci. U.S.A.*, 89, 10552, 1992.
127. Behan, P.O. and Behan W.M., Postviral fatigue syndrome, *Crit. Rev. Neurobiol.*, 4, 157, 1988.
128. Hyde, B., The clinical investigation of acute onset of ME/CFS and MS following recombinant hepatitis B immunization, *Proc. 2nd World Congr. Chronic Fatigue Syndrome and Related Disorders*, Brussels, 1999, p. 71.
129. Nicolson, G.L., Nasralla, M.Y., Franco, A.R., et al., Role of mycoplasmal infections in fatigue illnesses: chronic fatigue and fibromyalgia syndromes, Gulf War illness, and rheumatoid arthritis, *J. Chronic Fatigue Syndrome*, 6, 23, 2000.
130. Baseman, J.B. and Tully, J.G., Mycoplasmas: sophisticated, re-emerging, and burdened by their notoriety, *Emerg. Infect. Dis.*, 3, 21, 1997.
131. Nicolson, G.L., Nicolson, N.L., and Nasralla, M., Mycoplasmal infections and chronic fatigue illness (Gulf War illness) associated with deployment to Operation Desert Storm, *Intern. J. Med.*, 1, 80, 1998.

132. Nasralla, M., Haier, J., and Nicolson, G.L., Multiple Mycoplasmal infections detected in blood of patients with chronic fatigue syndrome and/or fibromyalgia, *Eur. J. Clin. Microbiol. Infect. Dis.*, 18, 859, 1999.
133. Vojdani, A. et al., Detection of Mycoplasma genus and *Mycoplasma fermentans* by PCR in patients with chronic fatigue syndrome, *FEMS Immunol. Med. Microbiol.*, 22, 355, 1998.
134. Choppa, P.C. et al., Multiplex PCR for the detection of *Mycoplasma fermentans, M. hominis,* and *M. penetrans* in cell cultures and blood samples of patients with chronic fatigue syndrome, *Mol. Cell. Probes*, 12, 301, 1998.
135. Nijs. J. et al., Prevalence of Mycoplasma infections among Belgian CFS-patients, submitted for publication, 2001.
136. Simecka, J.W. et al., Interactions of Mycoplasmas with B cells: antibody production and nonspecific effects, *Clin. Infect. Dis.*, 17 (suppl. 1), S176, 1993.
137. Blanchard, A. and Montangier, L., AIDS-associated mycoplasmas, *Annu. Rev. Microbiol.,* 48, 687, 1994.
138. Dallo, S.F. et al., Biofunctional domains of the *Mycoplasma pneumoniae* P30 adhesion, *Infect. Immun.*, 64, 2595, 1996.
139. Cassell, G.H., Infectious causes of chronic inflammatory diseases and cancer, *Emerg. Infect. Dis.*, 4, 475, 1998.
140. Kraft, M. et al., Detection of Mycoplasma pneumoniae in the airways of adults with chronic asthma, *Am. J. Resp. Crit. Care Med.*, 158, 998, 1998.
141. Gil, J.C. et al., Isolation of *Mycoplasma pneumoniae* from asthmatic patients, *Annu. Allergy*, 70, 23, 1993.
142. Sokolova, I.A., Vaughan, A.T., and Khodarev, N.N., Mycoplasma infection can sensitize host cells to apoptosis through contribution of apoptotic-like endonuclease(s), *Immunol. Cell. Biol.*, 76, 526, 1998.
143. Grayston, J.T., Chlamydial diseases: 159: *Chlamydia Pneumoniae (TWAR).* vol. 2, part III, section C, in *Principles and Practice of Infectious Diseases,* Mandell, G.L., Bennet, J.E., and Dolin, R., Eds., New York, Churchill Livingstone, 1995, pp. 1696–1701.
144. Rottenberg, M.E. et al., Regulation and role of IFN-γ in the innate resistance to infection with *Chlamydia pneumoniae, J. Immunol.,* 164, 4812, 2000.
145. Penttila, J.M. et al., Depletion of CD8+ cells abolishes memory in acquired immunity against *Chlamydia pneumoniae* in BALB/c mice, *Immunology*, 97, 490, 1999.
146. Airenne, S. et al., *Chlamydia pneumoniae* infection in human monocytes, *Infect. Immun.,* 67(3), 1445, 1999.
147. Rödel, J. et al., Production of basic fibroblast growth factor and interleukin 6 by human smooth muscle cells following infection with *Chlamydia pneumoniae, Infect. Immunol.,* 68(6), 3635, 2000.
148. Matchey, I., *Chlamydia pneumoniae* antibodies in myalgia of unknown cause (including fibromyalgia), *Br. J. Rheumatol.*, 36 (10), 1134, 1997.
149. Rottenberg, M.E. et al., Role of innate and adaptive immunity in the outcome of primary infection with *Chlamydia pneumoniae,* as analysed in genetically modified mice, *J. Immunol.*, 162, 2829, 1999.
150. Davies, D.H. et al., *Infection and Immunity,* London: Taylor & Francis, 1999, pp. 55–57.
151. Halme, S. et al., Cell-mediated immune response during primary *Chlamydia pneumoniae* infection, *Infect. Immun.,* 68(12), 7156, 2000.
152. Verbinnen, T., Masters Thesis Biomedical Science, Faculty of Medicine and Pharmacy, Vrije Universiteit Brussel, 2001.

153. Geng, Y. et al., *Chlamydia pneumoniae* inhibits apoptosis in human peripheral blood mononuclear cells through induction of IL-10, *J. Immunol.,* 164, 5522, 2000.
154. Moazed, T.C. et al., Evidence of systemic dissemination of *Chlamydia pneumoniae* via macrophages in the mouse, *J. Infect. Dis.,* 177, 1322, 1998.
155. Jadin, C.L., Common clinical and biological windows on CFS and rickettsial diseases, *J. Chronic Fatigue Syndrome*, 6, 133, 2000.
156. Daniel, V. et al., Impaired *in vitro* lymphocyte responses in patients with elevated pentachlorophenol (PCP) blood levels, *Arch. Environ. Health*, 50, 287, 1995.
157. McConnachie, P.R. and Zahalsky, A.C., Immunological consequences of exposure to pentachlorophenol, *Arch. Environ. Health*, 46, 249, 1991.
158. Lang, D. and Mueller-Rucholtz, W., Human lymphocyte reactivity after *in vitro* exposure to technical and analytical grade pentachlorophenol, *Toxicology*, 70, 271, 1991.
159. Petanova, J., Pucikova, T., and Bencko, V., Influence of cadmium and zinc sulphates on the function of human T lymphocytes *in vitro, Cent. Eur. J. Public Health*, 8(3), 137, 2000.
160. Salgueiro, M.J. et al., Zinc status and immune system relationship: a review, *Biol. Trace Elem. Res.*, 76(3), 193, 2000.
161. Kumar, P., Rai, G.P., and Flora, S.J., Immunomodulation following zinc supplementation during chelation of lead in male rats, *Biometals,* 7(1), 41, 1994.
162. Brousseau, P. et al., Flow cytometry as a tool to monitor the disturbance of phagocytosis in the clam *Mya arenaria* hemocytes following *in vitro* exposure to heavy metals, *Toxicology*, 142(2), 145, 2000.
163. Fugère, N. et al., Heavy metal-specific inhibition of phagocytosis and different *in vitro* sensitivity of heterogeneous coelomocytes from *Lumbricus terrestris* (Oligochaeta), *Toxicology,* 109, 157, 1996.
164. Leibbrandt, M.E. and Koropatnick, J., Activation of human monocytes with lipopolysaccharide induces metallothionein expression and is diminished by zinc, *Toxicol. Appl. Pharmacol.*, 124(1), 72, 1994.
165. Stacey, N.H., Effects of cadmium and zinc on spontaneous and antibody-dependent cell-mediated cytotoxicity, *J. Toxicol. Environ. Health*, 18, 293, 1986.
166. Schedle, A. et al., Response of L-929 fibroblasts, human gingival fibroblasts, and human tissue mast cells to various metal cations, *J. Dent. Res.*, 74(8), 1513, 1995.
167. Ohsawa, M. et al., Strain differences in cadmium-mediated suppression of lymphocyte proliferation in mice, *Toxicol. Appl. Pharmacol.*, 84, 379, 1986.
168. Whitacre, C.C., Reingold, S.C., and O'Looney, P.A., A gender gap in autoimmunity, *Science*, 283, 1277, 1999.
169. Scuri, M. et al., Inhaled porcine pancreatic elastase causes bronchoconstriction via a bradykinin-mediated mechanism, *J. Appl. Physiol.*, 89, 1397, 2000.
169. Vojdani, A., Choppa, P.C, and Lapp., C.W., Down-regulation of RNase L inhibitor correlates with up-regulation of interferon-induced proteins (2-5A synthetase and Rnase L) in patients with chronic fatigue immune dysfunction syndrome, *J. Clin. Lab. Immunol.*, 50, 1, 1998.
171. Cowan, K. et al., Complete reversal of fatal pulmonary hypertension in rats by a serine elastase inhibitor, *Nature Med.*, 6(6), 698, 2000.
172. Silverthorn, D., *Human Physiology, an Integrated Approach,* 2nd ed., Prentice Hall, New York, 2001.
173. Levine, P.H. et al., Chronic fatigue syndrome and cancer, *J. Chronic Fatigue Syndrome*, 7, 29, 2000.
175. Skornik, W.A. and Brain, J.D., Relative toxicity of inhaled metal sulfate salts for pulmonary macrophages, *Am. Rev. Respir Dis.,* 128(2), 297, 1983.

176. Lutton, J.D. et al., Zinc porpyrins: potent inhibitors of hematopoieses in animal and human bone marrow, *Proc. Natl. Acad. Sci. U.S.A.*, 94, 1432, 1997.
177. Nath, R. et al., Molecular basis of cadmium toxicity, *Prog. Food Nutr. Sci.,* 8(1–2), 109, 1984.
178. Boscolo, P. et al., Lymphocyte subpopulations, cytokines, and trace elements in asymptomatic atopic women exposed to an urban environment, *Life Sci.,* 67(10), 1119, 2000.
179. Fischbein, A. et al., Phenotypic aberrations of CD3+ and CD4+ cells and functional impairments of lymphocytes at low-level occupational exposure to lead, *Clin. Immunol. Immunopathol.*, 66(2), 163, 1993.
179. Hughes, M.F., Del Razo, L.M., and Kenyon, E.M., Dose-dependent effects on tissue distribution and metabolism of dimethylarsinic acid in the mouse after intravenous administration, *Toxicology*, 143(2), 155, 2000.
180. Vega, L. et al., Sodium arsenite reduces proliferation of human activated T-cells by inhibition of the secretion interleukin-2, *Immunopharmacol. Imunotoxicol.*, 21(2), 203, 1999.
181. Yu, H.S. et al., Defective IL-2 receptor expression in lymphocytes of patients with arsenic-induced Bowen's disease, *Arch. Dermatol. Res.*, 290(12), 681, 1998.
182. Gonsebatt, M.E. et al., Inorganic arsenic effects on human lymphocyte stimulation and proliferation, *Mutation Res.*, 283(2), 91, 1992.

9 Current Advances in CFS Therapy

Pascale De Becker, Neil R. McGregor, Karen De Smet, and Kenny De Meirleir

CONTENTS

0-8493-1046-6/02/$0.00+$1.50

9.1 INTRODUCTION

Clinical trials in CFS are difficult to perform and evaluate because of the heterogeneity of patient populations and the complexity of illness origin. Therefore, it has been suggested that data should be analyzed by comparing different groups of patients (e.g., gradual vs. sudden onset, symptom severity, etc.).[1] Despite anecdotal case reports, and limited clinical trials claiming the benefit of various therapeutic regimens, their value is limited since most have been uncontrolled, involved small numbers of patients, used variable case definitions, have had only short follow-up, and most importantly lacked reproducibility.[2] Proper assessment of treatment effectiveness is hampered by several factors, e.g., the cyclic relapsing and remitting pattern of the disorder, small numbers of patients involved in clinical trials, and response often assessed on subjective parameters.[3]

Several drug therapies have been used to treat CFS depending on the suspected mechanisms underlying the condition, e.g., persistent viral infection, immunologic or allergic disorders, neurally mediated hypotension, neuroendocrine disturbances, vitamin deficiency, or depression. Even though clinical benefit has been reported in some of these trials, concerns about study methods have precluded their general acceptance. Various medications are used in practice for treatment of headaches, insomnia, myalgia, or other symptoms of CFS, but no consistently effective, readily available, safe, and affordable pharmacological therapy has been identified for the disorder as a whole.

Chronic fatigue syndrome has a profound impact on the functional status of patients,[4,5] and while complete recovery does occur in a small subset, most patients experience functional limitations that persist for years.[6] In a recent review regarding the natural history of CFS, results showed that less than 10% of the patients returned to their pre-morbid levels of functioning and that the majority remained significantly impaired.[7] Very few patients will return to the active population, which has major socio-economic consequences. Thus, finding effective treatments for CFS should be considered a high priority.

When planning a treatment trial, several limitations should be considered. Epidemiological studies on the natural history of CFS show some spontaneous improvement, which suggests that investigation of treatment or management of the condition should include an untreated control group. We should also consider that, within the CFS population, subgroups exist that are different in symptomatology, illness duration, presence of opportunistic infections, immune status, psychiatric co-morbidity, etc. It has also been suggested that the illness takes a different course in those with CFS of sudden onset compared to those whose illness developed gradually, in children compared to adults, and in those with certain biomarkers, such as RNase L.

Several types of study design have been used to assess the effectiveness of interventions used in the treatment and management of CFS. These range from randomized controlled trials to single subject (case studies). The assessment of their effectiveness as reviewed in this chapter will consequently be confronted to the type of study considered and to the limitations indicated above.

9.2 PHARMACOLOGICAL THERAPY

9.2.1 Psychotropic Drugs

Analysis of De Becker's study data strongly suggests that in CFS patients there is no correlation between mood change and psychiatric symptoms and alterations in RNase L, but that there is an association between these symptoms and cytokine alterations (Chapter 7). Therefore depression and anxiety appear to be secondary phenomena that occur in a subgroup of CFS patients and should be treated as a co-morbid condition. Cytokine-mediated disturbances in brain neurochemistry may serve as a basis for the effectiveness of some antidepressants in CFS.[8] A therapeutic response may, however, occur at doses lower than those used in major depression.[9]

9.2.1.1 Hypnotics, Sedatives, and Anxiolytics

Benzodiazepines may be used in the treatment of the anxiety associated with CFS;[10] they cannot be used to correct the sleep abnormalities of CFS.[11,12] Modafinil has been used successfully in narcolepsy and could possibly have some benefit in CFS patients.[13]

9.2.1.2 Tricyclic and Related Antidepressants and SSRIs

The use of antidepressant medications in patients with CFS has achieved widespread theoretical and clinical support, despite the lack of controlled studies and data suggesting that they are only part of the clinical response (Chapter 7). The prevalence of depressive symptoms and anxiety in patients with CFS is high.[14-16] Moreover, a substantial number of patients with CFS suffer from sleep disturbances.[3,17,18] Some of the antidepressants, particularly tricyclic antidepressants (TCAs), can modulate the sleep pattern and have been shown to benefit some CFS patients.[11] Reducing depression and CFS symptoms could be obtained by using nortriptyline, a tricyclic antidepressant which was tested in a double-blind study.[8] A 60 mg per day dose significantly reduced Beck depression scores and CFS symptom scores. The clinically observed efficacy of low or nonpsychiatrically therapeutic doses of tricyclic antidepressants in CFS suggests that symptom reduction is not based on an antidepressant effect. However, CFS patients may experience significant side effects, including sedation and fatigue exacerbation from first generation TCAs.[9]

The selection of an antidepressant should be based on tailoring the tolerability profile of the drug with relief of particular CFS symptoms. If insomnia is a severe problem, a sedating tricyclic antidepressant (e.g., amitriptyline) may offer alleviation of symptoms, whereas when hypersomnia is the major complaint, a neutral (e.g., desipramine) or a serotonin reuptake inhibitor (e.g., fluoxetine) may offer some relief.[13]

Despite encouraging results from two preliminary case series using fluoxetine in CFS patients,[8] a recent study has shown that, at a 20 mg daily dose, it does not have a beneficial effect on any characteristic of CFS, including fatigue severity, depression, functional limitations, sleep disturbances, and cognitive functioning.[19] The authors suggest that the selective serotonin reuptake blocking mechanism of fluoxetine is not effective in CFS because depressed CFS patients do not show disturbed serotoninergic processes.[20] This is consistent with the neuroendocrinological evidence that CFS and major depression are discrete conditions, and also suggests that the pathophysiological basis of depressive symptomatology found in CFS differs from that in an isolated depressive disorder.[21] This is consistent with the data showing that the catecholamine precursors are reduced in CFS patients and not the serotonin precursor, tryptophan.[22]

Doxepin, a tricyclic antidepressant, had a 70% success rate in an uncontrolled trial in patients with CFS.[23] Finally, terfenadine is unlikely to be of clinical benefit in CFS.[24]

9.2.1.3 MAO-Inhibitors

In an open study with moclobemide, a monoamine oxidase inhibitor, significant but small reductions in fatigue, depression, and somatic complaints, as well as a modest overall improvement, were noted. The greatest improvement occurred in those individuals who had a co-morbid major depressive illness.[25] The authors concluded that moclobemide may be beneficial for CFS patients with co-morbid depressive illness. As moclobemide is a tricyclic that influences catecholamine levels, its effects might be attributed to changes in the tyrosine and catecholamine availability.[22] Natelson and colleagues hypothesized that CFS is a disorder of reduced central sympathetic drive. They performed a small study with phenelzine in which they were able to show some improvement of CFS symptoms, illness severity, and mood and functional status.[26] A more recent study also found improvement in the key symptoms of CFS after treatment with moclobemide; in this study the greatest reduction in clinician-rated disability was in patients with concurrent immune dysfunction.[27] Those antidepressants which influence catecholamine availability seem to have the greatest efficacy. This is consistent with our observation that a disturbance in tyrosine metabolism is present in CFS patients.[22]

9.2.1.4 5-Hydroxytryptamine (5-HT)

The pattern of fatigue in CFS suggests a central rather than a peripheral origin; therefore, functional studies of brain neurotransmitters such as serotonin (5-HT) are important since it can play a role in the expression of central fatigue.[28] Serotonin is also one of the neurotransmitters involved in the regulation of sleep, the modulation of pain sensation, temperature regulation, cardiovascular response, and mood.[12] Abnormalities in all of these are characteristic features of CFS.

A study by Yatham et al. revealed no alteration in overall central 5-HT activity in patients with CFS,[20] which would suggest that the depressive symptoms in CFS patients are not due to alteration in 5-HT function. Yet, Bakheit et al. reported increased prolactin responses to buspirone challenge in CFS patients compared to depressed patients and healthy controls, which was interpreted as indicating an up-regulation of 5-HT_{1A} receptors in CFS.[29] This could be explained by the fact that patients with CFS may have increased 5-HT_{1A} receptor sensitivity without any alteration in overall 5-HT activity. Another study showed a significant rise in prolactin release to administration of d-fenfluramine, a serotonin reuptake inhibitor, and of buspirone, a serotonin agonist.[30,31] These results are in line with the hypothesis of a "postsynaptic hyperresponsiveness" in CFS, which was tested recently in a study involving treatment with 5-HT_3 receptor antagonists.[32] Despite the limitations of the trial (small sample size, not double-blind, not placebo-controlled), the preliminary results indicate a benefit at least for a subgroup of CFS patients.[32] This is consistent with the observations that changes in catecholamine metabolism are more important than in serotonin metabolism.[22]

9.2.1.5 Galantamine

Galantamine is a tertiary alkaloid originally isolated from bulbs of the Caucasian snowdrop. When administered orally as the hydrobromide, it is a reversible and selective inhibitor of acetylcholinesterase and crosses the blood–brain barrier.[33] To test the hypothesis that the three main symptoms of CFS, namely fatigue, sleep disturbances, and myalgia are due to a cholinergic deficit, an acetylcholinesterase inhibitor, galantamine hydrobromide, was used in a few studies. Galantamine has been shown to increase plasma levels of cortisol, but it is not clear whether this is due to an inhibition of acetylcholinesterase. Galantamine, as a cholinesterase inhibitor, has also been shown to decrease REM sleep latency and increase REM sleep density, and also has some analgesic effects, perhaps by increasing the release of endorphins.[33]

Promising results were obtained with galantamine hydrobromide by Snorrason in an initial study.[33] Forty-three of the patients involved reported 50% improvement in fatigue, myalgia, and sleep, while 70% reported 30% improvement compared with only a 10% improvement in the same parameters in the placebo group. The most significant improvement was in the reporting of sleep disturbances. There were some methodological limitations in this trial, e.g., the study was originally organized as a double-blind, placebo-controlled trial, but was changed to an optional crossover after 2 weeks of treatment. Also the adverse effects of the active drug in 30% of patients could compromise the double-blind and a relatively small number of patients were involved (n = 39). These results could not be reproduced in a larger study. In a five-arm, randomized, double-blind, placebo-controlled study, the effects of different dosages of galantamine hydrobromide were evaluated. The patients were randomly allocated either placebo or galantamine hydrobromide, at one of four doses (total daily dose 7.5, 15, 22.5, or 30 mg) for 16 weeks. In this study, galantamine was not found to be an effective treatment for CFS.[34]

9.2.2 Drugs Targeting Endocrine Dysfunctions

9.2.2.1 Growth Hormone

Growth hormone (GH) studies are of particular interest because of the known link between GH deficiency states and fatigue.[35] GH, either acting directly or through its major mediator, insulin-like growth factor-I (IGF-I), is an important anabolic agent and a prerequisite for normal muscle homeostasis.[36] In CFS patients, both low and high serum IGF-I levels have been reported.[36,37] Low nocturnal GH secretion has been observed in CFS patients,[38,39] but given the high prevalence of sleep disturbance in CFS, it was not possible to ascribe this change as a cause or effect in CFS.[38] Results of such studies are not always consistent; in a recent study no differences between patients and controls in basal levels of IGF/IGFBP or in urinary GH excretion were observed.[35] In a small double-blind study, GH treatment during 12 weeks did not result in an improvement of quality of life, although four patients were able to resume work after a long period of sick leave.[40]

9.2.2.2 Hydrocortisone

Given the fact that in a subset of patients a mild glucocorticoid insufficiency is present,[41] and there is evidence of underactivity of the hypothalamic–pituitary–adrenocortical (HPA) axis,[35] studies with steroid supplementation have been conducted in CFS. In some patients with CFS, low-dose hydrocortisone (5 mg was considered to be sufficient, since there was no evidence of a dose-response effect) reduces self-rated fatigue levels and disability in the short term.[42] Treatment with hydrocortisone seems to produce some benefit, but this effect is rapidly attenuated when treatment is resumed.[43] Since the long-term effect is not known and because of the known side effects of corticosteroid treatment, the widespread use of hydrocortisone as a treatment strategy is not recommended.[42] In a recent study, low-dose hydrocortisone treatment appeared to have little impact on GH function.[35]

9.2.2.3 Fludrocortisone

The rationale behind treatment with fludrocortisone is that patients with CFS are more likely than healthy persons to develop neurally mediated hypotension (NMH) in response to prolonged orthostatic stress.[44-46] Therefore, it is reasonable to suppose that therapeutic strategies intended either to increase adrenergic activity or increase plasma volume could significantly improve orthostatic intolerance.[47] One such strategy consists of repleting volume either by increasing sodium intake or by treatment with fludrocortisone.[44,46] Other drugs such as disopyramide and atenolol have also been used successfully to treat neurally mediated hypotension.[44] However, in two recent studies monotherapy with fludrocortisone was no more efficacious than placebo in terms of symptoms improvement.[48,49] Moreover, it has been suggested that the degree of adrenal suppression after hydrocortisone therapy precludes its practical use for CFS.[50]

9.2.2.4 Dehydroepiandrosterone (DHEA)

Dehydroepiandrosterone (DHEA) tends to promote a Th1-type response pattern. It can restore immune functions through correction of deregulated cytokine release.[51] CFS patients have reduced basal DHEA and DHEA-sulfate levels,[52] and display a blunted serum DHEA response to i.v. ACTH injection.[53] An impaired activation of the HPA axis could account for several CFS symptoms, e.g., neuropsychiatric symptoms.[52] Dehydroepiandrosterone was reported in preliminary studies to improve symptoms in some patients;[54,55] however, larger and double-blind studies are necessary to clarify possible efficacy in the treatment of CFS-related symptoms.

9.2.3 Immunomodulating and Antiviral Drugs

Variation in cytokines and reactivation of viruses have significant effects upon symptom presentation in CFS. Changes in virus antibody levels were associated with alteration in immunophenotyping, suggesting that the changes in immune responses were virus specific. These changes were not consistent across the entire CFS population (Chapter 7) and these co-morbid disease entities may therefore explain the considerable variability in treatment results using immunomodulating and antiviral drugs. Studies should be designed to evaluate the effects of these drugs upon those subgroups of CFS patients with various viral and immune changes.

In a recent review it was mentioned that the two basic problems with immune function documented by most research groups are: 1) immune activation, as demonstrated by elevation of activated T-lymphocytes, cytotoxic T-cells, and circulating cytokines; and 2) poor cellular function, with low natural killer cytotoxicity, poor lymphocyte response to mitogens in culture, and frequent immunoglobulin deficiencies, most often IgG1 and IgG3.[56] These findings have a waxing and waning temporal pattern consistent with episodic immune dysfunction (with predominance of Th2 type cells, proinflammatory cytokines, and low NK cytotoxicity and lymphoproliferation) that can be associated as cause or effect, for example, of activation of latent viruses.[56]

Patients with CFS frequently report that their condition was preceded by an acute infective illness.[57,58] Immunologic abnormalities including reduced or absent delayed-type hypersensitivity (DTH) skin responses, impaired lymphocyte proliferation in response to mitogenic stimulation, and depressed natural killer cell cytotoxicity have been reported in CFS patients.[59-61] An increasing amount of evidence points toward an involvement of an immunoregulatory defect in CFS.[62,63] Collectively, these findings point to a disturbance in cell-mediated immunity and suggest that an immunoregulatory defect may play an important role in the etiology and maintenance of symptoms in CFS.[64,65] As a consequence, the efficacy of immunoglobulins and other antiviral and immunomodulating drugs has been evaluated in patients with CFS. Immunosuppressive therapy (such as cyclophosphamide, azathioprine, and corticosteroids) has been found ineffective.[3]

9.2.3.1 Ampligen, a Synthetic, Mismatched, Double-Stranded (ds) RNA

The investigational drug ampligen (Poly I:Poly $C_{12}U$ or PIPCU) is a specific form of mismatched, double-stranded ribonucleic acid (dsRNA) in which uridylic acid (U) substitution in the polycytidylic acid chain creates periodic regions of non-hydrogen bonding in the molecular configuration. Its chemical name is polyriboinosinic:polyribocytidylic (12:1) uridylic acid. Double-stranded RNAs act as biological response modifiers with interrelated activities: immunomodulatory activity, antiviral activity against RNA and DNA viruses, and tumor cell antiproliferative (antineoplastic) activity. Like natural dsRNA, ampligen directly activates intracellular pathways associated with antiviral and immune enhanced states.[66-69] It modulates 2',5'-oligoadenylate synthetase (2-5OAS) and the production of 2',5'-oligoadenylates (2-5A), which control RNase L. The immunomodulatory and antiviral effects of ampligen have been documented in a large number of studies.[67-69,70,71]

Ampligen possesses *in vitro* and *in vivo* antiviral activity against many viruses in a variety of model systems.[72-74] These viruses include: double-stranded DNA viruses such as human herpes virus type 6 (HHV-6), herpes simplex virus type 2, duck hepatitis, human cytomegalovirus, single-stranded (+) RNA viruses such as Coxsackie virus B5, poliovirus type 1, single-stranded (–) RNA viruses such as vesicular stomatitis virus, respiratory syncytial virus, and Caraparu bunyavirus, as well as retroviruses such as Rauscher leukemia virus and HIV. Thus, ampligen inhibits the replication of a wide variety of viruses in cell culture and *in vivo*. It modulates 2-5OAS and the subsequent production of 2-5A, which control RNase L. Ampligen also activates p68 protein kinase. Additionally, the drug fights viral infections by modulating the production of lymphokines, including interferons, IL2, IL6, and other components of the cellular immune system (macrophages, NK-cells, T-cells, B-cells).[75,76] It is a biological response modifier endowed with activity on both humoral and cellular components of immune response; it also exerts antitumoral effects.[77,78]

In man, the immunomodulatory role of ampligen was demonstrated in particular by the increase in NK-cell activity in response to treatment.[79] Zarling et al. initially demonstrated that pretreatment of human peripheral blood mononuclear cells from normal adults with ampligen significantly augmented NK-cell activity. In a study using a variety of dsRNAs including ampligen, the increase in cytotoxicity was associated with the interferon-inducing ability of the dsRNA. Because of its biological properties, ampligen has been evaluated in CFS. In CFS, the 2-5A pathway has been shown to be dysregulated in up to 70 to 75% of subjects.[62,80] Levels of 2-5OAS, bioactive 2-5A, and RNase L were studied by Suhadolnik et al. in 1990 in 15 individuals with CFS.[80] Compared to healthy controls, patients with CFS had lower mean basal levels of latent 2-5OAS ($p = 0.0001$), elevated levels of 2-5A ($p = 0.002$), and higher levels of RNase L activity ($p = 0.0001$), when measured in extracts of peripheral blood mononuclear cells (PBMC). Therapy with ampligen resulted in a downregulation of the 2-5A oligoadenylate synthetase/RNase L pathway toward normal and was associated with clinical improvement related to exercise tolerance and neurocognitive impairments.[80] These initial findings were confirmed in a second study including 92 CFS individuals entered into a controlled trial of ampligen (AMP502).[81]

In line with the study mentioned above, the mean levels of 2-5A and RNase L activity were significantly elevated at baseline compared to controls ($p < 0.0001$ and $p = 0.0001$, respectively).[82] In addition, in individuals that presented with elevated RNase L activity at baseline, therapy with ampligen resulted in a decrease in both 2-5A and RNase L activity ($p = 0.09$ and $p = 0.005$, respectively). Moreover, decreases in RNase L activity in individuals treated with ampligen correlated with cognitive improvements ($p = 0.007$). More recently, a relationship between improvement in physical performance (Karnofsky Performance Score) following Ampligen treatment and RNase L cleavage was demonstrated for patients with CFS (De Meirleir, unpublished data). As discussed in Chapters 2, 3, 6, and 10, the increase in RNase L activity observed in CFS PBMC results from a pathological cleavage of the enzyme to the inability of the 2-5A molecules produced to induce its normal regulation by homodimerization. Homodimerization of the protein protects from the cleavage. Some enzyme fragments generated by the cleavage are catalytically active *per se* and escape regulation by 2-5A. The therapeutic success of ampligen results from different biologic effects inducible by dsRNAs. Thus, ampligen can directly activate and regulate two important dsRNA-dependent enzymes, 2-5OAS and the p68 protein kinase (PKR), associated with antiviral[83] and antitumor[77,78] activity. Activation and regulation of the 2-5OAS results in the production of bioactive 2-5A capable of activating RNase L by regulating homodimerization, which then degrades viral RNAs. Activation of the p68 protein kinase also results in the phosphorylation of the protein translation initiation factor eIF2, which leads to the inhibition of protein synthesis.[66]

After administration by i.v. infusion, ampligen binds to surface receptors on target cells and activates latent 2-5OAS within the cells; ampligen-activated 2-5OAS produces bioactive 2-5A, which activates RNase L and destroys RNAs including viral-dependent transcripts necessary for viral multiplication.[84]

TABLE 9.1
Overview of Four Phase II Clinical Trials of Ampligen Therapy in CFS

	Study A	Study B	Study C	Study D
Study design	Open-label	Open-label	Randomized, placebo-controlled	Open-label, cost-recovery
# of patients	15	45	92 (19)	30
% female	73%	70%	75% (74%)	70%
Age (mean)	44	35	35 (40)	41
Duration (weeks)	24+	24+	24+	24+
Dose (mg)	200–400	200–400	200–400	200–400
Dosing frequency	Twice weekly	Twice weekly	Twice weekly (thrice weekly)	Twice weekly
Location	U.S.	Belgium	U.S.	U.S.

In the main clinical studies (Study A, B, C, D), ampligen patients received 200 mg for the first doses. After 2 weeks at 200 mg twice weekly, the patients were increased to 400 mg twice weekly for the duration of the 6-month study or stayed at 200 mg (low body weight). Table 9.1 provides a summary of the four phase II clinical trials of ampligen therapy in CFS. In all four studies, after 6 months of treatment, the KPS score increased significantly, and activities of daily living (ADL) scores showed a general progressive improvement within all activity modules. The patients' KPS as determined by the principal investigators and the ADL scores as determined by the patients were found to correlate. In all four studies, after 24 weeks of treatment, the patients receiving ampligen experienced a significant improvement in their cognitive ability as measured by SCL-90-R testing. Exercise testing (treadmill and bicycle ergometer) showed a ± 10% increase in exercise work and exercise duration after 24 weeks of ampligen therapy. Patients on ampligen also used much less concomitant medication to suppress morbidity. In study A, ampligen therapy resulted in a significant decrease in 2-5A and RNase L activity with decreased giant cell formation, in agreement with clinical and neuropsychological components.[80,85,86]

This drug, is now in phase III of clinical testing, seems to be an effective treatment for the severely debilitating form of CFS.[86]

9.2.3.2 Immunoglobulins

Immunological abnormalities observed in CFS, most notably IgG subclass deficiencies amounting to 60% in some studies,[87-92] prompted several investigators to study immunoglobulin therapy in CFS. It has been argued that intravenous immunoglobulins could provide potential therapeutic benefit to CFS in two possible ways: 1) by providing neutralizing antibodies against persistent viral antigen, or 2) by analogy with its efficacy in autoimmune disorders by correcting immunoregulatory disturbances.[65]

Four placebo-controlled studies have been reported: two studies showed improvement in the symptoms and restitution of premorbid activities,[92,93] but the two other studies did not show beneficial effect of immunoglobulin over placebo.[65,91] Another uncontrolled trial did not reveal beneficial effect.[2] However, in contrast to adults, intravenous immunoglobulin therapy was found to be effective in treating adolescents with CFS.[94]

Currently, intravenous immunoglobulin therapy is not recommended in CFS, partly based on expense, tolerability, available data showing limited efficacy, and transient nature of the benefits.[65] However, if one could identify a cohort of genuine postinfective cases with CFS and immunologic abnormalities, then it might be appropriate to re-evaluate intravenous immunoglobulin therapy in that specific group.[65]

From these data, the IgG subclass deficiencies would appear to be merely associated with susceptibility to the development of a persistent viral infection rather than with the alteration in symptom expression once CFS has developed.

9.2.3.3 Transfer Factor

Dialyzable leukocyte extract (DLE), also known as transfer factor, is extracted from white blood cells of normal healthy donors and has been used extensively in the treatment of cancer, immunodeficiency disorders, and a wide variety of other diseases. Transfer factor is an extract of T lymphocytes and contains fragments of CD4 and CD8 cells, nonspecific suppressor substances, IL-2, fragments of interferon, and other immune modulators. DLE has immunomodulatory properties; it has been shown to be a safe and effective prophylaxis and therapy for bacterial, viral, and protozoal infections.

DLE has been evaluated in a double-blind, placebo-controlled study conducted in conjunction with individual cognitive-behavioral therapy (CBT). No specific effect was shown with DLE therapy, either alone or in combination with CBT.[95]

9.2.3.4 Isoprinosine

Isoprinosine, an immunomodulator that stimulates the immune system, gave considerable improvement in a small study with CFS patients.[96] Larger double-blind studies should be undertaken to observe the clinical and immunological effects of this drug. From our clinical experience, about one-third of CFS patients treated with 1.5 g Isoprinosine per day report an improvement after a few months.

9.2.3.5 Amantadine

Amantadine is an antiviral agent which acts by inhibiting penetration of the virus into the host cell, e.g., it is known to prevent uncoating and uptake into host cells of Influenza A virus. Furthermore, it may elevate dopaminergic activity since it releases norepinephrine and dopamine from storage sites in the central nervous system (CNS) and retards reuptake of these neurotransmitters into neurons.[97] Interestingly, amantadine was found to be a very effective treatment of fatigue and pain in multiple sclerosis.[98] Although the precise mechanism of the action of amantadine on the CNS is unknown, it could be a nonspecific, general CNS stimulator.[98]

One anecdotal study reported a beneficial effect of amantadine in CFS,[2,3] but in a crossover study comparing L-carnitine and amantadine, no significant beneficial effect of amantadine could be observed.[97] Moreover, amantadine was not well tolerated: only 15 out of 28 patients were able to complete the 8-week treatment period;

the other 13 had to discontinue due to side effects. The development of the side effects in this study may be the result of the effects of the drug upon catecholamine metabolism, which appears defective in CFS patients.

9.2.3.6 Lentinan

Lentinan is a β1-3, β1-6 D-glucan extract from the mushroom *Lentinus edodes*. Open-label studies indicate that lentinan can prolong life in cancer patients, and it has proven to potentiate human immunity.[99] Although the mechanism of the anti-tumor action is still not completely clear, it is suggested that lentinan enhances cell-mediated immune responses *in vivo* and *in vitro* and acts as a biological response modifier; it is also considered to be a multi-cytokine inducer able to induce gene expression of various immunomodulatory cytokines and cytokine receptors.[100] Moreover, lentinan is able to inhibit the Th2-dominant condition in patients with digestive cancers and may improve the balance between Th1 and Th2.[101] Because Japanese researchers report that lentinan can cancel a Th2-dominant condition and may correct the balance between Th1 and Th2, clinical studies in CFS should be warranted.

9.2.3.7 Kutapressin

Kutapressin is a porcine liver extract with bradykinin-potentiating effects, but devoid of vitamin B12 activity, which has been shown to have potent antiviral effects. It inhibits *in vitro* viral replication of HHV-6, probably through inhibition of viral attachment to cellular receptors and inhibition of intracellular maturation of the virus.[102] It has also been tested against EBV, and the authors concluded that kutapressin blocks infection with EBV in mononuclear cells. The mechanism whereby kutapressin inhibits EBV immortalization remains to be determined.[103]

9.2.3.8 Acyclovir

CFS patients often report flu-like symptoms and there is considerable evidence that an acute viral infection initiates the illness in the majority of patients.[57] Moreover, some scientists suggest that viral infection persists in CFS patients;[104] consequently, antiviral therapy was tried in a small number of studies. Acyclovir is an antiviral agent that is active against herpes viruses including EBV and inhibits viral DNA replication.[3] Acyclovir was selected for inclusion in trials because of the suspected link between CFS and EBV, but the drug showed no beneficial effect over placebo in a prospective, double-blind, placebo-controlled trial.[105] Patients (n = 27) with symptoms of chronic fatigue and unusual EBV serological profiles were administered high-doses of acyclovir, but there was no correlation among the therapy, clinical improvement, and immunological profile.

9.2.3.9 Interferons

Interferons are a group of proteins produced early in viral infections and form a first line of defense against many different viruses (Chapter 1). They can induce resistance to viral infection in uninfected cells by causing the production of intracellular antiviral proteins and by inhibiting viral penetration and nucleic acid release. In addition, they have potent stimulatory effects on the immune system.[106] Interferon-α has a wide range of antiviral and immunomodulatory properties and has been successful in the treatment of chronic hepatitis B and C and other viral infections.[107] A proportion of CFS patients may also benefit from interferon-α therapy.[108,109] The rationale resides in the association between CFS and persistent viral infection.[110] A subset of CFS patients who had diminished NK function and normal lymphocyte proliferation benefited from this therapy.[109]

The use of this drug may need to be re-evaluated for its possible efficacy in specifically defined subgroups of CFS patients who have known viral-associated co-morbid illness. It is, however, noteworthy that patients suffering from hepatitis C treated with interferon-α often develop many of the symptoms of CFS including fatigue, pain, and mood change.[111,112]

9.2.4. Antibiotics

The treatment of co-morbid disease in patients with HIV is essential. A similar situation exists in CFS if the onset factor is an immune-depressing virus (e.g., CMV), or when subsequent opportunistic infections develop. Alternatively, microbial or viral interactions may exist in CFS, so that treatment of the microbial agent could have a significant effect upon the morbidity.

Nicolson et al. suggested that chronic mycoplasma infections are underlying events that may trigger CFS and Gulf War illness (GWI). They found several types of mycoplasma infections, which cannot be detected easily, but can be identified in blood leukocytes by a technique developed in their laboratory called PCR with gene tracking. This technique uses a very sensitive and specific DNA hybridization procedure to positively identify unique DNA sequences indicative of specific species of mycoplasma and other organisms. They found evidence of mycoplasma presence in about half of the GWI/CFS patients' blood leukocytes. The majority of the mycoplasmas present in the nuclear fractions prepared from blood leukocytes were identified as *Mycoplasma fermentans*. In 71% of CFS and fibromyalgia patients *Mycoplasma spp.* were detected, multiple infections being found in approximately one-half of the patient population.[113,114] In healthy individuals the incidence of mycoplasmal infections was only 6%. These results are similar to those reported by Choppa et al.[115] and Vodjani et al.[116] who showed increased PCR-detectable mycoplasmas in CFS patients.

In addition to activating B lymphocytes, mycoplasmas are capable of producing chemotactic factors and immunoglobulin proteases that may also be involved in lesion development and survival of the organisms.[117] Thus, both specific and nonspecific interactions of mycoplasmas with B cells can have important effects on disease progression, especially when one considers that many mycoplasma infections become chronic; the cumulative effect of these interactions may be substantial.[117] Mycoplasmas penetrate into nerve cells, synovial cells, and other cell types and are also very effective at evading the immune system. Synergy with other infectious agents can also occur.[114]

In our studies using PCR with gene tracking, which involved 272 Belgian CFS patients, one or more mycoplasma infections were identified in 68% of the patients. The following species were identified: *Mycoplasma fermentans* (25%), *Mycoplasma pneumoniae* (26%), and *Mycoplasma hominis* (36%).[118] Interestingly, *Mycoplasma spp.* were found to be associated with increased low molecular RNase L levels, suggesting that they more readily invade cells that have defects in this enzyme system or that they are able to induce changes in the RNase L pathway.[118]

Nicolson et al. report that four antibiotics are useful for the treatment of mycoplasma-positive GWI/CFS patients and suggest that these be used in multiple 6-week courses: doxycycline (200 to 300 mg/day), azithromycin (500 mg/day), minocyclin (200 to 300 mg/day), and ciproflaxin (1000 to 1500 mg/day). Most patients require multiple cycles (2 to 6) of antibiotic therapy to completely recover, and even then some of these patients relapse occasionally when they are physically stressed. However, their symptoms are almost always less severe than during their initial relapses after their first few cycles of antibiotic therapy. Multiple cycles are necessary, possibly because of the intracellular locations of mycoplasmas, the slow-growing nature of these microorganisms, and their relative insensitivity to drugs.[119,120] Alternatively the changes induced in apoptosis by these organisms may allow their greater persistence in patients with defects in the 2-5OAS/RNase L pathway.

Some patients develop significant side effects with specific antibiotics, such as doxycycline; however, they seem to tolerate others (azithromycin, ciprofloxacin), especially when administered intravenously. In addition, a number of natural remedies that boost the immune system are available and may be potentially useful, especially during antibiotic therapy or after therapy has been completed.[119,120] If patients have nutritional and vitamin deficiencies, these should be corrected. Because this subset of patients often suffers from poor intestinal absorption, high doses of some vitamins must be used, and others, such as vitamin B complex, must be given sublingual.[120] Long-term antibiotics, oxidative therapy, and nutritional supplement support do result in slow recovery of patients with chronic bacterial infection.[114] Recently, Chia et al. demonstrated the presence of chronic *Chlamydia pneumoniae* infection in CFS that can be successfully treated with azithromycin.[121]

Bombardier and Buchwald[122] and Buchwald et al.[5] have shown that many CFS patients have increased morbidity if they suffer from additional syndromes such as fibromyalgia. McGregor has identified CFS patients who show an increased occurrence of myofascial pain syndrome.[22,123] Myofascial pain occurs in the non-CFS population[124] and is associated with the carriage of toxin-producing coagulase-negative staphylococci. Up to 60% of CFS patients report this syndrome and they are found to carry the same organisms as those associated with myofascial pain syndrome in the control population.[22] Based upon these observations, a small systemic antibiotic pilot study was undertaken. Although 80% of the patients indicated a significant improvement in symptoms, there was a high level of relapse due to regrowth of toxin-producing staphylococci. As a more powerful alternative, a regimen has been designed which involves rotating topical nasal antibiotics (nasal Bactroban, Neomycin, Kenacomb), using each for one week and rotating back to the initial antibiotic.

These organisms are normal skin inhabitants and their elimination is not possible. Andersson et al.[125] conducted a staphylococcal toxoid vaccination study in CFS patients and found that there was a significant improvement in pain, sleeping problems, concentration, and memory difficulties, which supports the association between the staphylococcal toxins and the pain condition.

The major side effect of antibiotherapy is the alteration of the normal bowel microflora. A significant change in the bowel microbial flora was observed in CFS patients and changes in the microorganisms were correlated with changes in symptom expression (Chapter 7). Pilot data clearly show that alteration of the microbial flora with the use of antibiotic and probiotics combinations is of benefit in certain CFS patients. Thus, the use of prolonged antibiotic therapy should be associated with appropriate probiotics support. However, more studies are required to confirm and assess these observations.

9.2.5 Other Agents

9.2.5.1 H2-Antihistamines

Histamine H_2-receptor antagonists such as cimetidine and ranitidine produced major and even complete remission after only 1 or 2 days in a small cohort of 10 patients with chronic EBV infection.[126] Suppression of suppressor T lymphocytes has been postulated as a mechanism of action, but has not been substantiated. Studies need to be undertaken to assess these drugs more thoroughly in CFS patients.

9.2.5.2 Analgesics, Antipyretics, and Anti-Inflammatory Drugs

Nonsteroidal anti-inflammatory agents may be helpful in the treatment of arthralgia and myalgia.[2] Low doses of nonsteroidal anti-inflammatory drugs (NSAIDs) and mild analgesics, such as paracetamol or co-codamol, are useful when pain is a major feature of the condition.[127] In a controlled study, naproxen alone was no more effective than placebo. Low-dose amitryptiline was proved to be efficient in decreasing myalgia, improving sleep, and reducing fatigue. The combination of amytriptiline and naproxen was better than amitryptiline alone.[3]

9.2.5.3 Antifungal Agents

Because of the clinical similarities between CFS and the reported "candidiasis hypersensitivity syndrome," treatment with oral nystatin and ketoconazole has been suggested.[128] *Candida albicans* is a normal commensal in the human gut, but it can act as a systemic pathogen in immunosuppressed subjects or may cause localized candidiasis in the mouth or genitalia following broad-spectrum antibiotics, which disrupt the normal bacterial flora.[3] Interestingly, Friedberg et al.[129] published CFS patient rankings of treatment effectiveness for 29 listed medical and alternative therapies. Two relatively benign interventions, anti-allergy diet and anti-yeast diet, were more highly rated than were the vast majority of pharmacological therapies.[9] Analysis of the Newcastle data on over 1500 patients found no association between the prevalence or colony counts of candida species in the bowel of patients and any of the defining symptoms of CFS. However, the presence of *Candida* species in the bowel of CFS patients is associated with the reporting of increased neck and back pain symptoms, suggesting that these organisms may induce a co-morbid condition in some CFS patients as suggested by other investigators.[9] The use of systemic antifungal agents has resulted in toxic hepatitis in some CFS patients[127] and should not be used in CFS patients unless a co-morbid fungal condition is present.

9.2.5.4 Calcium Antagonists

A single case report with nifedipine, a calcium channel blocker, has been reported with beneficial effect.[129] In clinical practice, nimodipine and amlodipine are used to alleviate pain in CFS patients. Goldstein reports that dihydropyridines have the best effect, but that tolerance may develop to nimodipine.[130] Calcium channel blockers may have several effects in CFS patients, such as inhibition of the apoptotic enzymes, or relaxation of vascular smooth muscle, leading to a fall in blood pressure. How these effects influence symptom expression remains to be determined.

9.2.6 Minerals/Vitamins/Amino Acids and Other Nutritional Factors

Although a majority of patients report having tried self-directed or professionally prescribed vitamin therapies,[9] few studies of dietary intervention have been published.

9.2.6.1 Vitamins

As there is no known single cure for CFS, a large number of patients have tried other nonconventional therapies; one of the most used is vitamin supplementation.

9.2.6.1.1 Folic Acid

A subset of CFS patients appears to be deficient in folic acid,[131] which could cause impairment in brain function. The efficacy of folate supplementation in CFS patients has been studied in a small, double-blind study. After 1 week of intramuscular injections of 800 μg folate daily, no beneficial effects were observed.[132] It should be noted that the study was of very short duration. Moreover, the dosage was small when compared to the folate dose used in another study involving a group of patients successfully treated who, although they were not diagnosed with CFS, presented with easy fatigability and minor neurological signs. These patients received a minimum of 10,000 μg folate daily. It took 2 or 3 months for their fatigue to respond. Therefore, it could be deemed appropriate to use larger doses for a longer period of time in CFS.[133]

No association has been found between red blood cell folate levels and any of the CFS defining symptoms and therefore supplemenation should only be considered in those patients with a demonstrable deficiency.

9.2.6.1.2 Vitamin B12 (Cobalamin)

As in the case of folic acid, inadequate vitamin B12 nutriture could contribute to the clinical picture in a subset of patients. Regland et al. report that vitamin B12 levels are low in the central nervous system of CFS patients, as shown by low vitamin B12 and high homocysteine levels in the cerebrospinal fluid.[134] Cobalamin in the form of hydroxocobalamin or cyanocobalamin injections has been widely used to treat CFS. Some reports mention that the administration of 2500 to 5000 μg cyanocobalamin (subcutaneous or intramuscular) every 2 or 3 days results in an increase in energy, stamina, or well-being within 2 or 3 weeks of treatment.[133] An earlier placebo-controlled study found no beneficial results in the active group compared to the placebo group.[132] In the latter study, however, only 200 μg cyanocobalamin a day were administered during 1 week. Consequently, the question of whether higher doses are effective remains unanswered.

It seems that a high amount of vitamin B12 is required in order to relieve the symptoms of CFS when compared to the amount needed to correct a vitamin B12 deficiency. The improved feelings of well-being in CFS patients following vitamin B12 supplementation could be at least partly due to the analgesic effect of the vitamin at pharmacologic dosages.[133] Another theory regarding the possible mechanism by which vitamin B12 pharmacotherapy could reduce CFS symptoms is postulated by Mukherjee et al..[135] They found an increased percentage of abnormally shaped erythrocytes and suggest that this could result in a reduction in blood flow at the microcirculatory level, causing an oxygen deficit and an accumulation of byproducts of cellular respiration.[135] Based on this theory, Simpson administered 1000 μg cyanobalamin i.m. to ME/CFS patients with an increased percentage of abnormally shaped erythrocytes. The patients who responded to the therapy did so within 24 h and their improvement was correlated with a reduction in the abnormally shaped erythrocytes. Patients who did not improve showed no change in red cell shape. It was therefore concluded that vitamin B12 may relieve symptoms in CFS patients by reversing the erythrocyte abnormalities.[136]

Hydroxycobalamin can also be beneficial through its properties as nitric oxide scavenger, both *in vitro* and *in vivo*.[138] Elevated nitric oxide and its oxidant product peroxynitrite could be held responsible for generating CFS symptoms.[139] The most direct evidence supporting the view that nitric oxide levels are increased in CFS comes from the fact that neopterin, a marker for the induction of the inducible nitric oxide synthase, is elevated in CFS.[140-142] In addition, the levels of several inflammatory cytokines known to induce nitric oxide synthase (TNF-α, IL-1, IL-6, IFN-γ) are also reported to show elevation in CFS.[138,140,141,143] This proposed mechanism of CFS predicts that scavengers of nitric oxide, such as hydroxocobalamin, may be useful in CFS treatment.

9.2.6.1.3 Other B Vitamins

Riboflavin, thiamine, and pyridoxine might also be reduced in CFS patients.[137] Niacin nutriture has not been studied in CFS, but there is evidence that supplementation with nicotinamide adenine dinucleotide (NADH), which is the reduced form of the vitamin, may be beneficial in CFS patients. In a double-blind, placebo-controlled, crossover study, 10 mg of NADH reduced symptoms significantly.[144] These drug-associated improvements may be due to changes in oxidation or reduction and not simply to resolving the niacin deficiency.

9.2.6.1.4 *Vitamin C*

Ascorbic acid exerts substantial analgesic effects at pharmacological dosages, boosters immune responses, and has antiviral properties which may result, at least partly, from enhanced interferon activity.[133] Moreover, vitamin C in high oral dose is capable of enhancing NK activity up to tenfold.[145] Although an early report failed to find evidence of decreased serum ascorbate levels in CFS patients,[146] a preliminary report on vitamin C infusions in a large sample of CFS patients suggests that this may be helpful.[147,148] In a series of publications from their laboratory, Kodama et al. showed that the clinical control of CFS was obtained by vitamin C infusions in association with the enhancement of endogeneous glucocorticoids. Greater improvement was observed when DHEA was added.[147,148]

9.2.6.2 Minerals

9.2.6.2.1 *Magnesium*

Magnesium deficiency might play a role in CFS. Reduced red blood cell magnesium concentrations were reported in several studies.[146,149,150] In one study involving CFS patients with low erythrocyte magnesium levels, 100 mg magnesium or placebo were administered i.m. each week for 6 weeks. The patients treated with magnesium reported increased levels of energy, decreased pain, and improvement of emotional status.[149] Moreover, erythrocyte magnesium levels returned to normal in all of the patients receiving magnesium. A subsequent study with intravenous magnesium loading, however, did not find improvement in symptoms.[151] This could be due to the fact that those patients were not magnesium deficient.

If muscle pain is one of the key symptoms or if secondary fibromyalgia is present, magnesium supplementation can be combined with malic acid. The rationale behind such therapeutic approach is that malate plays an important role in energy metabolism, specifically in the generation of mitochondrial ATP. One study in fibromyalgia patients revealed significant decrease in the mean tender point index after treatment by 200 to 600 mg magnesium and 1200 to 2400 mg malate daily.[152] These studies need to be reproduced in CFS patients with specific pre- and post-treatment assessment of magnesium and malate levels.

9.2.6.2.2 *Sodium*

In a group of CFS patients, 61% reported that they usually or always try to avoid salt.[44] Symptoms associated with low sodium intake include fatigue, headache, sleeplessness, and inability to concentrate.[133] These symptoms are worse in females than in males. It is reasonable to assume that reduced sodium intake is likely to worsen symptoms of NMH. The ability of sodium intake to increase blood pressure through its effect on blood volume is well known. A recommendation to increase sodium intake is standard in those CFS patients with NMH. Indeed two studies have shown successful treatment with sodium loading.[153,154]

9.2.6.2.3 *Zinc and Iron*

Grant reported that the mean red cell zinc concentrations in CFS patients are lower than in healthy controls.[146] Zinc deficiency can cause immune depression, myalgia, and fatigue.[133] However, there are no reports available on zinc supplementation in CFS patients.

Because it is likely that CFS patients have multiple active infections, their daily needs in iron are increased. Most importantly, iron deficiency can lead to the development of fatigue when associated with anemia. The possibility of iron overload in patients who carry the hemochromatosis gene should be considered when planning iron supplementation. Assessment of iron metabolism is essential to eliminate these potential co-morbid conditions.

9.2.6.3 Amino Acids

Supplementation of individually tailored amino acid mixtures results in marked improvements in CFS symptomatology. The most commonly reported improvement is in self-perceived levels of mental functioning.[155] The aim of supplementation is to replace the amino acids lost from tissues, particularly muscles and the brain, as the result of a cytokine-mediated catabolic state. Myalgic muscles and muscles of CFS patients infected with enteroviruses contain low levels of protein and RNA. CFS patients have reduced excretion of amino and organic acids, the importance of which correlates with duration and severity of their illness.[22] Pilot study data show that amino acid supplementation of CFS patients results in improvements in myalgia and neurocognitive symptoms. Side effects may be greater in patients who have mood changes or hypertension. A controlled blinded study is currently being conducted, but results are not yet known. Individual amino acid supplements appear to be more appropriate than generalized supplements, as many different patterns of amino acid excretion exist within the normal population.[156]

9.2.6.3.1 L-Tryptophan

There is enhanced degradation of tryptophan in infectious diseases, possibly due to the increased production of interferon-γ during activation of cell-mediated immunity. However, it is not clear whether correcting a tryptophan deficiency will enhance cell-mediated immunity[157] in virally mediated illnesses. In CFS patients, serum L-tryptophan levels have been reported to be significantly decreased[155,158] or, alternatively, do not show any change.[22] In fibromyalgia patients, administration of 5-hydroxytryptophan, a metabolite of tryptophan and immediate precursor of serotonin, resulted in improvements of fatigue, number of tender points, pain intensity, anxiety, and quality of sleep.[159]

Supplementation with a specific amino acid is not recommended, as amino acids are transported through a number of amino acid transporters which competitively transport their amino acid substrates. Thus oversupply of an individual amino acid may result in a further deficiency of other amino acids transported by that transporter unless the amino acid supplementation is only given for a short period.

9.2.6.3.2 L-Carnitine

Carnitine plays a key role in muscle metabolism, and as such carnitine deficiency can impair mitochondrial function, with consequent symptoms such as myalgia, muscle weakness, and malaise following physical exertion.[160] L-carnitine administered orally is effective in treating lethargy and fatigue in patients with chronic neurologic conditions.[97] Studies have found significant decreases in serum acylcarnitine in CFS.[160,161] It might also be that insufficient carnitine is available for metabolic requirements, since Grant et al. found that the ratio of acylcarnitine to free carnitine is increased.[146] From a clinical perspective, the most important finding is that both total and free serum carnitine levels were inversely correlated with symptoms, and that serum carnitine levels were inversely correlated with capacity to function.[161] Another study found a similar relationship between serum acylcarnitine, symptoms, and functional capacity.[160]

The results of clinical trials with oral L-carnitine [up to 1 mg 3 to 4 times daily) are not always consistent.[97,161,162] L-carnitine gave significant clinical improvement in 12 of the 18 studied parameters after 8 weeks of treatment in a study comparing carnitine and amantadine in a crossover design.[97] Improvements were observed on indices of CFS severity and impairment, depression (BDI), and generalized distress (SCL-90-R). Only one-third of CFS patients seem to respond to carnitine supplementation, and baseline levels of L-carnitine fail to predict the responders.[97,161] CFS patients appear to have low mononuclear cell carnitine levels;[163] mononuclear cell carnitine would possibly be a better predictor of carnitine response similar to the observations made in AIDS patients. In AIDS patients with low mononuclear carnitine, 6 g of L-carnitine daily results in an improvement in metabolic and immunological parameters after 2 weeks.[164] L-carnitine plays an important role in facilitating the transport of essential fatty acids across cell membranes and in normalizing the intracellular lipid milieu. Because an optimal lipid milieu is required for lymphocyte proliferation and cytokine production, the enhancement of immune functions by L-carnitine is likely to depend on the correction of the impaired lipid metabolism at the lymphocyte level.[164]

9.2.6.3.3 Glutamine

The mean plasma glutamine concentration of CFS patients was found to be significantly lower than that of age- and sex-matched controls in one study[165] but not in another.[22] Glutamine supplementation with the consequent increases in plasma and muscle concentrations comparable to those of control subjects was not related to any change in the symptomatic status of the CFS subjects.[165] During periods where symptom exacerbation occurs, the excretion of the excitatory amino acids, such as glutamine and aspartic acid, is increased. The increase in excitatory amino acids appears to be implicated in the development of sleep disturbance and pain associated with spinal column hyperalgesia.[22] Once again, supplementation with a single amino acid cannot be recommended.

9.2.6.4 Other Nutritional Factors

9.2.6.4.1 Essential Fatty Acids

Another type of dietary treatment is based on the hypothesis that disordered metabolism of fatty acids might play a role in CFS. These substances are essential substrates for many cellular systems in the body, including muscle mitochondrial oxidative phosphorylation and neuronal metabolism.[166] Serum fatty acids are known to fall in several acute and chronic viral infections and may remain persistently low, correlating with the physical malaise, e.g., after acute Epstein–Barr virus infection. Studies on the serum fatty acid changes in CFS patients failed to reveal many significant differences.[167,168] None of the specific virally associated serum lipid alterations were found in CFS patients when assessed as a single group. A study of a large group of patients showed that increases in the n-6 fatty acids, which are the immediate precursors of the eicosanoids (including prostaglandins and leukotrienes), correlate with symptom severity.[169] This is consistent with the observations of Gray.[170] Finally, both unsaturated and saturated fatty acids may both activate and inactivate certain viruses *in vitro* and enhance or inhibit their replication *in vivo*.[166] Behan et al. showed evidence for persistently low levels of serum fatty acids following acute infections such as Epstein–Barr virus (EBV), and these observations were also noted in CFS patients with EBV early antigen, suggesting that activation of EBV is associated with certain changes in fatty acids (Chapter 7).

Fatty acids also play important roles in immunity. The decreased levels of essential fatty acids (EFAs) in EBV infections can be due to abnormalities in EFA metabolism. Indeed, it was found that the ratio of biologically active EFA metabolites was changed, which could in turn cause immune, endocrine, and sympathetic nervous system dysfunctions as those seen in CFS.[170] It has been postulated that viruses may reduce the ability of cells to make 6-desaturated EFAs, while interferon requires 6-desaturated EFAs in order to exert its antiviral effects.[171] In the Dunstan's study[169] (Chapter 7), increases in 6-fatty acid desaturase were associated with increases in the levels of general CFS factor scores, which are indicative of inflammatory activity and viral reactivation.

A prospective, randomized double-blind placebo-controlled treatment study using essential fatty acids (a mixture of 80% evening primrose oil and 20% concentrated fish oil), reported beneficial effects in 74% of the CFS population.[166] Muscle pain was reported significantly less at treatment termination, and a 3-month follow-up revealed significant improvements in fatigue, malaise, dizziness, concentration, and depression. In another small open study, similar results were obtained.[170] In the Dunstan's study,[169] increases in the levels of the major n-6 fatty acid constituent of evening primrose oil (di-homo-γ-linoleic acid) and the levels of this fatty acid were positively associated with symptom expression. This is likely to support the results of a recent study showing no benefits from essential fatty acids in CFS patients.[172] Unless there is a demonstrable anomaly in levels of any fatty acid, supplementation should not be considered.

9.2.6.4.2 Diets and Food Intolerance

A study evaluating the treatment of food intolerance in CFS patients revealed it to be a co-morbid condition.[173] Restriction of the diet intolerance factor results in a very significant reduction or even total loss of the gastrointestinal and food intolerance symptoms, with only small reductions in the severity of the defining CFS symptoms. Treatment of food intolerance as a co-morbid condition is thus recommended.

9.2.6.4.3 Coenzyme Q10

No double-blind studies regarding the effect of coenzyme Q10 in CFS are available. One open study found significant improvement in clinical symptoms, exercise tolerance, and post-exercise fatigue.[133]

9.2.6.4.4 LEFAC

No effect was observed after treatment with intramuscular liver extract-folic-acid-cyanobalamin (LEFAC) in a small study.[132]

9.2.7 Other Therapies

9.2.7.1 Lymph Node Extraction

Lymph node extraction, *ex vivo* cell culture followed by autologous cell reinfusion is a new technique recently tried in the U.S. This experimental therapy utilizes the cells of the lymph node, activated and grown in culture with defined media containing interleukin-2 and anti-CD3, to activate and enhance cellular immunological functions. This procedure was designed to change the cytokine pattern of the lymph node lymphocytes to favor expression of T-helper (Th1)-type over Th2-type cytokines. Once reinfused into the donor, the mixed populations of *ex vivo* immune-enhanced cells are likely to provide significant improvements in clinical status associated with a decrease in Th2-type cytokine production.[174]

9.2.7.2 Plasma Exchange

Following a combination of plasma exchange and intravenous injection of fresh frozen plasma, some patients with CFS described a mild improvement, but the effect wore off in 2 weeks.[3]

9.2.7.3 Staphylococcus Toxoid Vaccine

In a small placebo-controlled study, the effect of vaccination with a staphylococcus toxoid during 12 weeks was compared with the injections of sterile water. Significant improvements were observed regarding pain, sleeping problems, concentration, and memory difficulties.[125] The aim of such vaccination is to eradicate a co-morbid toxin-producing organism shown to be associated with symptom expression in CFS patients.

9.3 NONPHARMACOLOGICAL THERAPY

9.3.1 Cognitive Behavior Therapy (CBT)

Cognitive behavior therapy (CBT) focuses on the consequence of an initial trigger which begins a cycle of attributional and avoidant behavior. Cognitive behavior therapy offers a rational approach for dealing with symptoms of an illness whose pathogenesis is unknown or untreatable. CBT is based on a model of CFS that hypothesizes that certain cognitions and behavior may perpetuate symptoms and disability, that is, act as obstacles to recovery. Treatment emphasizes self-help and aims to help the patient recover by changing these unhelpful cognitions and behavior.[175]

In CFS patients the main targets of behavioral change are the level and pattern of activity, ability to relax, and pattern of breathing. Programs intended at increasing levels of activity are an important component of most rehabilitative programs for chronic illnesses. They seek small increases or changes in activity which allow the patient to achieve specific goals or to improve their independence and quality of life. Also commonly included are techniques of relaxation designed to improve the quality of rest, and breathing exercises to prevent hyperventilation.[176] This treatment approach does not address the organic basis of the disease process, but rather predominantly focuses on the psychological responses to the disease.

In an uncontrolled study, improvement was obtained in 70% of patients with CFS. The treatment focused on cognitive aspects by decreasing the perception of helplessness, increasing the predictability of symptoms associated with activity, and developing tolerance to these symptoms.[177] However, this study by Butler et al. had several shortcomings, e.g., a high dropout rate (31%), a nonstandardized treatment duration, and the absence of a control group. It is also plausible that this clinical sample represented a subset of CFS patients who have significant fatigue attributable to psychobehavioral factors. For instance, the Butler et al. sample was similar in therapeutic response to a clinically depressed group without CFS in a cognitive–behavioral treatment study.[9]

In a subsequent study, Lloyd and coworkers compared cognitive–behavioral treatment with immunological therapy in a double-blind placebo-controlled trial. They did not find any significant differences in measures of functional status or daily nonsedentary activity, nor in measures of depression, anxiety, or fatigue severity.[95] In a recent randomized controlled trial with cognitive behavior therapy in CFS patients, functional impairment and fatigue improved more in the group that received cognitive behavior therapy. At final follow-up, 70% of the completers in the cognitive behavior therapy group achieved good outcomes compared to 19% of those in the relaxation group. The improvements were sustained over 6 months of follow-up.[178]

In a randomized controlled trial of cognitive behavioral therapy vs. relaxation therapy, 70% of the CBT group vs. 19% in the relaxation group had a good outcome.[179] In a multi-group clinical study, cognitive–behavioral coping skills intervention was compared in CFS patients and patients with major depression. The outcome data revealed a trend toward reduced depression scores, but no reduction in self-reported stress symptoms or fatigue severity scores in the CFS group. When a median split of depression scores was analyzed in the CFS treated group, reductions in depression, stress, and fatigue severity were found in the more highly depressed subgroup of CFS participants. Although fatigue was significantly reduced in this subgroup, it still remained abnormally high compared with healthy individuals and depressed patients. These findings are important because they suggest that, for CFS patients, relieving depression with cognitive–behavioral treatment does not cure fatigue symptoms.[180]

One has to keep in mind that CBT is expensive and carries the risk of deterring patients fearful of contact with health workers.[181] This would suggest that cognitive behavior therapy is of benefit in facilitating patients' acceptance of the disease, particularly in those with mood disorders, but of little value in altering the organic basis of the disease.

9.3.2 Psychotherapy

Some authors recommend psychotherapy in certain circumstances; others feel it should be avoided since this type of therapy is insight-directed and many of these patients are already introspective.[15] Once again, if this therapy can be used to allow improved acceptance of the disease, it may benefit selected patients.

In a very recent study[182] CBT was found to be more effective for CFS patients than guided support groups or the natural course. In this study a lower proportion of patients improved compared to other CBT trials, probably because the therapists had no clinical experience with CFS patients.

9.3.3 Physical Rehabilitation and Exercise

Clinical findings and self-reported symptoms vary among patients with CFS, yet a worsening of symptoms, especially of fatigue, after previously well-tolerated levels of exercise continues to be the hallmark of the disorder.[4] Some investigators have suggested the onset of severe symptoms to occur 6 to 48 h following exercise[183] and to last from 2 days to 2 weeks.[184,185] Although physical deconditioning from inactivity has been suggested as a partial explanation for activity avoidance in CFS,[165,186] the results of these studies could also be influenced by the limited number of patients involved, which could bias the results considering the heterogeneity of CFS patients. In a recent large study,[187] it was proven that female CFS patients have a significantly decreased exercise capacity. Moreover, autonomic dysfunction could be why some of the subjects were unable to achieve a maximal effort. Other studies of graded exercise testing in CFS compared with age- and sex-matched controls failed to demonstrate any evidence for physical deconditioning.[188,189] One of the most interesting findings in the study of De Becker[187] was that the maximal workload achieved at the anaerobic threshold averaged 53% of normal in the CFS population, which is close to the 50% decrease in physical capabilities in the CDC's criteria for CFS.[190,191]

A group of CFS patients had evidence of physical and cardiovascular deconditioning, suggesting that a graded exercise program could lead to physical reconditioning and could increase their ability to perform physical activities.[192,193] Fulcher and White concluded that graded exercise treatment was more effective than relaxation and stretching exercises. The patients show sustained benefit 1 year after graded exercise treatment.[194] A recent study by the same authors confirms these results;[195] they found that treatment designed to reverse deconditioning helps to improve physical function.

Powell et al.[181] designed a randomized controlled trial of patient education to encourage graded exercise in CFS. Of the patients in the intervention group, 69% achieved a satisfactory outcome in physical functioning compared with 6% of controls, who received standardized medical care. Similar improvements were observed in fatigue, sleep, disability, and mood. In another study,[196] fluoxetine treatment was compared with graded exercise in a randomized, double-blind, placebo-controlled treatment trial. There was a high drop-out rate (29%); patients were more likely to drop out of the exercise group rather than nonexercise treatment. The study resulted in small improvements in depression in the fluoxetine group and in small but clinically significant improvements in case level fatigue and functional work capacity in the graded exercise group. However, the authors concluded that graded exercise may not be adhered to by CFS patients who are particularly impaired.[196]

In any case, one has to question the disability state of the population studied. For instance, in the study of Fulcher,[194] the patients had an average VO_2max of 31.8 ml/kg per min and 28.2 ml/kg per min, which places them within the range of sedentary control subjects according to the classification of MacAuley et al.[197] Therefore, the results of the graded exercise testing might only apply for those patients who are not that severely disabled. In a subgroup of patients from the De Becker's study group,[187] the RNase L ratios were assessed. The higher the RNase L ratio, the lower the exercise capacity. Thus the loss of intercellular protein and RNA was associated with a reduction in exercise capacity. Whether these parameters are improved with exercise remains to be assessed. This increase in the RNase L ratio may also suggest a reason why the more severely disabled patients are not able to perform the graded exercises.

9.3.4 Transcutaneous Electrical Nerve Stimulation (TENS)

A randomized controlled trial of massage therapy as compared with transcutaneous electrical nerve stimulation (TENS) for patients with CFS showed significant improvements in generalized distress, sleep, anxiety, pain, depression, fatigue, and somatic symptoms on self-report measures.[198] The massage group also showed evidence of decreased cortisol levels. Obviously, the stress-reducing effects of massage can have some positive effects in CFS patients.

9.3.5 Multidisciplinary Rehabilitation

In a group of 19 severely incapacitated CFS patients, a multidisciplinary rehabilitation program appeared very helpful.[199] The program involved three main areas: physical, psychological, and social rehabilitation. It was a combination of individually balanced cognitive behavior therapy, relaxation therapy, biofeedback, counseling, antidepressants, therapeutic massage, a graded activity or exercise program tailor-made for the individual, and social rehabilitation.[199] Eighty-nine percent of the patients involved had functionally improved by discharge, with the median KPS improvement 15 points. How these interventions influence the underlying biochemical changes is yet to be determined.

9.4 FUTURE DEVELOPMENTS

The appropriate treatment of any disease relies upon knowing the processes involved in the disease etiology and progression. In Chapter 7 we have assessed the potential mechanisms involved in the CFS disease process. These can be divided into three major areas: 1) etiological components; 2) host response components; and 3) co-morbid disease processes. These have been further discussed at the molecular level in other chapters. The development of therapy regimens is required that address each of these processes, such as antivirals, antibiotics, and antifungals for the etiological and co-morbid pathogen-associated events, antiinflammatories, apoptotic regulators, and supplements for the host response events, and psychological, graded exercises, and antidepressants for the outcomes of the disease processes. The interventions may range from methods of reduction of cytokine availability (drugs, essential fatty acid manipulation, etc.)[200,201] through alterations in ion channel activity or manipulation of the 2-5A synthetase RNase L pathway, ATP-binding cassette (ABC) proteins, or even double-stranded RNA aptamers.

9.5 CONCLUSIONS AND PROSPECTS

As yet, no single therapeutic agent has been found useful for all CFS patients. Given the multifaceted nature of CFS, it may be more productive to view the syndrome and its treatment within a biopsychosocial perspective. Perhaps promising biological therapies, such as ampligen, need to be combined with psychosocial interventions that focus on stress reduction and lifestyle adjustments in order to maximize positive clinical outcomes. Treatment of the co-morbid disease processes is also of paramount importance.

Treatment of CFS should begin with an initial clinical evaluation, i.e., a good history and physical examination by the physician. Since a number of chronic organic or psychiatric disorders may have similar clinical features, these diagnoses must be excluded. The treatment of CFS is to a large extent dependent on the underlying causes and co-morbid or perpetuating factors (e.g., opportunistic infections). The plethora of remedies tried so far appear to be based upon attempts to treat some small part of the disease process without any significant understanding of the underlying disease and, as such, none are universally effective. Many remedies help some patients to varying degrees, but none help all of them. The advances made in the understanding of CFS pathogenesis as described in the various chapters of this book allow us to view the therapeutic approaches under a new light and to consider the application of various regimens already approved for other diseases.

Nevertheless, the important therapeutic principles are to keep the therapeutic regimen as simple, effective, and inexpensive as possible and to understand that many ME/CFS patients are hypersensitive to medication, which may alter their ability to respond in a normal way to any therapy.

REFERENCES

1. Evengard, B., Schacterle, R.S., and Komaroff, A.L., Chronic fatigue syndrome: new insights and old ignorance, *J. Intern. Med.,* 246, 455, 1999.
2. Blondel-Hill, E. and Shafran, S., Treatment of the chronic fatigue syndrome. A review and practical guide, *Drugs*, 46, 639, 1993.
3. McBride, S. and McCluskey, D., Treatment of chronic fatigue syndrome, *Brit. Med. Bull.,* 47, 895, 1991.
4. Komaroff, A. et al., Health status in patients with chronic fatigue syndrome and in general population and disease comparison groups, *Am. J. Med.*, 101, 281, 1996.
5. Buchwald, D. et al., Functional status in patients with chronic fatigue syndrome, other fatiguing illnesses, and healthy individuals, *Am. J. Med.*, 101, 364, 1996.
6. Vercoulen, J.H.M.M. et al., Prognosis in chronic fatigue syndrome: a prospective study on the natural course, *J. Neurol. Neurosurg. Psychiatry*, 60, 489, 1996.

7. Joyce, J., Hotopf, M., and Wessely, S., The prognosis of chronic fatigue and chronic fatigue syndrome: a systematic review, *Q. J. Med.*, 90, 223, 1997.
8. Goodnick, P.J. and Sandoval, R., Psychotropic treatment of chronic fatigue syndrome and related disorders, *J. Clin. Psychiatry*, 54, 13, 1993.
9. Friedberg, F. and Jason, L.A., Medical and alternative therapies, in *Understanding Chronic Fatigue Syndrome*, Friedberg, F. and Jason, L.A., Eds., Am. Psych. Assoc., Washington, DC, 1998, 121–130.
10. Klonoff, D.C., Chronic fatigue syndrome, *Clin. Infect. Dis.*, 15, 812, 1992.
11. Goldenberg, D.L., Felson, D.T., and Dinerman, H., A randomized, controlled trial of amitriptyline and naproxen in the treatment of patients with fibromyalgia, *Arthritis Rheum.*, 29, 1371, 1986.
12. McCluskey, D.R., Pharmacological approaches to the therapy of chronic fatigue syndrome, in *Chronic Fatigue Syndrome*, Bock, G.R. and Whelan, J., Eds., John Wiley & Sons, Chichester, 1993, 280–297.
13. Moldofsky, H., Broughton, R.J., and Hill, J.D., A randomized trial of the long-term, continued efficacy and safety of modafinil in narcolepsy, *Sleep Medicine*, 1, 109, 2000.
14. Lane, T., Manu, P., Matthews, D., et al., Depression and somatization in the chronic fatigue syndrome, *Am. J. Med.*, 91, 335, 1991.
15. Wessely, S. and Powell, R., Fatigue syndromes: a comparison of chronic "postviral" fatigue with neuromuscular and affective disorders, *J. Neurol. Neurosurg. Psychiatry*, 52, 940, 1989.
16. Gold, P. et al., Chronic fatigue. A prospective clinical and virological study, *JAMA*, 302, 140, 1990.
17. Morriss, R. et al., Abnormalities of sleep in patients with the chronic fatigue syndrome, *BMJ*, 306, 1161, 1993.
18. Morriss, R., Insomnia in the chronic fatigue syndrome, *BMJ*, 307, 264, 1993.
19. Vercoulen, J. et al., Randomised, double-blind, placebo-controlled study of fluoxetine in chronic fatigue syndrome, *Lancet*, 347, 858, 1996.
20. Yatham, L.N. et al., Neuroendocrine assessment of serotonin (5-HT) function in chronic fatigue syndrome, *Can. J. Psychiatry*, 40, 93, 1995.
21. Scott, L.V. et al., Differences in adrenal steroid profile in chronic fatigue syndrome, in depression and in health, *J. Affect. Dis.*, 54, 129, 1999.
22. McGregor, N.R., An investigation of the association between toxin-producing staphylococcus, biochemical changes, and jaw muscle pain, Ph.D. thesis, University of Sydney, 2000.
23. Jones, J.F. and Straus, S.E., Chronic Epstein–Barr virus infection, *Annu. Rev. Med.*, 38, 195, 1987.
24. Steinberg, P., MCNutt, B., and Marshall, P., Double-blind placebo-controlled study of the efficacy of oral terfenadine in the treatment of chronic fatigue syndrome, *J. Allergy Clin. Immunol.*, 97, 119, 1996.
25. White, P.D. and Cleary, K.J., An open study of the efficacy and adverse effects of moclobemide in patients with the chronic fatigue syndrome, *Int. Clin. Psychopharmacol.*, 12, 47, 1997.
26. Natelson, B. et al., Randomized, double-blind, controlled placebo-phase in trial of low dose phenelzine in the chronic fatigue syndrome, *Psychopharmacology*, 124, 226, 1996.
27. Hickie, I.B. et al., A randomized, double-blind, placebo-controlled trial of moclobemide in patients with chronic fatigue syndrome, *J. Clin. Psychiatry*, 61, 643, 2000.
28. Sharpe, M. et al., Increased prolactin response to buspirone in chronic fatigue syndrome, *J. Affect. Dis.*, 41, 71, 1996.

29. Bakheit, A. et al., Possible up-regulation of hypothalamic 5-hydroxytryptamine receptors in patients with postviral fatigue syndrome, *BMJ*, 304, 1010, 1992.
30. Cleare, A.J. et al., Contrasting neuroendocrine responses in depression and chronic fatigue syndrome, *J. Affect. Disord.,* 34, 283, 1995.
31. Bearn, J. et al., Neuroendocrine responses to d-Fenfluramine and insulin-induced hypoglycemia in chronic fatigue syndrome, *Biol. Psychiatry,* 37, 245, 1995.
32. Spath, M., Welzel, D., and Farber, L., Treatment of chronic fatigue syndrome with 5-HT3 receptor antagonists — preliminary results, *Scand. J. Rheumatol.*, 29 (suppl. 113), 72, 2000.
33. Snorrason, E., Geirsson, A., and Stefansson, K., Trial of a selective acetylcholinesterase inhibitor, galantamine hydrobromide, in the treatment of chronic fatigue syndrome, *J. Chronic Fatigue Syndrome*, 2, 35, 1996.
34. Final Report Clinical Study. Study Title: A phase II, randomised, placebo-controlled study to assess the safety an efficacy of galantamine hydrobromide 2.5mg tid, 5 mg tid, 7.5 mg tid, and 10 mg tid taken for a period of 16 weeks in patients with a diagnosis of chronic fatigue syndrome (protocol GAL-IV-201), Shire Pharmaceutical Development Ltd., 1999.
35. Cleare, A.J. et al., Integrity of the growth hormone/insulin-like growth factor system is maintained in patients with chronic fatigue syndrome, *J. Clin. Endocrinol. Metab.*, 85, 1433, 2000.
36. Allain, T.J. et al., Changes in growth hormone, insulin, insulinlike growth factors (IGFs), and IGF-binding protein-1 in chronic fatigue syndrome, *Biol. Psychiatr.*, 41, 567, 1997.
37. Bennett, A.L. et al., SomatomedinC (insulin-like growth factor I) levels in patients with chronic fatigue syndrome, *J. Psychiatr. Res.*, 31, 91, 1997.
38. Berwaerts, J., Moorkens, G., and Abs, R., Secretion of growth hormone in patients with chronic fatigue syndrome, *Growth Hormone & IGF Research*, 8, 127, 1998.
39. Moorkens, G. et al., Characterization of pituitary function with emphasis on GH secretion in the chronic fatigue syndrome, *Clin. Endocrinol.,* 53, 99, 2000.
40. Moorkens, G., Wynants, H., and Abs, R., Effect of growth hormone treatment in patients with chronic fatigue syndrome: a preliminary study, *Growth Horm. IGF Res.*, 8 (suppl. B), 131, 1998.
41. Scott, L.V. and Dinan, T.G., The neuroendocrinology of chronic fatigue syndrome, focus on the hypothalamic–pituitary–adrenal axis, *Funct. Neurol.*, 14, 3, 1999.
42. Cleare, A.J. et al., Low-dose hydrocortisone in chronic fatigue syndrome: a randomised crossover trial, *Lancet*, 353, 455, 1999.
43. Reid, S. et al., Chronic fatigue syndrome, *BMJ,* 320, 292, 2000.
44. Bou-Holaigah, I. et al., The relationship between neurally mediated hypotension and the chronic fatigue syndrome, *JAMA*, 274, 961, 1995.
45. Schondorf, R., Wein, T., and Phaneuf, D., Orthostatic intolerance in patients with chronic fatigue syndrome, *Clin. Auton. Res.,* 5: 320, 1995.
46. De Lorenzo, F., Hargreaves, J., and Kakkar, V.V., Pathogenesis and management of delayed orthostatic hypotension in patients with chronic fatigue syndrome, *Clin. Auton. Res.*, 7, 185,1997.
47. Schondorf, R. and Freeman, R., The importance of orthtostatic intolerance in the chronic fatigue syndrome, *Am. J. Med. Sci.,* 317, 117, 1999.
48. Rowe, P.C. et al., Fludrocortisone acetate to treat neurally mediated hypotension in chronic fatigue syndrome, *JAMA*, 285, 52, 2001.
49. Peterson, P.K. et al., A preliminary placebo-controlled crossover trial of fludrocortisone for chronic fatigue syndrome, *Arch. Intern. Med.*,158, 908, 1998.

50. McKenzie, R. et al., Low-dose hydrocortisone for treatment of chronic fatigue syndrome, *JAMA,* 280, 1061, 1998.
51. Schiffito, G. et al., Autonomic performance and dehydroepiandrosterone sulfate levels in HIV-1-infected individuals: relationship to TH1 and TH2 cytokine profile, *Arch. Neurol.,* 57, 1027, 2000.
52. Kuratsune, H. et al., Dehydroepiandrosterone sulfate deficiency in chronic fatigue syndrome, *Int. J. Mol. Med.*, 1, 143, 1998.
53. De Becker, P. et al., Dehydroepiandrosterone (DHEA) response to i.v. ACTH in patients with chronic fatigue syndrome, *Horm. Metab. Res*., 31, 18, 1999.
54. Salvato, P.D. and Thompson, C., An open study of DHEA (dehydroepiandrosterone) in chronic fatigue syndrome, *AACFS*, San Francisco, 1996.
55. Himmel, P.B., Brown, S., and Coppolino, R., A pilot study employing dehydroepiandrosterone (DHEA) in the treatment of the chronic fatigue syndrome, *AACFS,* San Francisco, 1996.
56. Patarca-Montero, R. et al., Immunology of chronic fatigue syndrome, *J. Chronic Fatigue Syndrome,* 6, 69, 2000.
57. Schluederberg, A. et al., Chronic fatigue syndrome research, *Annu. Intern. Med*., 117, 325, 1992.
58. De Becker, P., McGregor, N., and De Meirleir, K., Possible triggers and mode of onset of chronic fatigue syndrome, *J. Chronic Fatigue Syndrome,* in press.
59. Lloyd, A. et al., Cell-mediated immunity in patients with chronic fatigue syndrome, healthy control subjects, and patients with major depression, *Clin. Exp. Immunol*., 87, 76, 1992.
60. Klimas, N. et al., Immunologic abnormalities in chronic fatigue syndrome, *J. Clin. Microbiol.,* 28, 1403, 1990.
61. Landay, A. et al., Chronic fatigue syndrome: clinical condition associated with immune activation, *Lancet*, 338, 707, 1991.
62. De Meirleir, K. et al., A 37-kDa 2-5A binding protein as a potential marker for chronic fatigue syndrome, *Am. J. Med*., 108, 99, 2000.
63. Patarca-Montero, R. et al. Cytokine and other immunologic markers in chronic fatigue syndrome and their relation to neuropsychological factors, *Appl. Neuropsychol.,* 8(1), 51, 2001.
64. Lloyd, A.R., Wakefield, D., and Hickie, I., Immunity and the pathophysiology of chronic fatigue syndrome, *Chronic Fatigue Syndrome,* Bock, G.R. and Whelan, J., Eds., John Wiley & Sons, Chichester, 1993, 176.
65. Vollmer-Conna, U. et al., Intravenous immunoglobulin is ineffective in the treatment of patients with chronic fatigue syndrome, *Am. J. Med*., 103, 38, 1997.
66. Carter W.A. et al., Preclinical studies with ampligen (mismatched double-stranded RNA), *J. Biol. Response Mod*., 4, 495, 1985.
67. Suhadolnik, R.J. et al., Ampligen treatment of renal cell carcinoma: changes in 2-5A synthetase, 2-5A oligomer size, and natural killer cell activity associated with antitumor response clinically, *Prog. Clin. Biol. Res.,* 202, 449, 1985.
68. Carter, W.A. et al., Clinical, immunological, and virological effects of ampligen, a mismatched double-stranded RNA, in patients with AIDS or AIDS-related complex, *Lancet*, 1(8545), 1286, 1987.
69. Haines D.S., Strauss K.I., and Gillespie D.H., Cellular response to double-stranded RNA, *J. Cell. Biochem*., 46, 9, 1991.
70. Carter W.A. et al., Mismatched double-stranded RNA, ampligen (poly(I): poly(C12U), demonstrates antiviral and immunostimulatory activities in HIV disease, *Int. J. Immunopharmacol*., 13 (suppl. 1), 69, 1991.

71. Hendrix, C.W. et al., Biologic effects after a single dose of poly(I):poly(C12U) in healthy volunteers, *Antimicrob. Agents Chemother.*, 37, 429, 1993.
72. Peterson, D.L. et al., Clinical improvements obtained with Ampligen in patients with severe CFS and associated encephalopathy, *The Clinical and Scientific Basis of ME/CFS*, The Nightingale Research Foundation, Ottawa, 1992, pp. 634.
73. Ablashi, D.V. et al., Ampligen inhibits human herpesvirus-6 *in vitro, In Vivo*, 8, 587, 1994.
74. Pyo, S., The mechanism of poly I:C-induced antiviral activity in peritoneal macrophage, *Arch. Pharm. Res.*, 17, 93, 1994.
75. Milhaud, P.G. et al., Free and liposome-encapsulated double-stranded RNAs as inducers of interferon, interleukin-6, and cellular toxicity, *J. Interferon Res.*, 11, 261, 1991.
76. Montefiori, D.C. and Mitchell W.M., Antiviral activity of mismatched double stranded RNA against human immunodeficiency virus *in vitro, Proc. Natl. Acad. Sci. U.S.A.*, 84(9), 2985, 1987.
77. Hirabayashi, K. et al., Inhibition of metastatic carcinoma cell growth in livers by poly(I):poly(C)/cationic liposome complex (LIC), *Oncol. Res.*, 11, 497, 1999.
78. Hirabayashi, K. et al., Inhibition of cancer cell growth by polyinosinic-polycytidylic acid/cationic liposome complex: a new biological activity, *Cancer Res.*, 59, 4325, 1999.
79. Zarling, J.M. et al., Augmentation of human natural killer cell activity by polyinosinic acid-polycytidylic acid and its nontoxic mismatched analogues, *J. Immunol.*, 124, 1852, 1980.
80. Suhadolnik, R.J. et al., Up-regulation of the 2-5A synthetase/RNase L antiviral pathway associated with chronic fatigue syndrome, *Clin. Infect. Dis.*, 18 (suppl. 1), 96, 1994.
81. Strayer, D.R. et al., A controlled clinical trial with a specifically configured RNA drug, poly(I).poly(C12U), in chronic fatigue syndrome, *Clin. Infect. Dis.*, 18 (suppl. 1), 88, 1994.
82. Suhadolnik, R.J. et al., Changes in the 2-5A synthetase/RNase L antiviral pathway in a controlled clinical trial with poly(I)-poly(C12U) in chronic fatigue syndrome, *In Vivo*, 8, 599, 1994.
83. Ushijima, H. et al., Mode of action of the anti-AIDS compound poly(I).poly(C12U) (Ampligen): activator of 2',5'-oligoadenylate synthetase and double-stranded RNA-dependent kinase, *J. Interferon Res.*, 13, 161, 1993.
84. Suhadolnik, R.J. et al., RNA drug therapy acting via the 2-5A synthetase/RNase L pathway, *Annu. N.Y. Acad. Sci.*, 23, 756, 1993.
85. Salahuddin, S.Z. et al., Isolation of a new virus, HBLV, in patients with lymphoproliferative disorders, *Science*, 234, 596, 1986.
86. Strayer, D., Carter W., Strauss, K., Long-term improvements in patients with chronic fatigue syndrome treated with ampligen, *J. Chronic Fatigue Syndrome*, 1, 35, 1995.
87. Komaroff, A.L., Geiger, A.M., and Wormsley, S., IgG subclass deficiencies in chronic fatigue syndrome, *Lancet*, 1, 1288, 1988.
88. Linde, A., Hammerstrom, L., and Smith, C.I., IgG subclass deficiency and chronic fatigue syndrome, *Lancet*, 1, 885, 1988.
89. Read, R. et al., IgG 1 subclass deficiency in patients with chronic fatigue syndrome, *Lancet*, 1, 241, 1988.
90. Straus, S.E., Intravenous immunoglobulin treatment for the chronic fatigue syndrome, *Am. J. Med.*, 89, 551, 1990.
91. Peterson, P. et al., A controlled trial of intravenous immunoglobulin G in chronic fatigue syndrome, *Am. J. Med.*, 89, 554, 1990.

92. Lloyd, A. et al., A double-blind, placebo-controlled trial of intravenous immunoglobulin therapy in patients with chronic fatigue syndrome, *Am. J. Med.*, 89, 561, 1990.
93. Dubois, R.E., Gamma globulin therapy for chronic mononucleosis syndrome, *AIDS Res.*, 2 (suppl. 1), 191, 1986.
94. Rowe, K.S., Double-blind randomized controlled trial to assess the efficacy of intravenous gammaglobulin for the management of chronic fatigue syndrome in adolescents, *J. Psychiatr. Res.*, 31, 133, 1997.
95. Lloyd, A. et al., Immunologic and psychologic therapy for patients with chronic fatigue syndrome: a double-blind, placebo-controlled trial, *Am. J. Med.*, 91, 197, 1993.
96. Hyde, B. and Mitoma, D., A seven-month, blinded-treatment trial with isoprinosine. *Proc. 2nd World Cong. Chronic Fatigue Syndrome and Related Disorders,* 1999, p. 79.
97. Plioplys, A. and Plioplys, S., Amantadine and L-carnitine treatment of chronic fatigue syndrome, *Neuropsychobiology*, 35, 16, 1997.
98. The Canadian MS Research Group, A randomised controlled trial of amantadine in fatigue associated with multiple sclerosis, *Can. J. Neurol. Sci.*, 14, 273, 1987.
99. Kidd, P.M., The use of mushroom glucans and proteoglycans in cancer treatment, *Altern. Med. Rev.,* 5, 4, 2000.
100. Ooi, V.E. and Liu, F., Immunomodulation and anti-cancer activity of polysaccharide-protein complexes, *Curr. Med. Chem.*, 7, 715, 2000.
101. Yoshino, S. et al., Immunoregulatory effects of the antitumor polysaccharide lentinanon Th1/Th2 balance in patients with digestive cancers, *Anticancer Res.*, 20, 4707, 2000.
102. Ablashi, D.V. et al., Antiviral activity *in vitro* of Kutapressin against human herpesvirus-6, *In Vivo*, 8, 581, 1994.
103. Rosenfeld, E. et al., Potential *in vitro* activity of Kutapressin against Epstein–Barr virus, *In Vivo*, 10, 313, 1996.
104. Yousef, G. et al., Chronic enterovirus infection in patients with postviral fatigue syndrome, *Lancet*, 1, 146, 1988.
105. Straus, S. et al., Acyclovir treatment of the chronic fatigue syndrome, *N. Engl. J. Med.*, 319, 1692, 1988.
106. Sen, G.C., *Cytokines in Health and Disease*, Remick, D.G. and Friedland, J.S., eds., Marcel Dekker, New York, 1997.
107. Jacyna, M. et al., Randomised controlled trial of interferon-alpha (lymphoblastoid interferon) in chronic non-A, non-B hepatitis, *BMJ*, 298, 80, 1989.
108. Brook, M., Bannister, B., and Weir, W., Interferon-alpha therapy for patients with chronic fatigue syndrome, *J. Infect. Dis.*, 168, 791, 1993.
109. See, D.M. and Tilles, J.G., Alpha interferon treatment of patients with chronic fatigue syndrome, *Immunol. Invest.*, 25, 153, 1996.
110. Levy, J., Viral studies of chronic fatigue syndrome. Introduction, *Clin. Infect. Dis.*, 18(Suppl 1), 117, 1994.
111. Dusheiko, G., Side effects of alpha interferon in chronic hepatitis C, *Hepatology,* 26,112S, 1997.
112. Valentine, A.D. et al., Mood and cognitive side effects of interferon-alpha therapy, *Semin. Oncol.*, 25, 39, 1998.
113. Nasralla, M., Haier, J., and Nicolson, G.L. Multiple mycoplasmal infections detected in blood of patients with chronic fatigue syndrome and/or fibromyalgia, *Eur. J. Clin. Microbiol. Infect. Dis.*, 18, 859, 1999.
114. Nicolson, G.L. et al., Role of mycoplasmal infections in fatigue illnesses: chronic fatigue and fibromyalgia syndromes, Gulf War illness, and rheumatoid arthritis, *J. Chronic Fatigue Syndrome*, 6, 23, 2000.

115. Choppa, P.C. et al., Multiplex PCR for the detection of *Mycoplasma fermentans, M. hominis,* and *M. penetrans* in cell cultures and blood samples of patients with chronic fatigue syndrome, *Mol. Cell. Probes* , 12, 301, 1998.
116. Vojdani, A. et al., Detection of mycoplasma genus and *Mycoplasma fermentans* by PCR in patients with chronic fatigue syndrome, *FEMS Immunol. Med. Micro.*, 22, 355, 1998.
117. Simecka, J.W. et al., Interactions of mycoplasmas with B cells: antibody production and nonspecific effects, *Clin. Infect. Dis.*, 17 (suppl. 1), S176, 1993.
118. Nijs, J. et al. Prevalence of mycoplasma infections among Belgian CFS-patients, submitted for publication, 2001.
119. Nicolson, G.L. and Nicolson N.L., Diagnosis and treatment of mycoplasmal infections in Persian Gulf War illness — CFIDS patients, *Int. J. Occup. Med. Immunol. Toxicol.*, 5, 69, 1996.
120. Nicolson G.L., Considerations when undergoing treatment for Gulf War illness/CFS/FMS/rheumatoid arthritis, *Intern. J. Med.*, 1, 123, 1998.
121. Chia, J.K. and Chia L.Y., Chronic chlamydia pneumonia infection: a treatable cause of chronic fatigue syndrome, *Clin. Infect. Dis.,* 29, 452, 1999.
122. Bombardier, C.H. and Buchwald, D., Chronic fatigue, chronic fatigue syndrome, and fibromyalgia. Disability and health-care use, *Med. Care*, 34, 924, 1996.
123. McGregor, N.R. et al., Preliminary determination of the association between symptom expression and urinary metabolites in subjects with chronic fatigue syndrome, *Biochem. Mol. Med.*, 58, 85, 1996.
124. McGregor, N.R. et al., Assessment of pain (distribution and onset), symptoms, SCL-90-R inventory responses and the association with infectious events in patients with chronic orofacial pain, *J. Orofacial Pain*, 10, 339, 1996.
125. Andersson, M. et al., Effects of *staphylococcus toxoid* vaccine on pain and fatigue in patients with fibromyalgia/chronic fatigue syndrome, *Eur. J. Pain*, 2, 133, 1998.
126. Goldstein, J.A., Cimetidine, ranitidine, and Epstein–Barr virus infection, *Annu. Intern. Med.*, 105, 139, 1986.
127. Gantz, N. and Holmes, G., Treatment of patients with chronic fatigue syndrome, *Drugs,* 38, 855, 1989.
128. Wessely, S. et al., Management of chronic (postviral) fatigue syndrome, *J. R. Coll. Gen. Pract.*, 39, 26, 1989.
129. Adolphe, A.B., Chronic fatigue syndrome: possible effective treatment with nifedipine, *Am. J. Med.*, 85, 892, 1988.
130. Goldstein, J.A., Treatment of chronic fatigue syndrome, in *Chronic Fatigue Syndrome: The Limbic Hypothesis*, Goldstein, J.A., Ed., The Haworth Medical Press, New York, 1993, 127–188.
131. Jacobson, W. et al., Serum folate and chronic fatigue syndrome, *Neurology*, 43, 2645, 1993.
132. Kaslow, J.E., Rucker, L., and Onishi, R., Liver extract-folic acid-cyanocobalamin vs. placebo for chronic fatigue syndrome, *Arch. Intern. Med.*, 149, 2501, 1989.
133. Werbach, M.R., Nutritional strategies for treating chronic fatigue syndrome, *Altern. Med. Rev.,* 5, 93, 2000.
134. Regland, B. et al., Increased concentrations of homocysteine in the cerebrospinal fluid in patients with fibromyalgia and chronic fatigue syndrome, *Scand. J. Rheumatol.*, 26, 301, 1997.
135. Mukherjee, T.M., Smith, K., and Maros, K., Abnormal red-blood cell morphology in myalgic encephalomyelitis, *Lancet*, 2, 328, 1987.
136. Simpson, L.O., Myalgic encephalomyelitis, *J. R. Soc. Med.*, 84, 633, 1991.

137. Heap, L.C., Peters, T.J., and Wessely, S., Vitamin B status in patients with chronic fatigue syndrome, *J. R. Soc. Med.*, 92, 183, 1999.
138. Pall, M.L., Cobalamin used in chronic fatigue syndrome therapy is a nitric oxide scavenger, *J. Chronic Fatigue Syndrome,* 8, 39, 2001.
139. Pall, M.L., Elevated peroxynitrite as the cause of chronic fatigue syndrome: other inducers and mechanisms of symptom generation, *J. Chronic Fatigue Syndrome,* 7, 45, 2000.
140. Matsuda, J., Gohchi, K., and Gotoh, N., Serum concentrations of 2',5'-oligo-adenylate synthetase, neopterin, and beta-glucan in patients with chronic fatigue syndrome and in patients with major depression, *J. Neurol. Neurosurg. Psychiatry,* 57, 1015, 1994.
141. Patarca, R. et al., Dysregulated expression of tumor necrosis factor in chronic fatigue syndrome: interrelations with cellular sources and patterns of soluble immune mediator expression, *Clin. Inf. Dis.,* 18 (suppl. 1), S147, 1994.
142. Linde, A. et al., Serum levels of lymphokines and soluble cellular receptors in primary Epstein–Barr virus infection and in patients with chronic fatigue syndrome, *J. Infect. Dis.,* 165, 994, 1992.
143. Moss, R.B., Mercandetti, A., and Vojdani, A., TNF-alpha and chronic fatigue syndrome, *J. Clin. Immunol.*, 19, 314, 1999.
144. Forsyth, L.M. et al., Therapeutic effect of oral NADH on the symptoms of patients with chronic fatigue syndrome, *Annu. Allergy Asthma Immunol.*, 82, 185, 1999.
145. Heuser, G. and Vojdani, A., Enhancement of natural killer cell activity and T- and B-cell function by buffered vitamin C in patients exposed to toxic chemicals: the role of protein kinase-C, *Immunopharmacol. Immunotoxicol.*, 19, 291, 1997.
146. Grant, J.E., Veldee, M.S., and Buchwald, D., Analysis of dietary intake and selected nutrient concentrations in patients with chronic fatigue syndrome, *J. Am. Diet Assoc.*, 96, 383, 1996.
147. Kodama, M., Kodama, T., and Murakami, M., The value of the dehydroepiandrosterone-annexed vitamin C infusion treatment in the clinical control of chronic fatigue syndrome. I. A pilot study of the new vitamin C infusion treatment with a volunteer CFS patient, *In Vivo,* 10, 575, 1996.
148. Kodama, M., Kodama, T., and Murakami, M., The value of the dehydroepiandrosterone-annexed vitamin C infusion treatment in the clinical control of chronic fatigue syndrome. II, Characterization of CFS patients with special reference to their response to a new vitamin C infusion treatment, *In Vivo,* 10, 585, 1996.
149. Cox, I., Campbell, M., and Dowson, D., Red blood cell magnesium and chronic fatigue syndrome patients, *Lancet,* 337, 757, 1991.
150. Moorkens, G. et al., Magnesium deficit in a sample of the Belgian population presenting with chronic fatigue, *Magnes. Res.*, 10, 329, 1997.
151. Clague, J.E., Edwards, R.H.T., and Jackson, M.J., Intravenous magnesium loading in chronic fatigue syndrome, *Lancet,* 340, 124, 1992.
152. Abraham, G.E. and Flechas, J.D., Management of fibromyalgia: rationale for the use of magnesium and malic acid, *J. Nutr. Med.*, 3, 49, 1992.
153. Streeten, D.H. and Anderson, G.H., Delayed orthostatic intolerance, *Arch. Intern. Med.*, 152, 1066, 1992.
154. Rosen, S.G. and Cryer, P.E., Postural tachycardia syndrome. Reversal of sympathetic hyperresponsiveness and clinical improvement during sodium loading, *Am. J. Med.,* 72: 847, 1982.
155. Bralley, J.A. and Lord, R.S., Treatment of chronic fatigue syndrome with specific amino acid supplementation, *J. Appl. Nutrition,* 46, 74, 1994.

156. Dunstan, R.H. et al., Characterisation of differential amino acid homeostasis amongst population subgroups: a basis for determining specific amino acid requirements, *J. Nutr. Environ. Med.,* 10, 211, 2000.
157. Fuchs, D., Weiss, G., and Wachter, H., Pathogenesis of chronic fatigue syndrome, *J. Clin. Psychiatry,* 53, 296, 1992.
158. Vassallo, C.M. et al., Decreased tryptophan availability but normal post-synaptic 5-HT2c receptor sensitivity in chronic fatigue syndrome, *Psychol. Med.*, 31, 585, 2001.
159. Puttini, P.S. and Caruso, I., Primary fibromyalgia syndrome and 5-hydroxy-L-tryptophan: a 90-day open study, *J. Int. Med. Res.*, 20, 182, 1992.
160. Kuratsune, H. et al., Acylcarnitine deficiency in chronic fatigue syndrome, *Clin. Infect. Dis.,* 18 (suppl. 1), S62, 1994.
161. Plioplys, S. and Plioplys, A., Chronic fatigue syndrome (myalgic encephalopathy), *South. Med. J.,* 88, 993, 1995.
162. Grau, J.M. et al., Chronic fatigue syndrome: studies on skeletal muscle, *Clin. Neuropathol.*, 11, 329, 1992.
163. Famularo, G. and De Simone, C., A new era for carnitine? *Immunol. Today,* 16, 211, 1995.
164. De Simone, C. et al., Carnitine depletion in peripheral blood mononuclear cells from patients with AIDS: effect of oral L-carnitine, *AIDS*, 8, 655, 1994.
165. Rowbottom, D. et al., The physiological response to exercise in chronic fatigue syndrome, *J. Chronic Fatigue Syndrome*, 4, 33, 1998.
166. Behan, P.O., Behan, W.M.H., and Horrobin, D., Effect of high doses of essential fatty acids on the postviral fatigue syndrome, *Acta Neurol. Scand.*, 82, 209, 1990.
167. Dunstan, R.H. et al., Changes in plasma lipid homeostasis observed in chronic fatigue syndrome patients, *J. Nutr. Environ. Med.,* 9, 267, 1999.
168. McGregor, N.R. et al., Assessment of plasma saponified lipid fractions and ratios in sudden and gradual onset chronic fatigue syndrome patients, *J. Nutr. Environ. Med.,*10, 13, 2000.
169. Dunstan, R.H. et al., Analysis of serum lipid changes associated with self-reported fatigue, muscle pain and the different chronic fatigue syndrome factor analysis symptom clusters, *AACFS Conf. Seattle,* Jan. 2001, Abstract 62.
170. Gray, J. and Martinovic, A., Eicosanoids and essential fatty acid modulation in chronic disease and the chronic fatigue syndrome, *Med. Hypotheses*, 43, 31, 1994.
171. Horrobin, D.F., Post-viral fatigue syndrome, viral infections in atopic eczema, and essential fatty acids, *Med. Hypotheses,* 32, 211, 1990.
172. Warren, G., McKendrick, M., and Peet, M., The role of essential fatty acids in chronic fatigue syndrome, *Acta Neurol. Scand.*, 99, 112, 1999.
173. Emms, T.M. et al., Food intolerance in chronic fatigue syndrome, *AACFS Conf. Seattle*, Jan. 2001, Abstract 15.
174. Klimas, N.G. et al., Clinical and immunological effects of autologous lymph node cell transplant in chronic fatigue syndrome, *J. Chronic Fatigue Syndrome*, 8, 39, 2001.
175. Sharpe, M., Cognitive behaviour therapy for chronic fatigue syndrome: efficacy and implications, *Am. J. Med.,* 105(3A), 104S, 1998.
176. Gardner, WN., Hyperventilation disorders, *J. R. Soc. Med.*, 83, 755, 1990.
177. Butler, S. et al., Cognitive behaviour therapy in chronic fatigue syndrome, *J. Neurol. Neurosurg. Psychiatry,* 54, 153, 1991.
178. Deale, A. et al., Cognitive behavior therapy for chronic fatigue syndrome: a randomized controlled trial, *Am. J. Psychiatry,* 154, 408, 1997.
179. Sharpe, M. et al.,. Chronic fatigue syndrome. A practical guide to assessment and management, *Gen. Hosp. Psychiatry,* 19, 185, 1997.

180. Friedberg, F. and Krupp, L., A comparison of cognitive behavioral treatment for chronic fatigue syndrome and primary depression, *Clin. Infect. Dis.*, 18 (suppl. 1), S105, 1994.
181. Powell, P. et al., Randomised controlled trial of patient education to encourage graded exercise in chronic fatigue syndrome, *BMJ*, 322, 387, 2001.
182. Prins, J.B. and Bleijenberg, G., Cognitive behavior therapy for chronic fatigue syndrome: a case study, *J. Behav. Ther. Exper. Psychiatry*, 30, 325, 1999.
183. McCully, K.K. et al., Reduced oxidative muscle metabolism in chronic fatigue syndrome, *Muscle Nerve*, 19, 621, 1996.
184. Komaroff, A.L., Clinical presentation of chronic fatigue syndrome, in *Chronic Fatigue Syndrome*, Bock, G.R. and Whelan, J., Eds., John Wiley & Sons, Chichester, 1993, 176–192.
185. Kent-Braun, J. et al., Central basis of muscle fatigue in chronic fatigue syndrome, *Neurology*, 43, 125, 1993.
186. Fischler, B. et al., Physical fatigability and exercise capacity in the chronic fatigue syndrome, association with disability, somatisation, and psychopathology, *J. Psychom. Res.*, 42, 369, 1997.
187. De Becker, P. et al., Exercise capacity in chronic fatigue syndrome, *Arch. Intern. Med.*, 160, 3270, 2000.
188. Montague, T. et al., Cardiac function at rest and with exercise in the chronic fatigue syndrome, *Chest,* 95, 779, 1989.
189. Bazelmans, E. et al., Is physical deconditioning a perpetuating factor in chronic fatigue syndrome? A controlled study on maximal exercise performance and relations with fatigue, impairment, and physical activity, *Psychol. Med.*, 31, 107, 2001.
190. Holmes, G. et al., Chronic fatigue syndrome: a working case definition, *Annu. Intern. Med.*, 108, 387, 1988.
191. Fukuda, K. et al., The chronic fatigue syndrome: a comprehensive approach to its definition and study, *Annu. Intern. Med.*, 121, 953, 1994.
192. De Lorenzo, F. et al., Chronic fatigue syndrome: physical and cardiovascular deconditioning, *Q. J. Med.*, 91, 475, 1998.
193. Sisto, S. et al., Metabolic and cardiovascular effects of a progressive exercise test in patients with chronic fatigue syndrome, *Am. J. Med.*, 100, 634, 1996.
194. Fulcher, K. and White, P., Randomised controlled trial of graded exercise in patients with the chronic fatigue syndrome, *BMJ,* 314, 1647, 1997.
195. Fulcher, K. and White, P., Strength and physiological response to exercise in patients with chronic fatigue syndrome, *J. Neurol. Neurosurg. Psychatry,* 69, 302, 2000.
196. Wearden, A.J. et al., Randomised, double-blind, placebo-controlled treatment trial of fluoxetine and graded exercise for chronic fatigue syndrome, *Br. J. Psychiatr.*, 172, 485, 1998.
197. Macauley, D. et al., Levels of physical activity, physical fitness, and their relationship in the Northern Ireland Health and Activity Survey, *Int. J. Sports Med.*, 19, 503, 1998.
198. Field, T.M. et al., I. Massage therapy effects on depression and somatic symptoms in chronic fatigue syndrome, *J. Chronic Fatigue Syndrome*, 3, 43, 1997.
199. Essame, C.S. et al., Pilot study of a multidisciplinary inpatient rehabilitation of severely incapacitated patients with the chronic fatigue syndrome, *J. Chronic Fatigue Syndrome*, 4, 51, 1998.
200. Sola, A. et al., Vilas. Inactivation and inhibition of African swine fever virus by monoolein, monolinolein, and gamma-linolenyl alcohol, *Arch. Virol.*, 88, 285, 1986.
201. Pottathil, R. et al., Establishment of the interferon-mediated antiviral state: role of fatty acid cyclooxygenase, *Proc. Natl. Acad. Sci. U.S.A.*, 77, 5437, 1980.

10 From Laboratory to Patient Care

Kenny De Meirleir, Daniel L. Peterson, Pascale De Becker, and Patrick Englebienne

CONTENTS

10.1 INTRODUCTION

Chronic fatigue syndrome is an illness defined and characterized exclusively by a group of symptoms, according to a case definition developed by the Centers for Disease Control.[1] The major characteristic of the syndrome is an extreme, long-lasting and severely disabling fatigue combined with a series of symptoms featuring impairments in concentration and short-term memory, sleep disturbances, and musculoskeletal pain. The diagnosis can be definitively made only if any possible known medical and psychiatric causes have been excluded. However, the symptoms on which the diagnosis is based occur in many illnesses, including depression.[2] It is therefore not surprising that a long-lasting controversy has developed over the years in medical circles as to whether CFS, along with its often associated fibromyalgia syndrome, has any somatic etiology.[3,4]

Furthermore, the nonspecificity of the symptoms, along with their self-reported ranking and importance, suggests that several subsets of sufferers can hardly be distinguished from the heterogenous population of CFS patients, unless sophisticated biological investigations are performed (Chapter 7). The psychosomatic origin of CFS is challenged by numerous reports in the scientific literature[2,5,6] that provide clear-cut evidences of physiological anomalies in the illness. However, the etiology and pathogenesis have remained obscure.

The observation made by Suhadolnik and colleages in 1994[7] that peripheral blood mononuclear cells (PBMC) of CFS patients were characterized by abnormalities in

0-8493-1046-6/02/$0.00+$1.50

the interferon 2-5A pathway has prompted further research in this area. The recent validation of the presence in PBMC of the 37-kDa truncated RNase L variant as a biological marker of CFS[8,9] has permitted a classification of the CFS subpopulations on grounds of an objective parameter[10] in contrast to the subjectively self-reported symptomatology used previously. The availability of the recombinant enzyme and the recent discovery that the 37-kDa protein, as well as other major cellular components, was undergoing proteolytic cleavage in CFS immune cells[11] prompted us to have a fresher and unprecedented look at the molecular biology level. Our observations are detailed and discussed in the previous chapters. These observations allow us to unravel not only the etiology and pathogenesis, but also the evolution of the syndrome. They have major implications in conducting the therapy from the onset to the established and ongoing illness.

10.2 ETIOLOGY, PATHOGENESIS, AND EVOLUTION OF CFS: A LABORATORY PERSPECTIVE

Our quest starts with a critical review of the abnormalities of the 2-5A pathway in the immune cells of CFS patients and continues with a confrontation with data from the literature and our molecular biology approach. The role of RNA and type I interferons (IFNs) in activating this cellular defense mechanism has been reviewed and the various molecular players presented in the first chapter. However, for a good understanding of the subsequent discussion, we present a rapid summary of the activation of this major battlefield (Figure 10.1). Upon infection of the cell, RNA

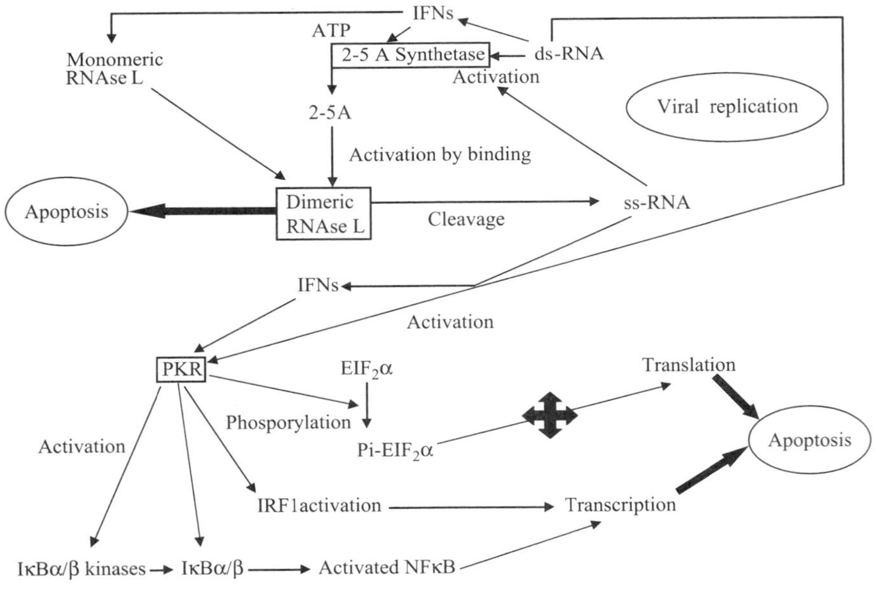

FIGURE 10.1 General overview of the 2-5A defense pathway.

(single- as well as double-stranded, ss- and ds-RNA respectively) activates the production of type I IFNs and the 2-5A synthetases (2-5OAS). IFNs induce the transcription of a set of responsive genes which, besides the 2-5OAS, include the ds-RNA-dependent protein kinase (PKR) and RNase L. The 2-5OAS polymerize ATP in 2',5'-oligoadenylates (2-5A), which in turn activate RNase L. Activation of this enzyme results in ss-RNA cleavage and leads to the activation of a first apoptotic program involving caspase activation (Chapter 6). The RNAs further activate PKR, which leads to a second apoptotic program mediated by a blockade of translation (phosphorylation of eIF-2α), along with the transcription of proapoptotic proteins through activation of IRF1 and NF-κB.

As recalled above,[7] CFS is characterized by abnormalities in the 2-5A pathway: 2-5OAS, RNase L as well as PKR[12] are upregulated. Recent evidence shows that RNase L is present as the native as well as the 37-kDa fragment in CFS PBMC,[8,9] the latter produced by proteolytic cleavage.[11] Activation of RNase L proceeds by 2-5A binding and dimerization. Therefore, we investigated (Chapter 2) the effect of dimerization on the susceptibility of the enzyme to proteolytic cleavage. Experiments with the recombinant enzyme show that dimerization protects the protein against cleavage. The susceptibility of RNase L to cleavage in the PBMC of CFS patients results, therefore, from its primary presence as the monomer instead of the dimer. Previous reports[7,13] indicate that biologically active 2-5A are upregulated in PBMC. By "biologically active" the authors mean "capable of binding to RNase L." For dimerization and full activation, RNase L requires 2-5A oligomers containing at least three adenylate repeats.[14] The 2-5A dimers bind to, but neither dimerize nor activate, the enzyme.[14,15] The concomitant presence of high levels of 2-5A and of RNase L fragments in CFS PBMC therefore indicates that 2-5A dimers are preferentially produced by the 2-5OAS.

The 2-5OAS protein family comprises three members of highly homologous proteins (Chapter 5) that have different catalytic properties, with the p100 protein synthesizing preferentially 2-5A dimers. The induction of 2-5OAS enzymes by type I IFNs is a sensitive, tightly regulated process. The pathway requires protein kinase C[16-18] and the promoter contains multiple regulatory elements[19,20] which are, respectively, either activated or repressed by interferon regulatory factors 1 and 2.[21] Moreover, upregulation of the 2-5OAS does not imply proper activation. Suitable activation of the enzymes to produce 2-5A oligomers longer than the dimer requires either ds-RNA of a given length (over 25bp),[22,23] or small ds- or ss-RNA aptamers with little secondary structure.[24] Our results show that RNase L dimer is protected from proteolytic cleavage by PBMC extracts (Chapter 2).

This implies that the native enzyme cleaved in CFS PBMC extracts is primarily present as the monomer. The 2-5A oligomers produced by 2-5OAS in these samples are therefore dimers and, consequently, the 2-5OAS are not properly activated to produce longer oligomers, as schematically depicted in Figure 10.2. This is further supported by the fact that poly(I).poly(C), one of the best activators of 2-5OAS *in vitro*[25,26] as well as *in vivo*[27] and provides good immunomodulatory activity in both CFS[7,28] and AIDS,[29] has been demonstrated to exert its activity *in vitro* by activating the 2-5OAS to produce longer 2-5A oligomers.[30,31] As shown in Figure 10.2, such

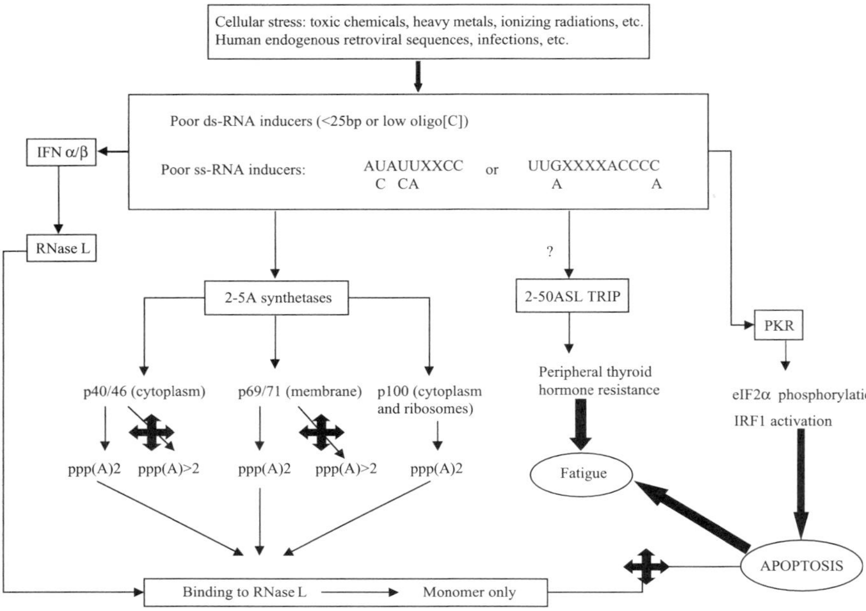

FIGURE 10.2 Suggested etiology of CFS. A poor activation of the 2-5OAS by small ds- or ss-RNAs generated by cellular stress (toxic chemicals, ionizing radiations), translation of human endogenous retroviruses or infection leads to the preferential production of 2-5A dimers which bind to, but do not activate, the RNase L. This blocks the RNase L apoptotic pathway. The simultaneous induction of 2-5OAS-like proteins (thyroid receptor interactors) leads to the extreme long-lasting fatigue. Finally, the activated PKR upregulates its apoptotic pathway.

upregulation, but poor activation, of the 2-5OAS can lead to the blockade of the RNase L apoptotic pathway. This does not preclude the simultaneous induction of the 2-5OAS-like proteins (thyroid-receptor interacting protein, TRIP, see Chapter 5), explaining the extreme fatigue. On the other hand, the ds-RNA-dependent protein kinase PKR has different nucleotide requirements for activation. Poor activators of the 2-5OAS, including ss-polynucleotides, can be repressors or, conversely, good activators of PKR.[32,33] The PKR apoptotic pathway thus becomes unbalanced with respect to the 2-5OAS pathway and we face an intracellular signal discrepancy.

Besides infection by, or reactivation of, the different viruses detected in CFS patients,[34] including the herpes viridae,[35] other mechanisms may be the source of an abnormal triggering of both 2-5OAS/PKR apoptotic pathways in immune cells: the transcription or even expression of endogenous retrovirus (HERV) sequences present in our genome, as well as of the short interspersed elements (SINEs) typified by the highly repetitive Alu sequences present in the human chromosomes (for recent reviews, see References 36 through 38). A large body of evidence is now available indicating that transcription and expression of HERVs and SINEs are induced by cellular stress and damage by ionizing radiations and toxic chemicals, both *in vitro* and *in vivo*.[39-47]

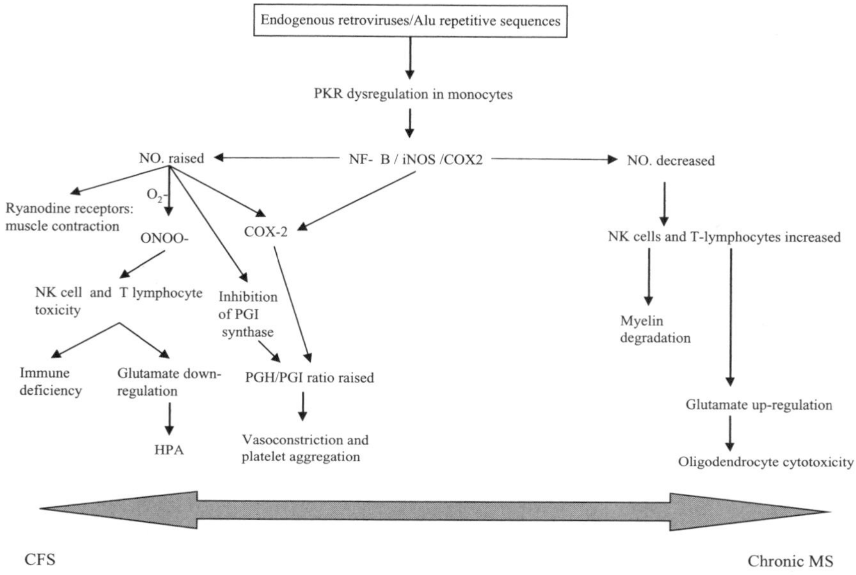

FIGURE 10.3 Potential effects of PKR dysregulation by HERV and Alu sequences through nitric oxide production. Activation of PKR (CFS case, left) induces iNOS and COX2 through NF-κB. Increased NO produces toxicity toward NK cells and macrophages which leads to immune deficiency, T_H2 shift, glutamate downregulation, and decreased CRH release. COX2 upregulation further induces vasoconstriction and platelet aggregation. PKR downregulation (chronic MS case, right) decreases nitric oxide which leads to T_H1 shift, glutamate up-regulation, and oligodendrocyte toxicity.

The importance of their respective transcription or expression in the etiology of autoimmune and malignant diseases begins only now to be unraveled.[37,48] Specific HTLV-II endogenous retroviral sequences have been noted in chronic fatigue syndrome,[49] although other studies have not confirmed this observation.[50] Alu sequences have been retrieved in Gulf War syndrome patients.[51] The presence of HERV sequences[52] and of antibodies to HERV proteins[53] has been demonstrated in multiple sclerosis (MS), a disease with which CFS shares some features,[34] and HERVs are progressively considered as an etiologic agent for MS.[54] Interestingly, we found the 2-5OAS dysregulation and RNase L cleavage in the PBMC of MS and CFS patients alike, particularly during active episodes in the former (unpublished data). As displayed in Figure 10.3, depending on the sequences transcribed (HERV, Alu, SINE), one is led to a large possible array of cellular situations where PKR activation or deactivation parallels the dysregulation of the 2-5OAS. Some Alu sequences, for instance, have been identified as negative regulators of PKR activity by binding to the enzyme and impairing its dimerization,[55] while other specific RNA sequences are most probably activating the enzyme.[56] At the extremes of the array are either full PKR hyperactivation or deactivation, with their corollary consequences.

For the sake of clarity, we will illustrate the effects of the extreme situations in the array. On the one hand (Figure 10.3, left-hand side), hyperactivation of PKR leads to NF-κB activation. This activation may proceed either through a ds-RNA-activated PKR[57] or a catalytically inactive enzyme,[58] which suggests that the enzyme bound by a given ss-RNA HERV or Alu sequence is still capable of activating NF-κB. NF-κB activation stimulates the expression of enzymes whose products contribute to the pathogenesis of the inflammatory process of CFS, including the inducible form of nitric oxide synthase (iNOS), which generates nitric oxide (NO·), and the inducible cyclooxygenase (COX2), which generates prostanoids.[59] NO· regulates numerous physiological processes, including neurotransmission, smooth muscle contractility, platelet reactivity, and the cytotoxic activity of immune cells.[60] NO· interacts with the cardiac and skeletal muscle ryanodine receptors (high-conductance Ca^{2+} release channels), dysregulating their contractile function.[61,62]

NO· further exerts a positive effect on COX2 activity or expression.[63] In presence of reactive oxygen species produced by monocytes and activated phagocytes, NO· is oxydized to the peroxynitrite ion ($ONOO^-$), which is extremely toxic to NK and lymphokine-activated killer (LAK) cells *in vitro* and *in vivo*.[64-67] This toxic process can be regulated by histaminergic control of the monocytes at H2 receptor level.[68,69] Interestingly, the extent of oxidative damage is correlated with symptoms in CFS (Chapter 7). Such deleterious effects on NK function, along with its associated 2-5OAS dysregulation, have reportedly been abrogated in breast cancer patients (a disease where HERVs are present and supposedly involved[37]) by treatment with the synthetic RNA poly(A)-poly(C),[70] which further illustrates the pertinence of our conclusions. Therefore, activated PKR is likely to play a central role in the T_H1/T_H2 balance, not only in inducing IgE class-switching in B cells,[71] but also in reducing NK and LAK cell activity.

Besides their direct immune functions, activated immune cells produce high quantities of glutamate. In the brain, glutamate exerts excitatory effects on neurons through the N-methyl-D-aspartate (NMDA) receptor.[72] A reduction in activated immune cells consequently impairs such cross-talk. Furthermore, NO·, which might be present in excess in the extraneuronal space, plays a role as feedback regulator of glutamatergic excitation, especially in the hypothalamus where it downregulates the secretion of corticotropin-releasing hormone (CRH).[73] These effects can be held responsible for the dysregulations in the central nervous system and hypothalamo-pituitary–adrenal axis (HPA),[74] as observed in CFS patients.[34] Finally, the simultaneous induction of iNOS (NO· blocks prostaglandin I synthase) and COX2 expression causes a dramatic increase in the ratio of prostaglandin (PG) H_2 (a vasoconstrictor) to PGI_2 (prostacyclin, a relaxant and the most potent platelet anti-aggregatory agent), which leads to vasoconstriction and activation of the coagulation system,[75] both symptoms observed in CFS.[76,77]

On the other hand (Figure 10.3, right-hand side), at the other extreme of the array, a deactivated PKR leads to a decreased NO· production in chronic MS and the subsequent T_H1-driven immune balance[78] which, through the autoreactive T-cells, leads to myelin degradation.[79] Furthermore, the activated immune cells produce glutamate in excess, which exerts in the brain a toxic activity on oligodendrocytes. These cells express exclusively the AMPA (α-amino-3-hydroxy-5-methyl-4-isoxazolepropionic acid)/kainate type of glutamate receptor, which renders them exquisitely sensitive to glutamate excitotoxicity.[80,81] Inducible NOS has been detected in monocytes of MS patients,[82] but the presence of the enzyme has been demonstrated to be restricted to the acute, not chronic, disease cases.[83] This has been confirmed by the excretion of NO· metabolites in urine.[84] Accordingly, iNOS induction and expression has been detected by *in situ* hybridization and immunocytochemistry in astrocytes of postmortem tissues from acute, not chronic, patients.[85] Furthermore, therapies aimed at inhibiting reactive nitrogen species in MS have provided variable results, sometimes exacerbating the disease.[86] The model presented in Figure 10.3 is thus consistent with our laboratory data and the *in vitro* and *in vivo* results presented in the literature. Between the two extreme situations presented for the worst CFS and chronic MS cases, several graded autoimmune pathological conditions may fit, including acute MS, lupus erythematosus that shares with CFS the presence in serum of antinuclear antibodies,[87] and type 1 diabetes, in which the abnormalities in the PKR/2-5A pathway receive sustained attention.[88]

The activation of PKR can also occur independently of RNA through the activation of PACT (PKR-activating protein),[89] the human cellular activator of PKR, homologous to the mouse RAX.[90] The 35-kDa protein is a ds-RNA-binding protein ubiquitously expressed which contains three ds-RNA binding domains (DRBD).[89] PKR contains two such domains which mediate protein–protein interactions between ds-RNA-binding proteins. PACT binds PKR through these DRBDs and activates PKR through a ds-RNA-independent mechanism.[91] Interestingly, the mouse homologue RAX is activated by the sphingolipid ceramide, an important second-signal molecule inducing apoptosis in response to every known stress stimuli, including reactive oxygen species (see Chapter 5 and Reference 92). Consequently, the production of oxydative species by activated phagocytes and macrophages in CFS might further enhance the PKR dysregulated activation through PACT.

Activated PKR induces apoptosis, not only by the inhibition of translation caused by eIF2-α phosphorylation, but also by another mechanism involving the activation of caspase 8 mediated by the Fas-associated death domain.[93,94] This latter involvement is in complete agreement with our observations in the PBMCs of CFS patients (Chapter 6), i.e., activation of caspases 8 and 2 in samples characterized with moderate to medium cleavage of RNase L. As depicted in Figure 10.4, these caspases activate caspase 3, which in turn activates caspase 6. This apoptotic induction is accompanied by a calcium release from the endoplasmic reticulum, which further activates PKR.[95] The caspases[96] and the increase in intracellular calcium further activate m-calpain (Figure 10.5), which in turn activates caspase 12,[97] inhibits caspase 9,[98] and cleaves its inhibitor calpastatin[99] and the monomeric RNase L generating the truncated 37-kDa fragment. By cleaving and inactivating calpastatin, the natural calpain inhibitor, caspases further upregulate calpain activity.[100] This sequence of events is in line with our observations of caspase activation as a function of RNase L cleavage. Priming of caspases 8, 2, and 3 increases in intracellular calcium, followed by inhibition of caspase 2 and 9. The cell receives contradictory signals: apoptotic induction through PKR and apoptotic inhibition through RNase L, which is impaired to homodimerize and activate. Our results further indicate that other important proteins are cleaved in this process, including G-actin, STAT1, RLI, and p53. These proteolytic cleavages are exerted by either calpain, the caspases, or both; they all lead to further apoptotic blockade.

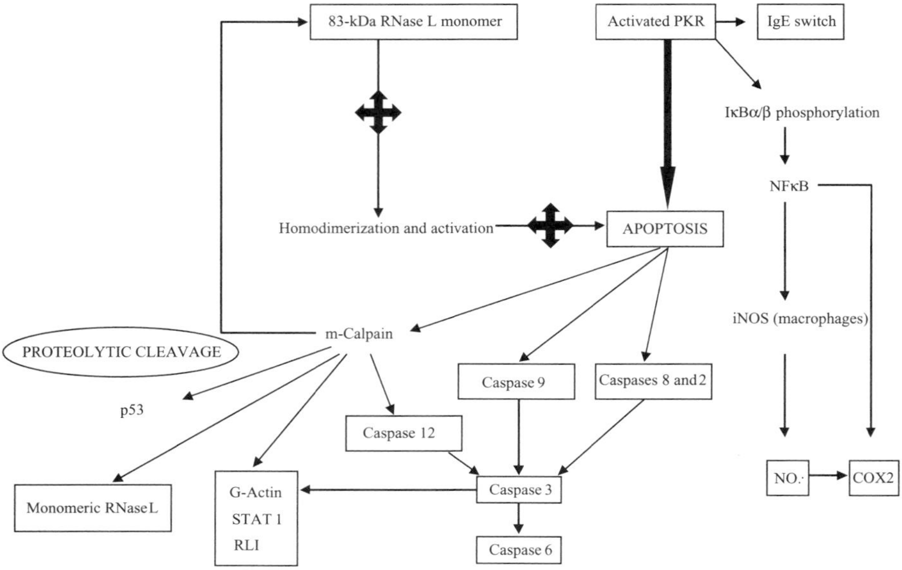

FIGURE 10.4 Apoptotic dysregulation sequence of events induced by PKR activation and RNase L inhibition in CFS. Activated PKR activates the caspase cascades and induces calcium release from the endoplasmic reticulum, which further activates m-calpain. These proteases cleave important cellular proteins. Due to lack of activation by proper 2-5A oligomers, RNase L does not send the expected parallel apoptotic signals.

The cleavage of p53 by calpain is particularly important since it might explain the propensity of some CFS patients to develop cancer, particularly lymphomas.[101] The cleavage of caspase 9 results in its inactivation. Cleavage of RLI results in the downregulation of this protein in CFS[102] and impairs its regulatory role toward the catalytically active truncated 37-kDa RNase L. The cleavage of STAT1 results in the incapacity of IFNs to further induce PKR and 2-5A synthetase expression. The cleavage of G-actin in monocytes leads to a dysregulation of the antigen presentation and proper activation of T-lymphocytes, which plays a role in the amplification of the T_H2 switch.[103] Finally, we suggest that the ankyrin fragment released from the cleavage of RNase L is capable of interaction with ABC transporters,[104] leading to the channelopathies associated with CFS and a further activation of m-calpain through calcium influx.

As further shown in Figure 10.6, the alternating agonistic and antagonistic apoptotic signals induced in immune cells of CFS patients open the gate for any opportunistic infection or viral reactivation.[105-107] The chronic inflammation that results induces the expression of elastase, which further cleaves RNase L and produces the unregulated catalytically active fragment. The opportunists, particularly mycoplasmas, are capable of producing proteases which may play a further role in the pathologic process.[108] The 37-kDa fragment of RNase L in turn cleaves the mRNa of interferon-stimulated gene 15, which codes for a ubiquitin-like protein that targets proteins for degradation by the proteasome.[109] The cleavage of STAT1

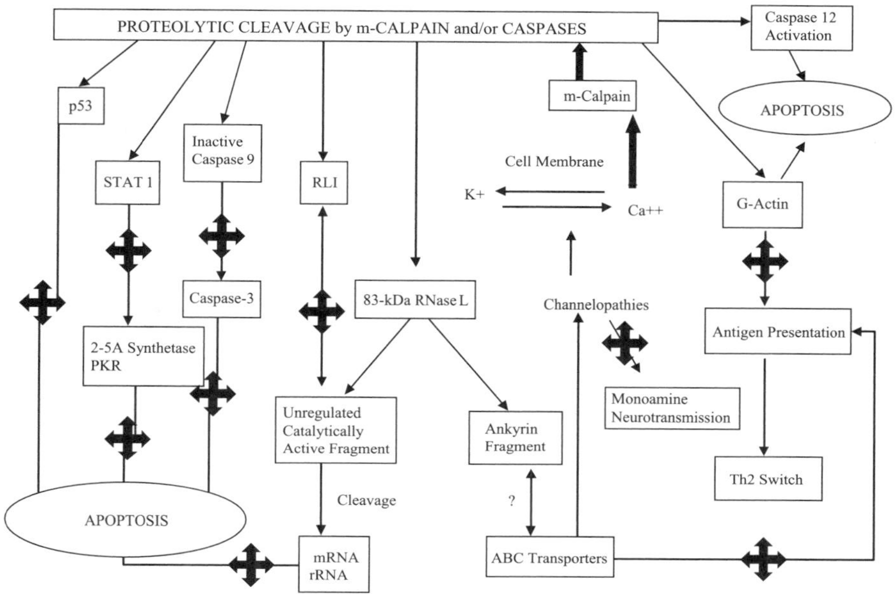

FIGURE 10.5 Further apoptotic dysregulation induced by calpain activation in CFS PBMC. Cleavage of p53 and caspase 9 block apoptosis. Cleavage of STAT1 further impairs the induction of 2-5A synthetase by the IFNs. Cleavage of RLI impairs the proper regulation of the 37-kDa RNase L fragment, which is allowed to cleave mRNA of proapoptotic genes. It is also possible that the ankyrin fragment released upon RNase L cleavage interacts with membrane ion channels and induces the channelopathies.

further blocks the induction by type I IFNs of the proapoptotic genes, and impairs IFN-γ to induce T-cell cytotoxicity.

The results gained by our laboratories shed a new light on the 2-5A/RNase L/PKR pathway in CFS. They convey the view of a vicious innate immune dysregulation induced by external stress stimuli and environmental factors. The dysregulation, depending on its gradation, leads to and explains the various accessory symptomatologies, pathologic expressions, and co-morbidities as expressed to different extents, which makes CFS patient populations as heterogeneous as those of cancer patients.[110]

10.3 DIAGNOSTIC STRATEGIES IN CFS

10.3.1 Medical History

Most patients who complained of CFS symptoms were active and healthy before they acquired the illness. Many became ill after an unusual or prolonged stress of the neuro-immuno-endocrine regulatory system. In a large study group including 1546 patients, we observed that, in their majority, an infectious event concomitant to a noninfectious event occurred simultaneously at the onset of the disorder.[111] These latter are capable of altering immune competence.

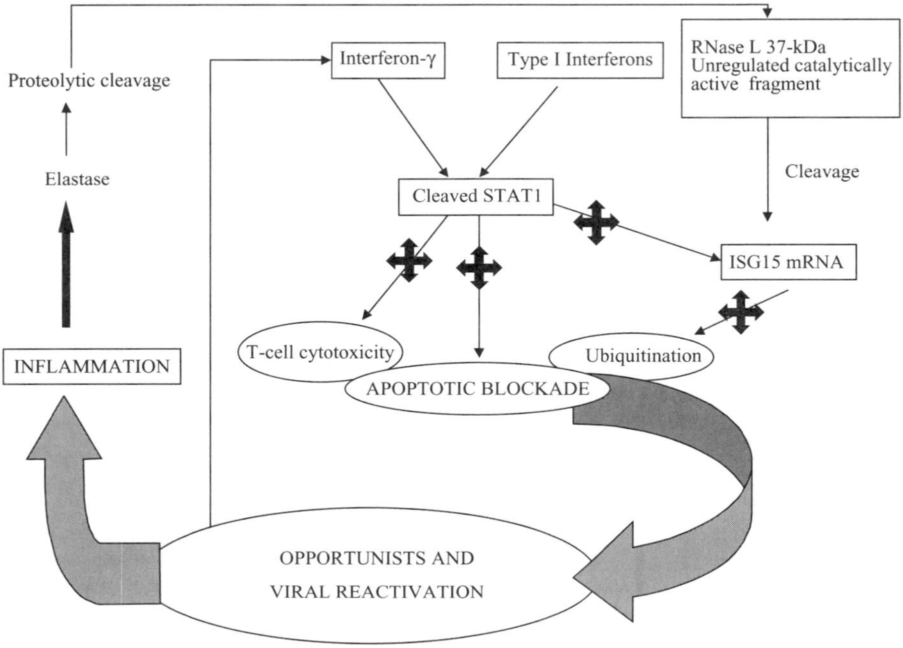

FIGURE 10.6 Apoptotic dysregulation in CFS PBMC opens the way for opportunist infections and viral reactivation to take place. The inflammatory reactions that result induce elastase production, which in turn enhances the cleavage of RNase L and the dysregulation process.

The questions that the physician may ask about the onset-related events are critical because they allow us to assign to them a chronological order. The questions that we address include (but are not limited to): date of onset? Did you receive a blood transfusion and when? Did you undergo surgery at that time (appendectomy, tonsillectomy, etc.)? Did you suffer from a flu-like illness at that period, gastro-intestinal infection, intestinal parasitosis, or intestinal bleeding? Is there a history of tick bite or insect bites, chemical or toxic exposure? Have you been subjected to long-standing stress (physical or mental)? Did you experience whiplash? Is there a history of repetitive upper respiratory tract infections like sinusitis or penumonia? Did you become ill in a foreign country or upon returning from one?

Male patients should be further interrogated about urethritis and prostatitis. Female patients should be asked to report about vaginal infections, cystitis, menstrual cycle abnormalities, abnormal vaginal bleeding, pregnancy, and possible iron deficiency. Questions regarding sleeping patterns, dental extractions and apectomy, history of dental-sinusal fistulae, thyroid problems, hepatitis B immunization, multiple vaccinations, and specific occupational health hazards should also be addressed. We also ask questions regarding the familial situation, about pets and birds (pigeons, parrots, etc.).[112] Finally, the patient's profession is an important factor because there is a higher prevalence of CFS among teachers and healthcare professionals.[113]

Once this anamnesis is complete, we ask ourselves whether the patient meets the clinical case definition criteria for CFS (Holmes et al. and Fukuda et al.). We match the symptoms to the most recent clinical case definition established by the Canadian Expert Consensus Panel[114] summarized in Table 10.1. According to this case definition, the symptoms must have begun, or have undergone significant changes, after the onset of the illness. According to our own experience, an improved

TABLE 10.1
Most Recent Clinical Case Definition of Chronic Fatigue Syndrome

1. **Post-exertional malaise and fatigue:** There is a loss of physical and mental stamina, rapid muscular and cognitive fatigability, post-exertional fatigue, malaise and/or pain, and a tendency for other symptoms to worsen. There is a pathologically slow recovery period, i.e., it takes more than 24 h to recover. Symptoms are usually exacerbated by stress of any kind. **Fatigue:** Patient must have a marked degree of new onset, unexplained, persistent, or recurrent physical and mental fatigue that substantially reduces activity level.
2. **Sleep dysfunction:** There is unrefreshed sleep or sleep quality and rhythm disturbance.
3. **Pain:** There is a significant degree of arthralgia and/or myalgia without clinical evidence of inflammatory responses of joint swelling or redness. The pain can be experienced in the muscles, joints, or neck and is often migratory in nature. Often, there are significant headaches of new type, pattern, or severity.
4. **Neurological/Cognitive manifestations:** Two or more of the following difficulties should be present: confusion, impairment of concentration and short-term memory consolidation, difficulty with information processing, categorizing, and word retrieval, intermittent dyslexia, perceptual and sensory disturbances, disorientation, and ataxia. There may be overload phenomena: informational, cognitive, and sensory and/or emotional overload which may lead to "crash" periods and/or anxiety.
5. **At least ONE symptom out of TWO of the following categories:**
   a. **Autonomic manifestations:** orthostatic intolerance such as neurally mediated hypotension (NMH), postural orthostatic tachycardia syndrome (POTS), delayed postural hypotension, vertigo, light-headedness, extreme pallor, intestinal or bladder disturbances with or without irritable bowel syndrome (IBS) or bladder dysfunction, cardiac arrhythmia, vasomotor instability, and respiratory irregularities
   b. **Neuro-endocrine manifestations:** loss of thermostatic stability, heat/cold intolerance, anorexia or abnormal appetite, marked weight change, hypoglycemia, loss of adaptability and tolerance for stress, worsening of symptoms with stress and slow recovery, and emotional lability
   c. **Immune manifestations:** tender lymph nodes, sore throat, flu-like symptoms, general malaise, development of new allergies or changes in status of old ones, and hypersensitivity to medications and/or chemicals
6. **The illness persists for at least 6 months. It usually has a distinct onset, although it may be gradual.** Preliminary diagnosis may be possible earlier. Three months is appropriate for children.

To be included, the symptoms must have begun or have been significantly altered after the onset of the illness. It is unlikely that a patient will suffer from all symptoms in criteria 4 and 5. The disturbances tend to form symptom clusters that are often unique to a particular patient. The manifestations fluctuate and may change over time. Symptoms can exacerbate with stress, resulting in a slow rate of recovery. Children may have numerous symptoms of similar severity but their hierarchy of symptom severity tends to vary from day to day.

With permission from Canadian Consensus Conference, 2001.

TABLE 10.2
Somatic Work-Up: First Round — Laboratory (General Practice or Internal Medicine)

- Complete blood count (CBC)
- Comprehensive metabolic panel (14)
- γ-glutamyl transferase (γ-GT)
- Intracellular (red blood cell) magnesium
- Immunoglobulin E (IgE)
- Rheumatoid factor (RF), antinuclear antibodies (ANA), fibrinogen, sedimentation rate
- Creatine phosphokinase (CPK), lactate dehydrogenase (LDH), cholesterol, triglycerides
- Ferritine and thyroid stimulating hormone (TSH)
- Infectious serology: toxoplasmosis, cytomegalovirus, mycoplasma, chlamydia, Epstein-Barr virus, herpes simplex virus, human herpes virus 6, human immunodeficiency virus
- Depending on specific symptoms or elements in the medical history: serology for rickettsia, dengue, hepatitis B/C, syphilis, yersinia, babesia

clinical case definition should include the minor criteria of the Holmes et al. definition, to which ten symptoms should be added: hot flushes (instead of low-grade fever), paralysis, new sensitivity to food and drugs, cold extremities, gastrointestinal symptoms, difficulties with words, exertional dyspnea, attention deficit, urinary frequency, muscle fasciculations, and light-headedness.[115] We believe that the establishment of symptom severity score requirements for the clinical case definition of CFS is an important tool in categorizing CFS patients. It is unlikely that a single disease model will account for every case of CFS, yet there is a common cluster of symptoms that allows a positive diagnosis.

10.3.2 The Somatic Work-Up

The somatic work-up plays a major role in excluding or confirming the presence of active diseases that fully explain all or most of the CFS-related symptoms: fatigue, sleep disturbance, pain, and cognitive and endocrine dysfunction. The physical examination may yield findings that provide an exclusion to the clinical case definition.

In a first round, we recommend the performance of the laboratory tests listed in Table 10.2 as well as of the other diagnostic procedures listed in Table 10.3. In particular, cardiopulmonary evaluation may provide information characteristic of the illness. In a recent study, we demonstrated that female CFS patients have a markedly reduced exercise capacity, an evaluation that can be used as an objective tool to document physical disability.[116] The ability of CFS patients to perform physical activity should therefore be assessed as a guide to the determination of the level of impairment. The differential diagnosis of chronic fatigue is crucial, and it is critical not to miss Addison's disease, Cushing's syndrome, hypo- or hyperthyroidism, iron deficiency or overload (hemachromatosis), anemia, diabetes mellitus, and cancer. Rheumatological disorders should also be excluded, such as rheumatoid arthritis, lupus, polymyositis, and polymyalgia rheumatica. As in CFS, infectious diseases such as HIV, tuberculosis, chronic hepatitis, Lyme disease, etc. can show a

TABLE 10.3
Somatic Work-Up: First Round — Other

- Electrocardiogram (U+T wave)/Exercise testing/Exercise electrocardiogram
- Chest X-ray
- Abdominal ultrasound
- Depending on specific symptoms:
  1. Computerized tomography, sinuses
  2. Echocardiogram
  3. Panoramic x-ray of dental roots
  4. Magnetic resonance imaging
  5. Gastroscopy, colonoscopy
  6. Pulmonary function tests

symptomatology dominated by overwhelming fatigue which is not alleviated by bedrest. The sleep apnea syndrome can be suspected from the medical history and, in such a case, it should be documented. Finally, indications of neurological diseases such as multiple sclerosis (MS), myasthenia gravis, and Parkinson's disease are most often detected on physical examination.

In a second round, we recommend more specific laboratory tests (Table 10.4), which should be selected on an individual basis. In men over 35 years of age who complain of hypersomnia during the day, a polysomnogram should be performed. A sleep study can also be helpful in other patients with a poor sleep architecture. Other tests that are sometimes indicated include the tilt table test intended at confirming the neurally mediated hypotension, magnetic resonance imaging of the brain

TABLE 10.4
Somatic Work-Up: Second Round — Specific Laboratory Tests (Internal Medicine)

- Low molecular weight RNase L + RNase L activity/actin fragments
- Immunophenotyping, with special emphasis on:
  1. Total lymphocytes (CD25+ CD14– CD3+) (CD = cluster determinant)
  2. Activated T-cells (CD25+ HLADR+ CD8+ CD38+)
  3. CD4+/CD8+ cells and ratio
  4. Ratio of memory/virgin CD4+ cells
  5. Natural killer (NK) subsets and NK cytotoxicity
  6. B lymphocytes (CD19+ CD5+)
- Antithyroid antibodies
- Immunoglobulins and IgG subclasses (IgG1, IgG3)
- T_H (T-helper)1/T_H2 profile
- Other: tumor necrosis factor (TNF)-α, interleukin (IL)-1
- Polymerase chain reaction (PCR) for the detection of chlamydia, brucella, mycoplasma, rickettsia
- Stool: search for parasites (amoebiasis)
- Urine: 24 h cortisoluria

TABLE 10.5
Neuropsychological Evaluation

Psychiatric evaluation	Neurocognitive evaluation
• Semi-structured interview (DSM-IV: diagnostic and statistical manual IV) for depression, anxiety disorders, and PTSD (post-traumatic stress disorders)	• Intelligence
• MMPI-II (Minnesota multiple personality inventory)	• Memory (verbal and visual)
• Coping checklist	• Reaction time
• MOS SF-36 (medical outcome survey short form 36)	• Attention
• SCL-90-R (symptom checklist 90, revised)	• Mental flexibility
• GHQ-30 (general health questionnaire)	• Interference
	• Index of general body strength
	• Simple motor speed
	• Word fluency

or cervical spine in order to detect MS, Arnold–Chiari malformation, and disc bulges. EEG, SPECT, and PET scans could become important diagnostic tools if patient subgroups can be defined using such techniques with improved sensitivity and specificity. Finally, a neurocognitive evaluation must be performed in order to evaluate the cognitive dysfunction and its severity; it should include several neuropsychological tests as listed in Table 10.5.

10.4 CONCLUSIONS AND PROSPECTS FOR FUTURE THERAPEUTIC STRATEGIES

The abnormalities observed so far at the molecular biology level have two major impacts. First, they allow us to assign the development of CFS to a major innate immunity dysfunction. Second, they permit us to relate CFS to other autoimmune diseases, including MS. These preliminary observations certainly warrant further investigation. The 2-5A/RNase L pathway dysregulations indicate that the assessment of RNase L cleavage by the measurement of the 37-kDa 2-5A-binding protein might not only provide a biological marker for CFS, but could also help in staging and subgrouping the various patient populations. Beside this protein evaluation, other cellular parameters may become relevant in evaluating and staging CFS dysfunctions.

The understanding of the abnormalities occurring at the molecular level in CFS is essential in order to identify and develop new therapies intended for their correction. It is clear from our work and from that of others that we report here that the use of immunotherapy, immunomodulators, and antiviral- and antibio-therapy will gain a prominent place in the therapeutic arsenal and strategies aimed at fighting the syndrome. The advances gained so far have already allowed us to identify several therapeutic regimens and new lead compounds that show efficacy *in vitro* as well as *in vivo*. We very much look forward to the not-so-distant time when these will become accepted standards.

REFERENCES

1. Fukuda, K. et al., The chronic fatigue syndrome: a comprehensive approach to its definition and study, *Annu. Intern. Med.,* 1994; 121: 953–9.
2. Komaroff, A. L., The biology of chronic fatigue syndrome, *Am. J. Med.,* 2000; 108: 169–71.
3. Reid, S. et al., Chronic fatigue syndrome, *Brit. Med. J.,* 2000; 320: 292–6.
4. Goldenberg, D. L., Fibromyalgia syndrome a decade later, *Arch. Intern. Med.,* 1999; 159: 777–85.
5. Komaroff, A. L. and Buchwald, D. S., Chronic fatigue syndrome: an update, *Annu. Rev. Med.,* 1998; 49: 1–13.
6. Evengard, B., Schacterle, R. S., and Komaroff, A. L., Chronic fatigue syndrome: new insights and old ignorance, *J. Intern. Med.,* 1999; 246: 455–69.
7. Suhadolnik, R. J. et al., Up-regulation of the 2-5A synthetase/RNase L antiviral pathway associated with chronic fatigue syndrome, *Clin. Infect. Dis.,* 1994; 18: S96–104.
8. De Meirleir, K. et al., A 37-kDa 2-5A binding protein as a potential biochemical marker for chronic fatigue syndrome, *J. Interferon Cytokine Res.,* 1999; 19: S94.
9. De Meirleir, K. et al., A 37-kDa 2-5A binding protein as a potential biochemical marker for chronic fatigue syndrome, *Am. J. Med.,* 2000; 108: 99–105.
10. Mc Gregor, N. R. et al., The biochemistry of chronic pain and fatigue, *J. Chronic Fatigue Syndrome,* 2000; 7: 3–21.
11. Roelens, S. et al., G-actin cleavage parallels 2-5A-dependent RNase L cleavage in peripheral blood mononuclear cells. Relevance to a possible serum-based screening test for dysregulations in the 2-5A pathway, *J. Chronic Fatigue Syndrome,* 2001, in press.
12. Vojdani, A. et al., Elevated apoptotic cell population in patients with chronic fatigue syndrome: the pivotal role of protein kinase RNA, *J. Intern. Med.,* 1997; 242: 465–78.
13. Suhadolnik, R. J. et al., Changes in the 2-5A synthetase/RNase L antiviral pathway in a controlled clinical trial with poly(I)-Poly(C_{12}U) in chronic fatigue syndrome, *In Vivo,* 1994; 8: 599–604.
14. Dong, B. and Silverman, R. H., 2-5A-dependent RNase molecules dimerize during activation, *J. Biol. Chem.,* 1995; 270: 4133–7.
15. Naik, S., Paranjape, J. M., and Silverman, R. H., RNase L dimerization in a mammalian two-hybrid system in response to 2',5'-oligoadenylates, *Nucleic Acids Res.,* 1998; 26: 1522–7.
16. Yu, F. and Floyd-Smith, G., A protein kinase C-dependent signaling pathway is required for transcriptional induction of the p69-71 isoform of the human 2-5A synthetase by interferon-alpha, *J. Interferon Cytokine Res.,* 1997; 17: S77.
17. Yu, F. and Floyd-Smith, G., Induction of the p40-46 and p69-71 isoforms of the human 2-5A synthetase by interferon-alpha requires protein kinase C, *FASEB J.,* 1997; 11: A1347.
18. Faltynek, C. R. et al., A functional protein kinase C is required for induction of 2-5A synthetase by recombinant interferon-alpha A in Daudi cells, *J. Biol. Chem.,* 1989; 264: 14305–11.
19. Wang, Q. and Floyd-Smith, G., The promoter of the p69-71 isoform of the human 2-5A synthetase contains multiple regulatory elements required for interferon-inducible and constitutive expression, *FASEB J.,* 1997; 11: A1203.
20. Wang, Q. and Floyd-Smith, G., The p69/71 2-5A synthetase promoter contains multiple regulatory elements required for interferon-alpha-induced expression, *DNA Cell Biol.,* 1997; 16: 1385–94.

21. Coccia, E. M. et al., Activation and repression of the 2-5A synthetase gene and p21 promoters by IRF-1 and IRF-2, *Oncogene*, 1999; 18: 2129–37.
22. Desai, S. Y. and Sen, G. C., Effects of varying lengths of double-stranded RNA on binding and activation of 2'-5'-oligoadenylate synthetase, *J. Interferon Cytokine Res.*, 1997; 17: 531–6.
23. Sarkar, S. N. et al., Enzymatic characteristics of recombinant medium isozyme of 2'-5' oligoadenylate synthetase, *J. Biol. Chem.*, 1999; 274: 1848–55.
24. Hartmann, R. et al., Activation of 2'-5' oligoadenylate synthetase by single-stranded and double-stranded aptamers, *J. Biol. Chem.*, 1998; 273: 3236–46.
25. Marié, I., Svab, J., and Hovanessian, A. G., The binding of the 69- and 100-kD forms of 2',5'-oligoadenylate synthetase to different nucleotides, *J. Interferon Res.*, 1990; 10: 571–8.
26. Pyo, S., The mechanism of poly I:C-induced antiviral activity in peritoneal macrophage, *Arch. Pharm. Res.*, 1994; 17: 93–9.
27. Hendrix, C. W. et al., Biologic effects after a single dose of poly(I):poly(C12U) in healthy volunteers, *Antimicrob. Agents Chemother.*, 1993; 37: 429–35.
28. Suhadolnik, R. J. et al., Changes in the 2-5A synthetase/RNase L antiviral pathway in a controlled clinical trial with poly(I)-poly(C12U) in chronic fatigue syndrome, *In Vivo*, 1994; 8: 599–604.
29. Ushijima, H. et al., Mode of action of the anti-AIDS compound poly(I)-poly(C12U) (Ampligen): activator of 2',5'-oligoadenylate synthetase and double-stranded RNA-dependent kinase, *J. Interferon Res.*, 1993; 13: 161–71.
30. Bonnevie-Nielsen, V., Husum, G., and Kristiansen, K., Lymphocytic 2',5'-oligoadenylate synthetase is insensitive to dsRNA and interferon stimulation in autoimmune BB rats, *J. Interferon Res.*, 1991; 11: 351–6.
31. Schröder, H. C. et al., Inhibition of DNA topoisomerase I activity by 2',5'-oligoadenylates and mismatched double-stranded RNA in uninfected and HIV-1-infected H9 cells, *Chem. Biol. Interact.*, 1994; 90: 169–83.
32. Baglioni, C., Minks, M. A., and De Clercq, E., Structural requirements of polynucleotides for the activation of (2'-5')An polymerase and protein kinase, *Nucleic Acids Res.*, 1981; 9: 4939–50.
33. Bevilacqua, P. C. et al., Binding of the protein kinase PKR to RNAs with secondary structure defects: role of the tandem A-G mismatch and noncontiguous helixes, *Biochemistry*, 1998; 37: 6303–16.
34. Komaroff A. L. and Buchwald, D. S., Chronic fatigue syndrome: an update, *Annu. Rev. Med.*, 1998; 49: 1–13.
35. Campadelli-Fiume, G., Mirandola, P., and Menotti, L., Human herpesvirus 6: an emerging pathogen, *Emerging Infect. Dis.*, 1999; 5: 353–66.
36. Löwer, R., Löwer, J., and Kurth, R., The viruses in all of us: characteristics and biological significance of human endogenous retrovirus sequences, *Proc. Natl. Acad. Sci. U.S.A.*, 1996; 93: 5177–84.
37. Urnowitz, H. B. and Murphy, W. H., Human endogenous retroviruses: nature, occurrence, and clinical implications in human diseases, *Clin. Microbiol. Rev.*, 1996; 9: 72–99
38. Schmid, C. W., Does SINE evolution preclude Alu function? *Nucleic Acids Res.*, 1998; 26: 4541–50.
39. Morris, T. and Thacker, J., Formation of large deletions by illegitimate recombination in the HPRT gene of primary human fibroblasts, *Proc. Natl. Acad. Sci. U.S.A.*, 1993; 90: 1392–6.
40. Igusheva, O. A., Bil'din, V. N., and Zhestianikov, V. D., The repair of gamma-induced single-stranded breaks in the transcribed and nontranscribed DNA of HeLa cells, *Tsitologiia*, 1993; 35: 54–63.

41. Schmidt, J. et al., Osteosarcomagenic doses of radium (224Ra) and infectious endogenous retroviruses enhance proliferation and osteogenic differentiation of skeletal tissue differentiation *in vitro, Radiat. Environ. Biophys.,* 1994; 33: 69–79.
42. Liu, W. M. et al., Cell stress and translational inhibitors transiently increase the abundance of mammalian SINE transcripts, *Nucleic Acids Res.,* A995; 23: 1758–65.
43. Thacker, J. et al., Localization to chromosome 7q36.1 of the human XRCC2 gene, determining sensitivity to DNA-damaging agents, *Hum. Mol. Genet.,* 1995; 4: 113–20.
44. Li, T. H. et al., K562 cells implicate increased chromatin accessibility in Alu transcriptional activation, *Nucleic Acids Res.,* 2000; 28: 3031–9.
45. Khodarev, N. N. et al., LINE L1 retrotransposable element is targeted during the initial stages of apoptotic DNA fragmentation, *J. Cell Biochem.,* 2000; 79: 486–95.
46. Botezatu, I. et al., Genetic analysis of DNA excreted in urine: a new approach for detecting specific genomic DNA sequences from cells dying in an organism, *Clin. Chem.,* 2000; 46: 1078–84.
47. Rudin, C. M. and Thompson, C. B., Transcriptional activation of short interspersed elements by DNA-damaging agents, *Genes Chromosomes Cancer,* 2001; 30: 64–71.
48. Mangeney, M. and Heidmann, T., Tumor cells expressing a retroviral envelope escape immune rejection *in vivo, Proc. Natl. Acad. Sci. U.S.A.,* 1998; 95: 14920–25.
49. De Freitas, E. et al., Retroviral sequences related to human T-lymphotropic virus type II in patients with chronic immune dysfunction syndrome, *Proc. Natl. Acad. Sci. U.S.A.,* 1991; 88: 2922–6.
50. Gelman, I. H. et al., Chronic fatigue syndrome is not associated with expression of endogenous retroviral p15E, *Mol. Diagn.,* 2000; 5: 155–6.
51. Urnowitz, H. B. et al., RNAs in the sera of Persian Gulf war veterans have segments homologous to chromosome 22q11.2, *Clin. Diagn. Lab. Immunol.,* 1999; 330–5.
52. Haahr, S. and Munch, M., The association between multiple sclerosis and infection with Epstein–Barr virus and retrovirus, *J. Neurol.,* 2000; 6, suppl. 2: S76–9.
53. Clerici, M. et al., Immune responses to antigens of human endogenous retroviruses in patients with acute or stable multiple sclerosis, *J. Neuroimmunol.,* 1999; 99: 173–82.
54. Rieger, F. et al., New perspectives in multiple sclerosis: retroviral involvement and glial cell death, *Pathol. Biol.* (Paris), 2000; 48: 15–24.
55. Chu, W. M. et al., Potential Alu function: regulation of the activity of double-stranded RNA-activated kinase PKR, *Mol. Cell. Biol.,* 1998; 18: 58–68.
56. Kumar, M. and Carmichael, G. G., Antisense RNA: function and fate of duplex RNA in cells of higher eukaryotes, *Microbiol. Mol. Biol. Rev.,* 1998; 62: 1415–34.
57. Gil, J. et al., The catalytic activity of dsRNA-dependent protein kinase, PKR, is required for NF-κB activation, *Oncogene*, 2001; 20: 385–94.
58. Ishii, T. et al., Activation of the IκBα kinase (IKK) complex by double-stranded RNA-binding defective and catalytic inactive mutants of the interferon-inducible protein kinase PKR, *Oncogene*, 2001; 20: 1900–12.
59. Yamamoto, Y. and Gaynor, R. B., Therapeutic potential of inhibition of the NF-κB pathway in the treatment of inflammation and cancer, *J. Clin. Invest.,* 2001; 107: 135–142.
60. Hobbs, A. J., Higgs, A., and Moncada, S., Inhibition of nitric oxide synthetase as a potential therapeutic target, *Annu. Rev. Pharmacol. Toxicol.,* 1999; 39: 191–220.
61. Eu J. P. et al., Regulation of ryanodine receptors by reactive nitrogen species, *Biochem. Pharmacol.,* 1999; 57: 1079–84.
62. Salama, G., Menshikova, E. V., and Abramson, J. J., Molecular interaction between nitric oxide and ryanodine receptors of skeletal and cardiac sarcoplasmic reticulum, *Antioxid. Redox Signal*, 2000; 2: 5–16.

63. Perez-Sala, D. and Lamas, S., Regulation of cyclooxygenase-2 expression by nitric oxide in cells, *Antioxid. Redox Signal,* 2001; 3: 231–48.
64. Ito, M. et al., Inhibition of natural killer cell activity against cytomegalovirus-infected fibroblasts by nitric oxide-releasing agents, *Cell. Immunol.,* 1996; 174: 13–8.
65. Ito, M. et al., Inhibition of natural killer (NK) cell activity against varicella-zoster virus (VZV)-infected fibroblasts and lymphocyte activation in response to VZV antigen by nitric oxide-releasing agents, *Clin. Exp. Immunol.,* 1996; 106: 40–4.
66. Orucevic, A. and Lala, P. K., Effects of N(G)-nitro-L-arginine methyl ester, an inhibitor of nitric oxide synthesis, on IL-2-induced LAK cell generation *in vivo* and *in vitro* in healthy and tumor-bearing mice, *Cell. Immunol.,* 1996; 169: 125–32.
67. Salvucci, O. et al., The induction of nitric oxide by interleukin-12 and tumor necrosis factor-α in human natural killer cells: relationship with the regulation of lytic activity, *Blood,* 1998; 92: 2093–102.
68. Asea, A. et al., Histaminergic regulation of interferon-gamma (IFN-γ) production by human natural killer (NK) cells, *Clin. Exp. Immunol.,* 1996; 105: 376–82.
69. Hellstrand, K. et al., Histaminergic regulation of NK cells. Role of monocyte-derived reactive oxygen metabolites, *J. Immunol.,* 1994; 153: 4940–7.
70. Hovanessian, A. G. et al., Enhancement of natural killer cell activity and 2-5A synthetase in operable breast cancer patients treated with polyadenylic;polyuridylic acid, *Cancer,* 1985; 55: 357–62.
71. Rager, K. J. et al., Activation of antiviral protein kinase leads to immunoglobulin E class-switching in human B cells, *J. Virol.,* 1998; 72: 1171–6.
72. Kennedy, M. B., Signal-processing machines at the postsynaptic density, *Science,* 2000; 290: 750–4.
73. Riedel, W., Role of nitric oxide in the control of the hypothalamic–pituitary–adreno-cortical axis, *Z. Rheumatol.,* 2000; 59, suppl. 2, II: 36–42.
74. Pall, M. L., Elevated, sustained peroxynitrite levels as the cause of chronic fatigue syndrome, *Med. Hypotheses,* 2000; 54: 115–25.
75. Camacho, M., Lopez-Belmonte, J., and Vila, L., Rate of vasoconstrictor prostanoids released by endothelial cells depends on cyclooxygenase-2 expression and prostaglandin I synthase activity, *Circ. Res.,* 1998; 83: 353–65.
76. Weir, W. R. C., *Post-Viral Fatigue Syndrome,* John Wiley and Sons, London 1991: 246–53.
77. Hannan, K. L. et al., Activation of the coagulation system in Gulf War illness: a potential pathophysiologic link with chronic fatigue syndrome. A laboratory approach to diagnosis, *Blood Coagul. Fibrinolysis,* 2000; 11: 673–8.
78. Trabattoni, D. et al., Augmented type 1 cytokines and human endogenous retroviruses specific immune responses in patients with acute multiple sclerosis, *J. Neurovirol.,* 2000; 6, suppl. 2: S38–41.
79. Steinman, L., Multiple approaches to multiple sclerosis, *Nature Med.,* 2000; 6: 15–6.
80. Smith, T. et al., Autoimmune encephalomyelitis ameliorated by AMPA antagonists, *Nature Med.,* 2000; 6: 62–6.
81. Pitt, D., Werner, P., and Raine, C. S., Glutamate excitotoxicity in a model of multiple sclerosis, *Nature Med.,* 2000; 6: 67–70.
82. Lopez-Moratalla, N. et al., Monocyte inducible nitric oxide synthase in multiple sclerosis: regulatory role of nitric oxide, *Nitric Oxide,* 1997; 1: 95–104.
83. Oleszak, E. L. et al., Inducible nitric oxide synthase and nitrotyrosine are found in monocytes/macrophages and/or astrocytes in acute, but not in chronic, multiple sclerosis, *Clin. Diagn. Lab. Immunol.,* 1998; 5: 438–45.

84. Giovannoni, G. et al., Increased urinary nitric oxide metabolites in patients with multiple sclerosis correlates with early and relapsing disease, *Mult. Scler.,* 1999; 5: 335–41.
85. Liu, J. S. et al., Expression of inducible nitric oxide synthase and nitrotyrosine in multiple sclerosis lesions, *Am. J. Pathol.,* 2001; 158: 2057–66.
86. Smith, K. J., Kapoor, R., and Felts, P. A., Demyelination: the role of reactive oxygen and nitrogen species, *Brain Pathol.,* 1999; 9: 69–92.
87. Konstantinov, K. et al., Autoantibodies to nuclear envelope antigens in chronic fatigue syndrome, *J. Clin. Invest.,* 1996; 98: 1888–96.
88. Bonnevie-Nielsen, V. et al., The antiviral 2',5'-oligoadenylate synthetase is persistently activated in type 1 diabetes, *Clin. Immunol.,* 2000; 96: 11–8.
89. Patel, R. C. and Sen, G. C., PACT, a protein activator of the interferon-induced protein kinase PKR, *EMBO J.,* 1998; 17: 4379–90.
90. Ito, T., Yang, M., and May, W. S., RAX, a cellular activator for double-stranded RNA-dependent protein kinase during stress signaling, *J. Biol. Chem.,* 1999; 274: 15427–32.
91. Peters, G. A. et al., Modular structure of PACT: distinct domains for binding and activating PKR, *Mol. Cell. Biol.,* 2001; 21: 1908–20.
92. Ruvolo, P. P. et al., Ceramide regulates protein synthesis by a novel mechanism involving the cellular PKR activator RAX, *J. Biol. Chem.,* 2001; 276: 11754–8.
93. Balachandran, S. et al., Activation of the dsRNA-dependent protein kinase, PKR, induces apoptosis through FADD-mediated death signaling, *EMBO J.,* 1998; 17: 6888–902.
94. Williams, B. R. G., Signal integration via PKR, *Science's Stke.,* 2001: www.stke.org/cgi/content/full/OC_sigtrans; 2001/89/re2: 1–10.
95. Srivastava, S. K., Davies, M. V., and Kaufman, R. J., Calcium depletion from the endoplasmic reticulum activates the double-stranded RNA-dependent kinase (PKR) to inhibit protein synthesis, *J. Biol. Chem.,* 1995; 270: 16619–24.
96. Wood, D. E. and Newcomb, E. W. Caspase-dependent activation of calpain during drug-induced apoptosis, *J. Biol. Chem.,* 1999; 274: 8309–15.
97. Nakagawa, T. and Yuan, J., Cross-talk between two cysteine protease families: activation of caspase 12 by calpain in apoptosis, *J. Cell Biol.,* 2000; 150: 887–94.
98. Chua, B. T., Guo, K., and Li, P., Direct cleavage by the calcium-activated protease calpain can lead to inactivation of caspases, *J. Biol. Chem.,* 2000; 275: 5131–5.
99. Porn-Ares, M. I., Samali, A., and Orrenius, S., Cleavage of the calpain inhibitor, calpastatin, during apoptosis, *Cell Death Diff.,* 1998; 5: 1028–33.
100. Kato, M. et al., Caspases cleave the amino-terminal calpain inhibitory unit of calpastatin during apoptosis in human Jurkat T cells, *J. Biochem.* (Tokyo), 2000; 127: 297–305.
101. Levine, P. H. et al., Cancer and a fatiguing illness in Northern Nevada — a causal hypothesis, *Annu. Epidemiol.,* 1998; 8: 245–9.
102. Vojdani, A., Choppa, P. C., and Lapp, C. W., Down-regulation of RNase L inhibitor correlates with up-regulation of interferon-induced proteins (2-5A synthetase and RNase L) in patients with chronic fatigue immune dysfunction syndrome, *J. Clin. Lab. Immunol.,* 1998; 50: 1–16.
103. Montaner, L. et al., Type 1 and type 2 cytokine regulation of macrophage endocytosis: differential activation by IL-4/IL-13 as opposed to IFN-γ or IL-10, *J. Immunol.,* 1999; 162: 4606–13.
104. Englebienne, P. et al., Interactions between RNase L ankyrin-like domain and ABC transporters as a possible origin for pain, ion transport, CNS, and immune disorders of chronic fatigue immune dysfunction syndrome, *J. Chronic Fatigue Syndrome,* 2001; 8: 83–102.

105. Treib, J. et al., Chronic fatigue syndrome in patients with lyme borreliosis, *Eur. Neurol.,* 2000; 43: 107–9.
106. Nasralla, M., Haier, J., and Nicolson, G. L., Multiple mycoplasmal infections detected in blood of patients with chronic fatigue syndrome and/or fibromyalgia syndrome, *Eur. J. Clin. Microbiol. Infect. Dis.,* 1999; 18: 859–65.
107. Ablashi, D. V. et al., Frequent HHV-6 reactivation in multiple sclerosis (MS) and chronic fatigue syndrome (CFS) patients, *J. Clin. Virol.,* 2000; 16: 179–91.
108. Watanabe, T., Proteolytic activity of *Mycoplasma salivarium* and *Mycoplasma orale* 1, *Med. Microbiol. Immunol.* (Berl.), 1975; 161: 127–32.
109. Li, X.-L. et al., RNase-L-dependent destabilization of interferon-induced mRNAs, *J. Biol. Chem.,* 2000; 275: 8880–8.
110. Levine, P. H. et al., Chronic fatigue syndrome and cancer, *J. Chronic Fatigue Syndrome,* 2000; 7: 29–38.
111. De Becker, P., McGregor, N., and De Meirleir, K., Possible triggers and mode of onset of chronic fatigue syndrome, *J. Chronic Fatigue Syndrome,* in press.
112. Glass, R. T., The human/animal interaction in myalgic encephalomyelitis/chronic fatigue syndrome: a look at 127 patients, *J. Chronic Fatigue Syndrome,* 2000; 6: 65.
113. Levine, P. H., Summary and perspectives: epidemiology of chronic fatigue syndrome, *Clin. Infect. Dis.,* 1994;18: S65.
114. Corruthers, B. et al., Myalgic encephalomyelitis/chronic fatigue syndrome. Clinical case definition, diagnostic and treatment protocols, The National ME/FM Action Network (Canada), 2001.
115. De Becker, P., McGregor, N., and De Meirleir, K., A definition-based analysis of symptoms in a large cohort of patients with chronic fatigue syndrome, *J. Intern. Med.,* 2001; 250: 234–40.
116. De Becker, P. et al., Exercise capacity in chronic fatigue syndrome, *Arch. Intern. Med.,* 2000; 160: 3270–7.

Index

A

B

C

D

E

J

K

L

M

R

S

T

V

W

Y

Z